Mathematical Knowledge
and the Interplay of Practices

Mathematical Knowledge
and the Interplay of Practices

JOSÉ FERREIRÓS

PRINCETON UNIVERSITY PRESS
PRINCETON AND OXFORD

Contents

List of Illustrations   ix

Foreword   xi

1 On Knowledge and Practices: A Manifesto   1

2 The Web of Practices   17
2.1. Historical Work on Practices   18
2.2. Philosophers Working on Practices   22
2.3. What Is Mathematical Practice, Then?   28
2.4. The Multiplicity of Practices   34
2.5. The Interplay of Practices and Its Basis   39

3 Agents and Frameworks   44
3.1. Frameworks and Related Matters   45
3.2. Interlude on Examplars   55
3.3. On Agents   59
3.4. Counting Practices and Cognitive Abilities   65
3.5. Further Remarks on Mathematics
   and Cognition   74
3.6. Agents and "Metamathematical" Views   79
3.7. On Systematic Links   83

4 Complementarity in Mathematics   89
4.1. Formula and Meaning   89
4.2. Formal Systems and Intended Models   94
4.3. Meaning in Mathematics: A Tentative Approach   99
4.4. The Case of Complex Numbers   104

5

Ancient Greek Mathematics: A Role for Diagrams   112

5.1. From the Technical to the Mathematical   113
5.2. The *Elements*: Getting Started   117
5.3. On the Euclidean Postulates: Ruling Diagrams
    (and Their Reading)   127
5.4. Diagram-Based Mathematics and Proofs   131
5.5. Agents, Idealization, and Abstractness   137
5.6. A Look at the Future—Our Past   147

6

Advanced Math: The Hypothetical Conception   153

6.1. The Hypothetical Conception: An Introduction   154
6.2. On Certainty and Objectivity   159
6.3. Elementary vs. Advanced: Geometry and
    the Continuum   163
6.4. Talking about Objects   170
6.5. Working with Hypotheses: AC and the
    Riemann Conjecture   176

7

Arithmetic Certainty   182

7.1. Basic Arithmetic   182
7.2. Counting Practices, Again   184
7.3. The Certainty of Basic Arithmetic   189
7.4. Further Clarifications   195
7.5. Model Theory of Arithmetic   198
7.6. Logical Issues: Classical or Intuitionistic Math?   200

8

Mathematics Developed: The Case of the Reals   206

8.1. Inventing the Reals   207
8.2. "Tenths" to the Infinite: Lambert and Newton   215
8.3. The Number Continuum   221
8.4. The Reinvention of the Reals   227
8.5. Simple Infinity and Arbitrary Infinity   231
8.6. Developing Mathematics   236
8.7. Mathematical Hypotheses and Scientific Practices   241

9 Objectivity in Mathematical Knowledge   247

9.1. Objectivity and Mathematical Hypotheses:
   A Simple Case   249

9.2. Cantor's "Purely Arithmetical" Proofs   253

9.3. Objectivity and Hypotheses, II: The Case of $\wp(\mathbb{N})$ 257

9.4. Arbitrary Sets and Choice   261

9.5. What about Cantor's Ordinal Numbers?   265

9.6. Objectivity and the Continuum Problem   273

10 The Problem of Conceptual Understanding   281

10.1. The Universe of Sets   283

10.2. A "Web-of-Practices" Look at the
   Cumulative Picture   290

10.3. Conceptual Understanding   296

10.4. Justifying Set Theory: Arguments Based on
   the Real-Number Continuum   305

10.5. By Way of Conclusion   310

References   315

Index   331

List of Illustrations

Figure 1 A schematic picture of interrelated practices.   38

Figure 2 Two representations of a theory, idealized vs. concretely given in practice.   53

Figure 3 Diagram from a late edition of the *Zhou Bi* (an early text in astronomy), related to third-century commentaries of this work and of the *Jiu Zhang* or *Nine Chapters*.   121

Figure 4 Diagrams corresponding to the *Elements* Book I. Prop. 5, Book I. Prop. 32, and Book III. Prop. 2.   136

Figure 5 First page of the Paganini-Pacioli edition of the *Elements* (Venice, 1509), with diagrams of line, point, angle, circle, etc.   145

Figure 6 The cone representation of the world of sets   292

Foreword

The philosophy of mathematics has experienced a renewal in recent years due to a more open and interdisciplinary way of asking and answering questions. Traditional philosophical concerns about the nature of mathematical objects and the epistemology of mathematics are combined and fructified with the study of a wide variety of issues about the way mathematics is done, evaluated, and applied—and in connection therewith, about historical episodes or traditions, educational problems, cognitive questions. The outcome is a broad outward-looking approach to the philosophy of mathematics, which engages with mathematics in practice.[1] This book is an outcome of such initiatives.

Purists may perhaps feel that this is not philosophy anymore; but, in my opinion, such judgments come from an unduly narrow understanding of the philosophical enterprise. In any event, it is our hope that a broad, open-minded approach to the study of mathematical practice can only act for the good. The study of mathematical knowledge and how it is produced is an important topic, and it is certainly desirable that the work of philosophers be of interest to mathematicians, mathematical educationalists, and scientists who are mathematics users.

In spite of the rather new orientation, there is a sense in which this book deals with traditional topics in the philosophy of mathematics. There are many interesting questions about the practice of mathematics (explanation, fruitfulness, computers, cognitive roots, and more) that I don't discuss carefully enough, and in some cases not at all.[2] This is not because of a lack of interest in them, but simply because I believed I had something to say about more traditional problems—questions about mathematical knowledge and its objectivity—from the standpoint of

---

[1] Cf. the description of the aims and goals of the APMP, which I have merely made mine in these sentences: http:// institucional.us.es/apmp/index_about.htm.

[2] A very good introduction to this area is the collective volume Mancosu (2008).

an approach that emphasizes the web of intertwined knowledge and practices, and the role of agents.

The main topics discussed have to do with core mathematical theories, in fact, with those structures that practicing mathematicians tend to regard as reflecting categorical models. (An axiom system is said to be *categorical* when all its models are isomorphic; traditionally, the theory of real numbers has been regarded as categorical, although this is more a theoretical desideratum than a fact of mathematical logic; the same applies of course to the natural-number structure.) Not much will be said about the natural numbers, since the topic is abundantly treated in other places and I wanted to concentrate on something different. There is a sustained argument about how the mathematical tradition evolved from Euclidean geometry to the real numbers, and from them to set theory.

I know some readers may be disappointed that I don't concentrate on very advanced material (e.g., on the worlds of contemporary mathematics), but I do believe the philosopher still has to deal adequately with more basic stuff.

After Chapter 1, which is introductory and written as a manifesto, Chapters 2–4 form a philosophical introduction to the standpoint from which the whole book is written. Chapter 2 can be read as a general introduction to the current trend of studies of mathematical practice, focusing above all on historical and philosophical work; it proposes a preliminary explanation of the notion of mathematical practice. Chapter 3 goes to the heart of the matter by laying out some crucial ingredients for the historically oriented, agent-based philosophy of mathematics that I develop (a web of practices based always on agents; symbolic frameworks; theoretical frameworks). Chapter 4 presents a thesis about what I consider to be a crucial *complementarity* of elements in the practice of mathematics, and in the configuration of mathematical knowledge: both conceptual and formalistic ingredients enter the game, both necessary, and neither reducible to the other.

As should be obvious from this brief overlook, the first four chapters are heavily philosophical in orientation. But the reader should feel free to employ this book in several different ways. Readers interested more directly in the mathematical material might go to Chapter 5 directly, perhaps after having read the "manifesto" in Chapter 1. In fact, some readers might even have the impression that the philosophical content in the first four chapters is not used very much in the second half of the book. While it is true that I have not paused to make explicit how the analysis of mathematical knowledge in the second part depends on the philosophical perspective presented in the first part (mainly because this would have considerably lengthened the work), nonetheless that perspective is presupposed. Readers versed in philosophy would certainly find my discussion utterly incomplete if I had left out the material in Chapters 2–4.

Chapter 5 considers the ancient tradition of geometrical proof in the light of recent studies by Manders and others. This in my opinion is highly important, since our *leitfaden* throughout the book is the idea of the continuum, and this originates in geometry (well, to be more precise, in notions of space, time, and movement). Also, the analysis of old geometric practices is a paradigm of the philosophy of mathematical practice, showing most clearly how this is different from traditional philosophy and how it complements and expands twentieth-century conceptions of proof. And, not least, mathematicians who approach the matter with an open mind (both researchers and students) will be intrigued by the challenge that Euclid poses to their basic assumptions about math.

Chapter 6 presents a crucial tenet of this work, the idea that advanced mathematics is based on hypotheses—that far from being a priori, it is based on *hypothetical* assumptions. The fact that an axiom is well established and generally accepted does not detract from its being, as I state, hypothetical in nature.[3] I do not claim originality here, since one can find related ideas in a long and important tradition of reflection on mathematical knowledge (including Riemann, Peirce, Poincaré, Weyl, Quine, Putnam, and others). What is original in my study is the way I elaborate on the idea and its implications: the devil is in the details. As that standpoint—the way I present it—depends on a distinction between elementary and advanced math, Chapter 7 studies some aspects of the paradigm of elementary math, basic arithmetic. I elaborate on the idea

---

[3] To be a bit precise, examples I have in mind include the Axiom of Completeness for the real number system, the Axioms of Infinity and Powerset in set theory, and the Axiom of Choice in category-theoretic foundations.

that we have *certainty* in our basic arithmetic knowledge. This is unlike what happens in advanced mathematical topics, which means (or so I claim) almost all of mathematics, beginning with plane geometry and the theory of real numbers (studied in Chapter 8).

Perhaps the most interesting element in this work will be the argument I put forward for reconciling the hypothetical conception of mathematics with the traditional idea of the objectivity of mathematical knowledge. This is the purpose of Chapter 9, to which I direct the reader's attention. The basic notion is that, because new hypotheses are embedded in the web of mathematical practices, they become systematically linked with previous strata of mathematical knowledge, and this forces upon us agents (e.g., research mathematicians or students of math) certain results, be they principles or conclusions. Mathematical knowledge is neither invention nor discovery: in advanced math we rather find something like invention-*cum*-discovery,[4] which shows that the simple dichotomy (invention vs. discovery) is quite useless for the analysis of scientific knowledge.

Finally, Chapter 10 is a preliminary look into one of the most intriguing questions that a philosophy of mathematics in practice must, sooner or later, confront: how *understanding* of math is obtained.

Let me insist on the idea that the reader has several options. Mathematically inclined readers might want to concentrate on the second half of this work, reading sections of the early chapters along the way, if and as they feel the need to have a stronger basis for the philosophical aspects. Philosophically inclined readers will certainly want to start with the early chapters, but they are invited to judge the perspective presented in them by its fruits in the concrete analysis of mathematical material discussed in the last few chapters. This is not an academic book of the kind that tries to consider all the alternative viewpoints, arguing for or against each of them. It is rather an attempt to explore in depth the possibilities of one particular take on the issues; by their fruits ye shall know them.

As for requirements, neither advanced knowledge of mathematics nor of philosophy is presupposed—yet it's true that some things have

---

[4] In contrast, in the case of elementary math we essentially discover traits of the objective world (a world to which agents belong).

been presupposed. An effort has been made to write clearly, yet the book requires a certain level of maturity; as Frege suggested, the philosophy of mathematics is in a difficult position, since many mathematicians will regard it as too philosophical, and many philosophers will find it too mathematical. On the side of mathematics, it is expected that the reader will have studied a rigorous definition of the real numbers, but a careful reading of Dedekind's beautiful essay "Continuity and irrational numbers" should suffice.[5] Set theory is mentioned as an example in many passages, but the usual rather informal acquaintance with the theory should be enough for the reader. Acquaintance with the axiom system of Zermelo-Fraenkel, however, will certainly be an advantage.[6]

Other than that, there are passages in which I mention concepts such as the von Neumann ordinals, nonstandard models of arithmetic, or Goodstein's theorem. Since those are in no way essential to following the argument, I have taken the freedom to mention them without giving the details; this seems to be a viable option, particularly nowadays, when it is so easy to find basic information about any of those points in Wikipedia and other places.

On the side of philosophy, the concepts or principles required to understand the position advocated here have been given explicitly with sufficient clarity—I hope. Given that philosophy is, by nature, heavily argumentative, there are many passages devoted to answering possible worries or objections I anticipate. This may make things somewhat difficult for less expert readers, who may be unfamiliar with the points of view that I am taking into account. My counsel is simply to read through those passages without trying to fully grasp what is at stake. There's much to be gained from reading books of this kind, even when one does not attain full command of the discussion.

The ideas that I present in this work have long been under elaboration, and that makes it very difficult for me to acknowledge properly

---

[5] For a more advanced work, I can recommend the recent Stillwell 2013 (much of whose content deals with material that goes beyond what is treated here).

[6] See, e.g., Thomas Jech's entry on "Set Theory" in the *Stanford Encyclopedia of Philosophy*, edited by E. N. Zalta, and its supplement on Zermelo-Fraenkel Set Theory," http://plato.stanford.edu/archives/fall2014/entries/set-theory/ and http://plato.stanford.edu/entries/set-theory/ZF.html.

everyone who has influenced my thinking. In fact, I began to reflect seriously on this topic during my undergraduate studies in philosophy, decades ago. Thus, my readings in philosophy (Descartes, Berkeley, Leibniz, Kant, Russell, Wittgenstein, Piaget, Quine, Kuhn, Kitcher, etc.) have always been shaped by my interest in understanding mathematics. And then, my readings in mathematics (Gauss, Cantor, Dedekind, Riemann, Poincaré, Frege, Hilbert, Weyl, Gödel, etc.) have been shaped by my interest in philosophy. Of special importance to my way of thinking have been, as far as I can tell, at least three ingredients: the interdisciplinary orientation, which I have always followed (e.g., with excursions into psychology and cognitive science); the contrast between the areas of mathematics and logic on the one hand, and historical work on the other, a contrast that is full of methodological lessons; and, more generally, the contrasts between scientific and philosophical readings, e.g., between old theory of knowledge and issues in biology and cognitive science.

Work on this book has been made possible by a number of research projects funded by the Spanish Ministerio de Ciencia e Innovación and by the Junta de Andalucía.[7] Part of the work was done while staying at the Instituto de Filosofía (CSIC, Madrid) as a research professor in 2009–2011; I thank my colleagues there, especially Javier Moscoso and Concha Roldán, for their help and understanding. Also I acknowledge gratefully a travel grant from the Ministerio de Educación that allowed me to make a stay at the University of Berkeley as a visiting scholar at the Philosophy Department.

Special thanks must go to Leo Alonso, Javier Ordóñez, Ralf Haubrich, José F. Ruiz, Roberto Colom, Tino Blanco, Angel Rivière, Huberto Marraud, John Heilbron, Antonio Durán, Guillermo Curbera, Javier Aracil, Jessica Carter, Jeremy Gray, Hourya Benis, Ignacio Jané, Erhard Scholz, Leo Corry, Marcus Giaquinto, Paolo Mancosu, Marco Panza, Philippe Nabonnand, Catherine Goldstein, Dominique Flament, Karine Chemla, and the former équipe Rehseis (now Sphere), Jean Paul van Bendegem, Jamie Tappenden, and Abel Lassalle Casanave for their various contributions to the shaping of this work. Jeremy, in particular,

[7] Proyecto de Excelencia P07-HUM-02594, Junta de Andalucía. Proyecto FFI2009–10224, Ministerio. Proyecto de Excelencia P12-HUM-1216, Junta de Andalucía.

has regularly accompanied the writing of this book and read it with great care and understanding. I would like to thank many other people as well, including Luis Vega, Javier de Lorenzo, Sol Feferman, Ken Manders, Fernando Zalamea, Jens Høyrup, Sonja Brentjes, Valeria Giardino, Wilfried Sieg, Bill Lawvere, John Steel, Joan Bagaría, Roberto Torretti, Jesús Mosterín, Norbert Schappacher, Javier Legris, Luis Carlos Arboleda (and the group of Historia y Educación Matemática in Cali), André Porto, Rafael Núñez, Brendan Larvor, Bart van Kerkhove, the Logic Group at Berkeley, and others, for conversations and exchanges that have helpfully guided my thoughts on these topics. And I wouldn't like to forget the names of students who helped me clarify thoughts in lectures and discussions (some of them related to their doctoral dissertations), in particular, Nicasio Ledesma, Enrique Sarrión, Mario Bacelar, Elías Fuentes, and Andrés Chaves.

Apologies to those whose names I may be overlooking. And thanks also to the personnel at Princeton University Press, in particular to Vickie Kearn and Jenny Wolkowicki, for their careful work transforming the manuscript into a well-edited book.

Despite everyone's good efforts, the flaws and peculiarities readers will find in this work, are a natural result of my stubborn ways of thinking.

The book is dedicated to Dolores, Inés, Lucía, and Juana, for all their love and care.

José Ferreirós
Sevilla, January 2014

Mathematical Knowledge and

the Interplay of Practices

1

On Knowledge and Practices

A Manifesto

Although the idea of emphasizing them is relatively new, and there is still some disunity concerning how to focus our analysis, mathematical practices are in the agenda of every practicing philosopher of mathematics today. Mathematical knowledge, on the other hand, has always figured prominently among the mysteries of philosophy.[1] Can we shed light on the latter by paying attention to the former? My answer is yes. I believe the time is ripe for an ambitious research project that targets mathematical knowledge in a novel way, operating from a practice-oriented standpoint.

Let us begin by placing this kind of enterprise within the context of the philosophy of mathematics. During the twentieth century, we have seen several different broad currents in this field, which, simplifying a great deal, can be reduced to three main types: *foundational* approaches (logicism, intuitionism, formalism, finitism, and predicativism), *analytic* approaches (focused on questions of ontology and epistemology), and the so-called "*maverick*" approaches (to use Kitcher's colorful terminology), which have typically been anti-foundational and focused on history, methodology, and patterns of change. Mixed approaches have, of course, been present throughout the century, although one can say that they remained relatively uninfluential; early examples are the work of Jean Cavaillès in France during the late 1930s and that of Paul Bernays in Germany and Switzerland from the 1930s on. But in the 1980s and

---

[1] A mystery that some twentieth-century philosophers (Wittgenstein prominent among them) tried to dispel like a simple fog, but quite unconvincingly. For reasons why mathematics is not a game of tautologies, nor a mere calculus of symbol transformations, see especially Chapters 4 and 6.

1990s the situation—at least judged from an Anglo-American perspective—seemed to be one of confrontation between the anti-foundational maverick camp and the system-oriented camp.

It seems to be the case that a new generation of philosophers of mathematics has arisen whose work is superseding those distinctions. They follow upon the footsteps of Cavaillès, Bernays, Manders, and others. Examples of this new phenomenon are provided in some recent anthologies, such as Mancosu (2008), Ferreirós and Gray (2006), and van Kerkhove, de Vuyst, and van Bendegem (2010). These philosophers engage in an analysis of mathematical practices that incorporates key concerns of the "mavericks," without adopting their anti-foundational, anti-logical orientation. They are no longer obsessed with all-encompassing formal systems (e.g., axiomatic set theory) and the associated metalogical results, directing their attention instead to different branches and forms of mathematics—geometry ancient and modern; different ways of practicing analysis, algebra, topology, and so on. But thereby they do not imply—not at all, in my case—that there's nothing to learn about mathematics and its methodology from the crisp results of foundational studies. They keep considering the traditional questions of ontology and epistemology, but within a broader palette of issues concerning the evaluation of mathematical results (see the different aspects treated in Mancosu 2008) and the place of mathematics within human knowledge—one of my central concerns here.

All of that is meant when I say that I aim at providing a novel analysis of mathematical knowledge from a practice-oriented standpoint.

Notice that the new orientation in the philosophy of mathematics is highly interdisciplinary. Some authors emphasize knowledge of mathematics itself and logic, coupled with careful scrutiny of epistemological issues; some put an emphasis on combining philosophical issues with historical insight; some others stress the role of cognitive science (Giaquinto 2007) or sociological approaches (van Bendegem and van Kerkhove 2007); and the list goes on, with mathematics education, anthropology, biology, etcetera. My own approach, as will become clear, has a strong interdisciplinary bent. But one has to be quite clear and careful about the ways in which the different disciplines could or should contribute to the enterprise. Instead of trying to provide a principled discussion at this point, we shall clarify the matter as we go along. However, let me give an example of what I mean by "careful": it is highly relevant to establish contact with cognitive science and with the biological underpinnings of human knowledge, but I believe the time is not ripe for simply taking some 'established' theories or models from cognitive science and "applying" them to mathematics. While paying attention to what goes on in cognitive science and neuroscience, and aiming at convergence with that kind of research, a philosopher of mathematics can and perhaps should remain independent from the concrete current theories in those fields (see Chapter 3).

To briefly describe the crucial traits of the approach I shall put forward, I can say that this is a cognitive, pragmatist, historical approach:

- *agent-based* and cognitive, for it emphasizes a view of mathematics as knowledge produced by human agents, on the basis of their biological and cognitive abilities, the latter being mediated by culture;
- *pragmatist* or practice-oriented, as it places emphasis on the practical roots of math, i.e., its roots in everyday practices, technical practices, mathematical practices themselves, and scientific practices;
- *historical* and hypothetical, because it emphasizes the need to analyze math's historical development, and to accept the presence of hypothetical elements in advanced math.[2]

Our perspective on mathematics will thus stress the provenance of mathematical knowledge from particular kinds of interplays between cognitive resources and cultural practices, with agents at the center, making such interactions and the development of new practices possible.

There is an aspect in which such an approach goes against well-established habits of philosophers of mathematics. It was customary to focus one's attention on a single mathematical theory assumed to be sufficiently broad to embrace all of current mathematics; the chosen system was commonly a form of axiomatic set theory based on classical logic.[3]

---

[2] On this topic, see Chapter 6 and also the introduction to Ferreirós and Gray (2006).

[3] However, other systems have been considered in foundational work, e.g., constructivist systems, and in recent decades there has been much discussion of category theory as an alternative foundation (see, e.g., issues of *Philosophia Mathematica* for 2004 and 2005).

This tradition emerged as a result of foundational developments in the early twentieth century, but its roots lie deeper, in the centuries-old vision of Math as an Ideal Theory, given in some Platonic realm (often conceived as God's mind), fixed and static, offering a unified foundation which one may bring to light by excavating the imperfect glimpses of the "true" math that our current theories provide.[4] Of course, this old tendency has been heavily criticized, both by the "mavericks" and by more recent proponents of a philosophy of mathematical practice. One of the reasons for such a critique is that set-theoretic reductionism blinded philosophers of mathematics to important phenomena, such as the hybridization of branches of mathematics, progress through the expansion of mathematical domains, or the specificity of methods characteristic of this or that branch of math. As a general rule, it is not only salutary, but necessary, to avoid the excessive systematicity and reductionism that was characteristic of much philosophy of science and epistemology. But there is another reason to go against that old habit, one that is crucial to my whole approach.

It is a key thesis of this work that several different levels of knowledge and practice are coexistent, and that their links and interplay are crucial to mathematical knowledge. What I mean will be spelled out in the next chapter, but for a first approximation consider Philip Kitcher's idea that given historical periods in mathematics are each under the spell of what he called a "mathematical practice," a system or aggregate of linguistic statements, methods, questions, and results (Kitcher 1984) that guide mathematical work. Intentionally or not, Kitcher's very interesting work promoted a rather Kuhnian image of a "normal period" in which a discipline, math in our case, is governed by a single paradigm or disciplinary matrix;[5] it also seems to have been influenced by the dogma of systematicity that we have discussed. Contrary to this, I maintain that in any given historical period one can find more than one framework for mathematical practice. Hence different levels of

---

[4] On this topic and the Quinean dogma of systematicity, see Ferreirós (2005).

[5] The reference is to Kuhn's classic (1970) and especially the 1969 Epilogue. Kitcher's analysis of a "mathematical practice" is very different from Kuhn's notion of a "disciplinary matrix"; yet I mean to say that there seemed to be a common underlying assumption. A more detailed presentation and critique of Kitcher's idea can be found below, in Chapter 2.

knowledge and practice coexist historically during the same period, and even within the same agent, an individual mathematician.

The Kuhnian element in such an analysis is very misguided, as it actually makes it impossible to address key epistemological issues. Analysis of a certain mathematical framework requires us at the very least, in all cases that I can think of, to consider its connections with another level of knowledge and practice. Different practices are linked in a systematic fashion, which is learnable and reproducible, and their interplay is crucial for math as the peculiar kind of knowledge it is: crucial, that is, not only to the way individuals learn mathematics and to mathematical understanding, but also to the lack of arbitrariness that is typically found in mathematical changes, to the emergence of new practices, and to the objectivity of new results. All of these ideas will be developed and discussed in the body of the present work.

As already announced, I shall argue that our knowledge of mathematics cannot be understood without emphasizing the practical roots of math, including its roots in scientific practices and technical practices.[6] The general scheme of my non-reductionist way of analyzing the *interplay* of practices will do a crucial job in making it possible to substantiate this claim. The idea is akin to that of the pragmatist tradition, in particular to the elucidation of "meaning" in terms of "use" (although I must resist the *reduction* of "meaning" to "use," see Chapter 4). Presented in a simple form, the claim is that a mathematician confronted with an axiomatic system for, say, the real numbers, can only obtain *knowledge* of it by (re)establishing its links with other mathematical practices (e.g., and importantly in this case, geometry), and ultimately with the basic technical practices of measuring and counting. There is a sense of understanding, to be explained below, in which one can claim that mathematical understanding is only gained by this kind of process, i.e., by linking back to more elementary practices and to the cognitive roots. This is not to be conceived as a process of reduction, but rather as a form of substantiation.

In the course of history, scientific practices have also been poles of reference for mathematical practices. The study of geometry was not

---

[6] By *technical* practices I mean something specific (section 2.4), examples being the practices of counting, measuring, and drawing geometrical shapes.

stimulated only by basic technical needs of drawing geometrical figures (such as for architectural construction); it was also, and rather constantly throughout history, motivated by the use of geometrical models in astronomy. Likewise, if we take the paradigmatic case of the function concept, we have an instance in which a central mathematical idea has emerged in response to needs arising in the context of scientific modeling of natural processes. This crucial case can be turned, I believe, into an argument for two points: that mathematical theories and practices are not totally independent or autonomous, and that they can be more fully understood when they are regarded as part and parcel of the theoretical work mobilized by the enterprise of science.

Our strategy will be to understand mathematical results and theories as *knowledge*, contiguous with (and not essentially different from) other forms of knowledge; but we shall insist on the specificity of mathematics, in contrast with holistic strategies such as those of Quine (1951) and Kitcher (1984). What kind of knowledge does mathematics provide? What are its relations with other forms of knowledge? These are difficult questions that lie at the heart of the theory of knowledge, and all major epistemological approaches have attempted to solve them. It suffices to mention Plato and his views about the links between mathematics and the realm of Ideas, mathematical discovery, and reminiscence; Leibniz and his views about mathematical truths being analytic "verités de raison," so close to twentieth-century logical positivism; Mill and his ideas about mathematical truths being obtained inductively from an extraordinarily broad basis of evidence; or Kant and his views about mathematical truths being synthetic a priori, but totally irrelevant to the world of "things in themselves," relevant only to the phenomenal world.

From the time of the ancient Greeks, with their definition of knowledge as "justified true belief,"[7] to the twentieth century under the influence of denotational semantics, it has been common to consider mathematics as a body of *truths* (otherwise, it is felt, there couldn't be

---

[7] My beliefs can only be called knowledge when they happen to be true (normally, of states of affairs in the world) and I can provide justification for them. If I believe there is a king of France, this is false and hence not knowledge; if I believe there is a king of Spain, and I can provide justification, then I know it.

knowledge of them). Math has been regarded as the discipline that provides justifications of the strongest kind, namely deductive proofs. But what are mathematical truths true of? What do they refer to? Some of the main types of answers are given by the authors just mentioned. Mathematical truths have been taken to be truths about objects of a peculiar non-physical kind (platonism), truths about or deriving from concepts or linguistic conventions (analytic truths), truths about the empirical world as the mind or subject knows it perceptively (empiricism), or truths about the forms of pure intuition (Kantianism).[8] Yet all of these answers run into difficulties—some of which will be mentioned later—and in my view none of them is satisfactory.

Though I cannot attempt to provide here anything close to a careful analysis of this broad topic, I feel it will be helpful to offer readers some remarks that may help them locate my proposals within a general scheme. But it should be taken into account that my argument in the bulk of the book does not depend on the considerations offered here. It is in this spirit that the following paragraphs ought to be read; the numbering is intended to underscore their *non*-systematic character; I do not in the least aim at completeness.

1. Common to all those attempts at answering the question of referents is the assumption that the semantics for mathematics must be similar to the semantics of common language. When I say "Venus is the morning star," the word "Venus" is the proper name of an object, a planet, and the word "star" is the common name for objects in a class to which the sun and other brilliant celestial masses belong. Medieval logicians knew already that there are many terms of common language to which this kind of semantics does not apply, namely what they called "syncategorematic terms" (particles like "all," "and," and "is"). Yet, the idea is that whenever we find proper names in mathematics (1, *e*, the Riemann theta function), they should be taken to refer in the same way

---

[8] The empiricist view, or a view more or less akin to it, admits rather different versions. When mathematics was regarded as the "science of magnitudes," many understood this to mean that it referred to physical magnitudes (but notice that a Kantian interpretation was also possible). In a strong move of refinement, Gauss defended mathematics as the "science of relations," and this position can be given a (broadly speaking) empiricist reading.

as "Venus" does; and whenever we find common names (real number, analytic function, closed set), they should be taken to refer like "star" does. If we resist this idea, then we shall be providing a semantics for math and science that will deviate from our semantic conceptions in common life. Perhaps the reader will not be surprised if I say that this would be a salutary move.

Without entering into complexities at this point, we can suggest how the reorientation would proceed. The common ingredient in the previous attempts was to adopt a referential semantics as the basis for a correspondence theory of truth: there had to be a specific domain of objects, which provide the reference for proper and common names (or constant terms, variables, predicates, and relations). To avoid this, one must avoid the correspondence theory, adopting some kind of deflationary theory of truth, an account of how we can arrive at truths that does not invoke a domain of previously given objects. This would actually be the starting point for another way of presenting the project of this book. I shall try to provide a strong account of how we arrive at—and share—mathematical truths, without invoking mathematical objects in any strong ontological sense. Of course, the word "deflationary" by itself does not make any miracle; what is required is a detailed account, a robust theoretical explanation of the basis on which mathematical truths can be found and shared.

2. My use of the word 'truth' at this point must be relativized by implicit or explicit reference to a mathematical theory. This agrees with the practice of most mathematicians; hence it should not be perceived as a shortcoming. That use is such that one can say, for instance, that Theorem 29 of the *Elements*, Book I ("the angles in a triangle add up to two right angles," 180°), is a truth of Euclidean geometry and the theorem that "the angle sum in a triangle is less than 180°" is a truth of the geometry of Bolyai-Lobatchevskii—while everybody knows that they cannot both be true of physical space, since they contradict each other. Similarly, the Well-ordering theorem is a truth of the Zermelo-Fraenkel set theory ZFC.

For that reason, it is actually not very enlightening to speak of "mathematical truths," and in this book I shall normally not employ that phrase. What I aim to provide, more than anything else, is an account of our *knowledge* of mathematical results; and the core of the matter is accounting for the *objectivity* of mathematical results. The motto here, following Kreisel and Putnam, could be "objectivity without objects".[9] The phenomenon that we aim to explain is the following: mathematical results force themselves upon us in such a way that we cannot help accepting them (except perhaps by some strong revision of the basic axioms or the logical principles employed).[10] Let me give an example that Einstein praised for the beauty of its statement and its method of proof, namely, Cantor's theorem of the nondenumerability of the set of real numbers, proved by the diagonal method. This is a mathematical truth, i.e., an objective result, and our task is to explain why and how this is so. My answer will be found in Chapter 9. (A follower of Brouwer can, of course, reject Cantor's theorem, because he or she rejects the basic axioms of modern mathematics, and some of its logical principles. The simplest way of avoiding Cantor's result is by rejecting the concept of an infinite set altogether—but once we accept the sets of natural and real numbers, we are forced to accept the result; see Chapter 9.)

3. Since physics and other sciences (natural and social) employ mathematics, the account we give of mathematical truth—or the objectivity of mathematical results—will affect accounts of scientific truth. The kind of analysis that I shall propose here calls for a critical understanding of the mathematical aspects of scientific models of natural phenomena. From the standpoint we shall adopt, it makes very little sense to assume that the mathematical basis of an empirically adequate model should automatically be interpreted in a realistic fashion. Notice that this is often done, and sometimes even proposed explicitly, e.g., in the context of interpreting the conception of space-time that underlies relativistic models.

---

[9] The word "object" should be understood here in the metaphysical sense. There is no problem in speaking of objects within the context of a given mathematical framework, on the understanding that their mathematical existence (modulo a given framework) is not metaphysical existence. See Chapter 6.

[10] The phraseology is Gödel's (1947); he wrote that the axioms of set theory "force themselves upon us as being true" (this is found in the 'Supplement to the second edition', 1964, p. 271).

However, the approach I shall propose here cannot simply be called non-realist, fictionalist, or idealist. The question of realism cannot be posed in a "flat" way, so to speak, as if one could give a yes-or-no answer that applies uniformly to all mathematical theories and all mathematical practices. (Similarly, there is more to say about the relation between "math" and "truth" than found in the preceding paragraphs.) This is a consequence of the ideas about mathematical knowledge that I shall be presenting: in particular the ideas concerning the interplay of practices presented in Chapter 2 and developed subsequently, my views on arithmetic certainty (Chapter 7), and the hypothetical conception of advanced math (Chapter 6). Once again, it seems to me that the over-systematic, reductionistic attitudes that were prevalent in the past have impeded a nuanced approach to questions like this, which require it. We shall return to this issue in the last chapters.

4. Recently, Giaquinto (2007, 5) called attention to the fact that twentieth-century philosophy of mathematics has abandoned the traditional topic of how knowledge is gained by the individual. This was a result of the concentration on the problem of rigor that was characteristic of foundational studies, plus the focus on theories and their justification as the sole question for philosophy. It was also aided by a psychologistic-subjectivistic understanding of the question of how knowledge is gained by individuals, which in my view has been superseded by more recent conceptions in cognitive science. Those philosophical and foundational orientations overshadowed the traditional concern for the epistemology of discovery by individuals that we find in Plato, Kant, and Mill. Giaquinto has proposed reviving this concern, which is obviously of great interest in connection with cognitive science, and also of practical interest within the field of math education. As has been said before, the approach I am proposing is centered on the agents, hence cognitive, and thus resonates with Giaquinto's proposal.

The subjectivistic understanding of individual knowledge, which I have mentioned as superseded, was a close ally of another influential idea of the twentieth century: the call for an epistemology without a subject. The shortcomings of traditional conceptions in epistemology—based on an attempt to reconstruct human knowledge as if it were obtained by a single perceiving mind, solely on the basis of its observations of phenomena—were more than clear. It was probably because of those subjectivistic connotations that a need was felt to free epistemology within the general scheme of contrapositions between subjective/objective and descriptive/normative. In France, the call was to replace the "*philosophie du sujet*"; and Cavaillès proposed a "*philosophie du concept*" that was later turned into philosophical structuralism. In England, Popper raised the call for an epistemology without a subject. In my opinion, however, these were wrong solutions based on a correct perception of symptomatology; the right surgery should have proceeded differently. Nothing is gained by trying to study epistemology without a subject; in fact, without a community of subjects. But the notion of a "subject" has to be seriously reconceived.

5. The general direction of "pragmatists" (interpreting this label in a broad sense that may include the later Wittgenstein) was, in my opinion, much better oriented than the moves of Popper or Cavaillès. To be brief, we should employ the notion of an *agent*, rather than a subject, and specifically of a community of agents. As Dewey liked to emphasize, the tradition in epistemology has been to rely on a "spectator theory" of knowledge as merely based on perception (the ego portrayed as a passive spectator sitting in the theater, its perceptions as what happens in a scene, out of its reach and control). I dare say that the *oblivion of action* was the worst defect of traditional epistemology, its main shortcoming that, in my view, has made impossible a well-grounded account of human knowledge.[11] Even perception, which modern philosophers from Descartes and Locke onward understood as a passive reception of impressions, is in the light of neurobiology and cognitive science a complex system that results from the interplay of sensory input and motor output (not anything like a primitive faculty of the mind, as philosophers tended to think). Input and output, sensation and action—without this feedback, one cannot even make sense of perception, let alone the further complexities of knowledge production.

[11] This was the clearest lesson that I learned from studying epistemology, back in my undergraduate days, when I knew almost nothing about pragmatism—but was aware of the work of Piaget and several other scientists in biology and psychology.

A brief excursion into etymology may help hammer the point. The word "subject" comes from Latin *subjicere* (formed from *sub*, "under" + *jacere*, "to throw," also related to *subjectare*), and has a strong passive meaning, which accords very well with the traditional oblivion of action. Originally, it meant a person is subject to another (dependent on, submitted to, another, like a king's subjects in the ancien régime);[12] it seems that only in the seventeenth century did it come to be used for the thinking subject. Possibly this was meant to suggest the mind or soul that "lay beneath" the world as perceived. By contrast, the word "agent" comes from Latin *agere*, "to act," and is related to *actio*, "action"; and if that were not enough, it is formed from *agens*, the active participle of *agere*. When we stop talking about subjects in favor of agents, we should be constantly reminded of the role of action in all facets of human life. *Im Anfang war die Tat*: in the beginning was the deed (Goethe).

6. One last clarification concerning the general topic of knowledge is in order. Parallel to the centuries-old debate about the role of language in thinking (recall the Sapir-Whorf hypothesis, heavily criticized by some cognitive scientists, or the debate between Chomsky and Piaget) is the corresponding debate about knowledge. What is the role of linguistic knowledge, versus implicit or tacit forms of knowledge? In recent decades, the debate has found new forms of expression within the field of AI, in the contrast between traditional systems based on symbolic processing and neural network systems; the former are enormously successful in tasks such as calculation and chess, but reveal well-known shortcomings in pattern recognition, where neural networks have presented significant advantages.[13] In the twentieth century, both analytic and hermeneutic philosophers made a strong bet on language with the celebrated linguistic turn, but this far from ended the debate, and there

---

[12] It could also mean a subject matter, what lies beneath (subjicere) a certain theme.

[13] Notice that pattern recognition tasks bear relations with some important topics of twentieth-century philosophy, in particular with Wittgenstein's reflections on language (family resemblance) and Kuhn's views on the cognitive organization of scientific knowledge (his insistence on the role of exemplars).

has recently been a re-turn.[14] Meanwhile, Polanyi and Kuhn and others were arguing for the importance of tacit knowledge in scientific practices, which agrees with the views of cognitive linguists such as Lakoff. Although I do not intend to contribute to the debate here, I feel the need to clarify my position.

In my opinion, the roots of knowledge and thought lie deeper than language; thus I tend to take sides with Piaget, Kuhn, and Lakoff. Steven Pinker [1994], in discussing how people think, argues that language is not adequate to account for all of our thoughts, and concludes that, if there can be two thoughts corresponding to one word [indeed, one sentence], thoughts can't be words.[15] This is obviously not to say that language and symbolic systems do not play a crucial role in human knowledge generally, and mathematics in particular. I believe they do, and I am even tempted to think that the use of written symbols (ciphers, diagrams, formulas) is a defining trait of mathematical practices. The idea that language is the very source of thought appears to me quite obviously wrong, but this still leaves room for the more nuanced view that language does contribute to human thinking abilities in a crucial way—and, by the way, we should distinguish between the modalities of oral and written language. In fact, I do hope that the study of mathematical cognition can play the role of a privileged arena for sorting out this vexing question. But this is a task for future studies.

7. The idea that there is no methodological difference between math and science serves, no doubt, the purpose of making things simpler for the philosopher. But it lacks power of conviction; in fact it makes things too simple (see Chapter 6). On the other hand, the opposite move of making math and natural science essentially different, most often carried through by means of the distinction between formal and empirical sciences, is again simplistic and misleading. Carnap, e.g., considered mathematics as a formal science based on linguistic convention, in contrast with the empirical sciences, which were supposed to be based

---

[14] This includes authors such as Van Fraassen.

[15] He indicates problems of ambiguity, lack of logical explicitness, co-reference, deixis, and synonymy as examples.

on mere observation and induction.[16] This simple dichotomy has to be abandoned, but to conclude from this that there exists no difference in terms of methodology would be to throw the baby out with the bathwater. Although these are issues to which we shall be returning, let me advance two ideas at this point.

In an analysis of the natural sciences from a practice-oriented standpoint, it becomes necessary to distinguish between several kinds of practices: experimental practices, theoretical practices, practices of writing and communication, etcetera.[17] Several authors have made the point that all of these kinds of practices—and, most interestingly, experimental practices—can indeed be found within the field of mathematics (consider the role of empirical methods in number theory and many other branches of math). Conjectures and hypotheses, too, are no less present in this field than they are in the natural sciences. Nevertheless, it remains true that empirical methods in mathematics are of a different kind than those in the natural sciences: experiment and observation in the sciences depend on physical interactions with objects or phenomena under study, while mathematical experimentation is done only with symbols and diagrams (either on paper or on the computer). A way of clarifying this situation—and one that I favor—is by arguing that *mathematics is an essential part and parcel of the theoretical workings of science*: essential because it is crucial to the processes of concept-formation, model-formulation, and theory-formation, but also distinctive and peculiar, because here the theoretical work becomes autonomous. That is to say, the activity of mathematicians (at least since Archimedes) has typically involved work on problems emerging from the theoretical framework itself, from a concentration on conceptual and theoretical issues *independently* of their potential role as models for physical phenomena.

---

[16] As everyone knows, Quine devoted much of his work to criticizing this dichotomy, which in my view is untenable. See Quine (1951), Putnam (1965), and also Mancosu (2005).

[17] There is a large body of literature on these topics, as the philosophy of scientific practices has also flowered in recent times. I have contributed myself in joint work with J. Ordóñez and in Ferreirós (2009), where I emphasize the three kinds of practices mentioned above, in connection with Newton's optical work.

8. We have to resist the temptation of picturing math as *totally* independent from other activities. This has been common in recent times, partly due to the influence of the ideology of "pure mathematics" in the twentieth century, partly because of older philosophical trends represented by Leibniz and Kant, received and transformed by mathematicians like Gauss, Dedekind, Frege, Brouwer, and others. A typical example—quite representative because of the extreme form of his claims—is the French philosopher Jean Cavaillès, for whom mathematics is a fully autonomous kind of human activity, to the point that it can only be elucidated by doing mathematics; hence the philosophy of mathematics must grow "*a l'intérieur des mathématiques.*"[18] Many works in the philosophy of mathematics follow this orientation, often without the least explicit reflection, which of course helps that trend to reproduce itself dogmatically.

Notice how the proposed approach, by emphasizing the interplay between math and other kinds of practice, has the effect of shifting the celebrated problem of the "applicability" of mathematics.[19] In fact, the terminology of "pure" versus "applied" mathematics is already a bias. From the perspective of my practice-oriented approach, the problem of "applicability" ceases to be posed as external to mathematical knowledge, and becomes internal to its analysis. This is a move toward the dissolution of the problem, since in some respects mathematical frameworks (all mathematical frameworks) are designed to be applicable, and so there is little to explain—that's nothing unreasonable. At the same time, however, articulating a novel application of mathematics can be very difficult, and obtaining from a mathematical model the results one might desire may be impossible—hence the uncooperativeness.

Critics of my views may perhaps argue that such apparent progress by dissolution is only attained at the very high cost of complicating our

---

[18] Cavaillès was reflecting the influence of neokantianism and phenomenology, but also that of the German mathematical tradition that he knew well from Göttingen in the 1920s and 1930s.

[19] Be it the "unreasonable effectiveness" of math (Wigner 1960) or its "unreasonable uncooperativeness" (Wilson 2000). On this topic, see Ferreirós (2013).

analysis of mathematics, and of science in general. I agree that I am complicating those analyses, but in my view this is no disadvantage. We should abandon previous models of the workings of science—which were simplistic to the point of being of little use in understanding real cases—in favor of more complex models that can help us explain and understand.

2

The Web of Practices

What exactly are we to understand by mathematical practice or practices? An obvious answer would be along the lines of "what people do when they are doing mathematics." But such an answer is almost devoid of information, and it will describe different things, depending on who says that someone is doing math, i.e., on the person or community of reference ("doing mathematics" comes down to "said to be doing mathematics"). At one end, some people will say that only research mathematicians *really do* mathematics, hence we should pay attention to them and what they do when doing research;[1] at the other end, some will say that doing mathematics is present even in cultural practices that most of us would never regard as "mathematics" proper.[2] I find something interesting and commendable in each one of these proposals, but obviously when put in the context of epistemology they would point us in very different directions. Indeed, neither of those extreme options seems to be the most suitable one for an epistemological approach.

No doubt the question above is a difficult one, and the lack of a clear and uniform answer is one of the most serious shortcomings in recent philosophy of mathematics. I shall try to remedy this situation, partially at least. Without attempting to provide a rigid "definition" of mathematical practice, I shall work my way toward a serious characterization of that notion (a "definition" of sorts that, while quite inclusive, is not meant to be fully exhaustive, and hence not meant to establish a set of

---

[1] According to Dieudonné, you only qualify as a mathematician if you have published an important theorem. This again does not avoid relativism: who says which theorems are important?

[2] Examples can be musical patterns of basically any kind, or the beautiful shapes drawn on sand by the *malekula* in Oceania (ritual drawings, based on simple patterns that are reiterated and can be drawn without lifting the pen, which are relevant in connection with the myths and rites of this people).

necessary conditions). Such a characterization may help us understand recent work, fitting it into an overall picture, and also point the way toward future developments.

Let us begin by considering the work of historians and philosophers on mathematical practices, and let us try from there to work our way toward a general overview.

2.1. HISTORICAL WORK ON PRACTICES

As in other similar cases, it is wise to consider what people who pay attention to practices are reacting to. They are, in all cases, trying to open up the field of historical or philosophical studies, avoiding what they perceive to be an overly narrow focus on written texts. For reasons that range from relatively obvious facts about the study of history (which has been primarily based on texts) to relatively sophisticated motives having to do with concerns about rigor in mathematics itself and in the study of its foundations, traditional studies of mathematics have indeed been studies of the written text. In the case of the historian, the text in question may be a Babylonian cuneiform tablet such as Plimpton 322, some version of Euclid's *Elements*, Euler's *Introductio in analysin infinitorum* (1748), perhaps the collected works of Élie Cartan, or the books and papers by Alexander Grothendieck. In the case of an expert in foundations, the text will most likely be of a much tighter kind, namely, a formal system formulated in one of the current systems of mathematical logic (section 2).

Many cultures have established their own corpuses of reference texts, paradigmatic works that are admired and studied for generations: the *Elements* and the works of Archimedes in Europe, the *Nine Chapters* [*Jiu-zhang suanshu*] and its commentary by Liu Hui in China, and others. In the twentieth century, the corpus of reference texts became internationalized, and it was essentially formed by "serious," professional contributions to mathematics organized around the presentation of rigorous proofs: Hilbert's *Grundlagen der Geometrie*, van der Waerden's *Moderne Algebra*, Kolmogorov's *Foundations of the Theory of Probability*, and others.[3]

---

[3] See Corry (2008) and the anthology of *Landmark Writings* edited by Grattan-Guinness (2005).

Needless to say, the proof-oriented tradition of mathematical writing has not been universal, and one may feel the need to speak of mathematical knowledge even when the written texts are not organized around the presentation of rigorous proofs.

Historians have long been discussing the need to open up their field of studies, even if one's ultimate purpose is to understand the history of a written text, and even if one is interested only in paradigmatic texts (the quintessential example, for us Westerners at least, being the *Elements*):

A.   Where does the material recorded in the text come from? Not infrequently from oral traditions, from the cross-fertilization of cultures, or from activities that we usually classify under a label different from "mathematics" (such as cult and rites, as in the famous problem of the duplication of the cube; architecture and related pursuits, as is likely to have been the case with the so-called Pythagorean theorem; legal questions and accounting, as encountered in the Arabic tradition; and so on).

B.   How did a relevant text get prepared and perfected? This often requires us to study other related texts that the author may have known, but also available manuscript sources of all kinds, correspondence, information about the activities of the author, e.g., educational or professional activities.

C.   What were the processes that turned it into a reference text? This again requires us to inquire in a much larger context including the community of users of the text, the uses they put it to, how it was received by other authors, what attempts were there to transform or change it, and why the community returned to the original text, if indeed this happened.

All these questions call for the *study of the different activities that form "ecological" or paidological[4] niches for the writing of mathematics*, activities

---

[4] I allow myself here to coin a neologism: "culture" can be translated into Greek παιδεια, from which I derive "paidological." The added connotation of education or educational is actually welcome.

that have an influence on the emergence and the fortunes of written texts. Among these one may include problem traditions of different kinds; cultural pursuits and professional activities that have created the need for mathematical methods; educational, religious, and other institutions; the configuration of the community of practitioners of mathematics, including shifts in the idea of math itself and its cultural meaning; and different types of media employed for the communication of mathematical problems and results. Historians who have paid attention to practices have, in some cases, become interested in recreational mathematics; in ways landmark texts may reflect the influence of "lower" cultural traditions; in the emergence and transformation of styles of writing; etcetera.

A very good example of such concerns is the recent *Oxford Handbook of the History of Mathematics*, edited by Eleanor Robson and Jacqueline Stedall (Oxford University Press, 2009), which may very well serve as a general introduction to the kind of historical literature we are discussing. Incidentally, the book features several main sections entitled "People and Practices", where obvious examples of practices that create the need for mathematical methods appear, such as building, feeding, weaving, calculating, and drawing for all kinds of purposes. The historiography of mathematics in cultural contexts that are relatively distant from us has changed (and, I would say, improved) quite dramatically in recent decades: think, e.g., of ancient Babylonian or Egyptian mathematics, discussed (in the aforementioned book) by Robson and by A. Imhausen. In all these cases, attention to the cultural context from which the text emerged and to other practices that were closely associated with the practice of text-writing have been crucial to the changes.

Perhaps we can take as paradigmatic the work of Jens Høyrup. It is well known that the towering figure of Neugebauer stands behind a well-established tradition of interpreting the Babylonian cuneiform tablets as an extraordinarily forward-looking development of proto-algebraic techniques. Starting from a paper published in 1990, Høyrup has established an alternative interpretation of the tablets, understanding this sort of mathematics not as a theoretical pursuit, but as entirely practical, not as algebra, but linked to a concrete cut-and-paste geometry. In this case, careful attention to philological issues was crucial in offering hints about the referents of the cuneiform symbols and operations, which pointed to drawing practices in the immediate cultural (educational) niche of the "mathematical" tablets.[5] We could give other examples, from ancient China, for instance, showing how important it is to consider the role of instruments—such as the counting rod—that were employed with the text, and the practices of working with text and instrument in combination.[6] (Needless to say, instruments from the abacus to the differential analyzer and the computer have continued to figure prominently in mathematical practices, increasingly so in the age of the computer.)

In this connection, some historians like Chemla talk about different "elements of practice" that ought to be studied in their plurality, in their diverse configurations through time and cultures. Among these "elements" are different kinds of texts (from the canon to the commentary, from the monograph to the recreational narrative, to mention some extremes), figures used in different ways (diagrams, tables, etc.), and instruments such as those mentioned above. Different cultural contexts not only bring different ideas of math and its value, and different configurations of the agents of mathematics, but also peculiar styles of reasoning that may display very significant variety. The role of proofs and of algorithms, for instance, can be present or absent, articulated in very different forms; the emphasis placed on generality and abstraction can also adopt many nuances and sharp differences. Quite naturally, the historian's viewpoint has a tendency to emphasize diversity, to look for multiplicity.

Concurrently with all these new developments, there were other developments in historiography that reinforced the shift. Let me mention some very briefly and schematically. There was the emergence of the history of scientific institutions, prominent from the 1970s, and of the social history of science, which opened the field to the influence of sociology. The sociological perspective has been very influential in emphasizing the role of practitioners and their practices, well above the products (concepts, results, theories) that had been the almost-exclusive concern of historians and philosophers previously. There

[5] See Høyrup (1990); also his books (1994) and (2002).
[6] For example, as with the Chinese counting rod and table beside the written text (sometimes moving inside); see K. Chemla (2005).

was also, more recently, the move in cultural history and the history of science to consider not only "high culture" but also "low culture": the popular literature of science, and not just celebrated monographs; mathematical almanacs; journals for amateurs; and so on. And, finally, there was the more political move toward questioning the "hegemonic narrative" propagated from the West and including in the history of mathematics non-proof-oriented (but clearly mathematical) knowledge that originated in non-Western cultures.

Having mentioned the social history of science and scientific institutions—a topic associated with the issue of scientists as professionals—it is worthwhile adding a few words about the sociological perspective. Sociologists have a very clear conception of practice, at least for the case of modern, highly regulated societies: mathematics is a profession with a clear presence in society and institutions; hence mathematical practice would be what mathematicians do when they *act as* professional mathematicians. Mathematicians are experts in their discipline, officially recognised as such first through degrees (in particular the PhD) and through positions in universities or research institutions. Their acting as such experts of course includes their doing research, teaching, and going to congresses, but also working as heads of research projects, as directors of institutions, and as members of government committees. This scope is clearly wider than formerly, when the focus was on "doing mathematics;" and with recent sociology the implication has often been that one cannot separate neatly (even for advanced research) doing mathematics from acting as a mathematician in society.[7]

2.2. PHILOSOPHERS WORKING ON PRACTICES

Philosophers interested in practice come from a different angle, but their orientation also aims to go beyond the traditional corpus of written texts and their traditional interpretation, to pay attention to the

---

[7] This is often the case, but not always: contrast the case of Bloor and others, who insist on the idea that there is no distinction between the more "political" and the "scientific" aspects, with the work of Bettina Heintz: *Die Innenwelt der Mathematik* (2000). Although the work of sociologists is an important reference point for me, this book is not the place to enter into an explicit discussion.

activities of the community of mathematicians. Since twentieth-century philosophy of math was deeply impacted by foundational studies, the traditional way of conceiving the corpus was through different forms of logical reconstruction of mathematical theories, with a special emphasis on axiomatic theories and formal systems. A traditional approach to, e.g., set theory would focus on some axiomatic, formalized version of the theory—say the Zermelo-Fraenkel ZFC axiom system formulated in first-order logic—and ask a series of questions about its ontology and epistemology (e.g., how is it possible for us to come to know the axioms as true? What is the justification for the axioms? Or, to what extent is a certain conceptual account, e.g., the iterative conception of sets, successful in motivating the axiomatic system?).[8]

In connection with this, an exemplary case of the practical turn is the philosophy of Penelope Maddy—exemplary because of the neat way in which it associates with that tradition, and at the same time moves away from it. Maddy chose to concentrate on one of the topics that were regarded as crucial by the community of experts in set theory: propositions that were candidates to becoming axioms, such as the Continuum Hypothesis, large cardinal axioms, or determinacy principles. In due course, her attention was focused on the project of clarifying the methodologies that, in actual practice, guide and have guided the community of set theorists in their choice of some axioms (the axioms of Choice and Replacement), their reluctance to adopt others (e.g., the Continuum Hypothesis and its alternatives), and their rejection of still others (such as the axiom of Constructibility). Justifying or "believing" the axioms became mostly a matter of the methodologies applied in the practice of set theory, with a distinction between extrinsic or "regressive" methods and intrinsic or "intuitive" reasons (a distinction that goes back to Gödel and a few others).

The shift should be clear. Other philosophical work takes a static picture of a mathematical theory as the excuse or motivation for posing anew some traditional questions about ontology or epistemology; but the main aim of Maddy's work is to clarify some methodological issues that emerge in the research practice of a certain community of

[8] A fine example of such work is Potter (2004). We shall come back to this topic in the last chapter.

mathematicians. That makes it necessary to go beyond the canonical text, which in this case would be the formalized axiom system ZFC, because 1) the formal system is already the outcome of the methodological choices that one aims to investigate, and 2) a dynamic element enters the picture because the goal is to add new axioms to ZFC in a reasonable, principled way. At the same time, the way in which the formal system is employed by set theorists—adding to it some propositions, obtaining metatheoretical results from such enlarged systems about other simpler systems—is part of the object of investigation. Maddy's work is in clear continuity with traditional studies, but the *practical shift* is evident; the philosopher is not aiming to contribute to foundational studies, but to a methodological analysis of the practice of the community of experts in set theory. Faithfulness to actual practice, both presently and historically, is a key requirement for this kind of work.[9] In fact, although the main purpose is to illuminate current practice, the enterprise itself forces the philosopher to engage in careful study of the historical antecedents of that practice.

A central element—perhaps the defining element—in the philosopher's conception of his or her canonical texts has been the *idea of rigorous proof.* Clarifying what proof and rigor meant was the task of logical analysis, which led to the twentieth-century identification of "rigor" with respect for the rules of a logical calculus, and "proof" with certain sequences of strings of symbols in formal systems. Interestingly enough, this understanding had the effect of creating tensions that have moved the enterprise into the arena of practices. On the one hand, it was clear that mathematicians in the past did not live by this ideal of rigor (they didn't even know it), as was the case not only in the notorious example of Leonhard Euler but also in that of Euclid's *Elements,* which for centuries was regarded as the paradigm of rigor. On the other, for those who were conversant with the actual practice of mathematics, it was clear that there is much more to a proof than merely establishing step by step that a certain proposition follows from some given ones, in a perfectly unobjectionable way. The interest of practicing mathematicians in multiple proofs of one and the

---

[9] The "faithfulness constraint" has also been emphasized by S. Shapiro (1997).

same result, for instance,[10] is evidence that there is more methodology behind the conduct of proofs than a narrow-minded logical analysis reveals.

Those two observations make it possible to understand many of the concerns of recent philosophy of mathematics. Philosophers have acknowledged the limits of logical analysis in understanding mathematical knowledge, and they have set out to introduce new types of theoretical discourse that complement or supplement logic. Examples are in-depth studies of the conduct of proof in cases that clearly violate twentieth-century standards, as is the case of Euclid's *Elements*, where the diagrams play an essential role in proofs; or analyses of value judgments in mathematics, especially as they apply to the evaluation of proofs and theories. Thus, K. Manders has developed a theoretical model of Euclidean diagram-based proof (see his contributions to Mancosu 2008, and Chapter 5). Other authors have paid attention to instances when mathematicians value a certain proof as being more "explanatory" than others (Mancosu), to evaluations of concepts and theories according to their "fruitfulness" (Tappenden), or to the great efforts that some mathematicians have made to attain "purity of method" in the developments of certain results or theories (Hallett, Detlefsen).[11] The traditional concern with the idea of rigorous proof also explains why philosophers have been excited by the emergence of computer-assisted proofs, especially in cases when the proof cannot be checked by human experts.[12]

---

[10] Carl Gauss, a giant of modern mathematics, is a very clear example: during his lifetime he published four different proofs of the so-called Fundamental Theorem of Algebra (that a polynomial of nth degree has n complex roots) and six different proofs of the quadratic and higher reciprocity laws in number theory (see Lemmermeyer 2000 and Goldstein and Schappacher 2006). For an elementary example, there are more than 200 different proofs of the Pythagorean theorem (see Maor 2007).

[11] All of the above are covered in the recent volume Mancosu (2008), which has chapters devoted to Visualization (by Marcus Giaquinto), Diagrammatic Reasoning and Representational Systems (Kenneth Manders), Explanation (Paolo Mancosu and Johannes Hafner), Purity of Methods (Mic Detlefsen and Michael Hallett), Mathematical Concepts (James Tappenden), the Philosophical Relevance of Category Theory (Colin McLarty), the Philosophical Relevance of Computers in Mathematics (Jeremy Avigad), and the Philosophical Relevance of the Interaction between Mathematical Physics and Pure Mathematics (Alasdair Urquhart).

[12] A notorious recent example is Thomas Hales's proof of the Kepler conjecture about sphere packing in three-dimensional space.

Another line of development has been the rise of interest in the cognitive resources that go into knowledge of mathematics. This is an active area of research on the side of experimental cognitive science, although its integration with philosophy of mathematics is still problematic. A careful attempt to achieve this integration is provided by the work of Marcus Giaquinto (2007) on the role and the epistemic reach, so to speak, of visual thinking. The case is very interesting, because the rise of modern mathematics around 1900 (with careful reconstructions of algebra, analysis, and even geometry along the lines first of so-called "arithmetization" and then of set-theoretic approaches, achieving "full rigor," as it was then said, at the price of eliminating intuition) and the subsequent development of foundational studies have both led to an anti-Kantian orientation that reduced visual thinking to a merely heuristic role, emphasizing its lack of rigor. And yet, visual thinking is clearly important in the practice of mathematics—an already mentioned, notorious example is diagram-based proofs; it is all the more important in the early stages of development of mathematical knowledge, hence also in math education.

Finally, and returning to the topic of historical changes in standards of rigor, historically minded philosophers of mathematics have acknowledged the multiplicity of traditions leading to proofs, and have tried to analyze their relations. This tendency resonates with the themes of moving away from foundationalism and reductionism. A prominent example was the work of Kitcher (1984) on patterns of change in the development of mathematics. His analysis was based on what he called a "mathematical practice," that is to say, a quintuple $<L, M, S, R, Q>$ formed by the set of statements S, methods of reasoning R, and questions Q, stated in a given language L, which (together with certain "metamathematical" views M about rigor, the "map" of the mathematical disciplines, and so on) regulate the discipline of mathematics at a given time. Kitcher's approach enabled him to provide an interesting analysis of the state of mathematics at a given time, to work toward a classification of patterns of change, and to provide arguments about the rationality and progressiveness of such changes.

Let me add a few more words on Kitcher's *The Nature of Mathematical Knowledge* (1984), which I use below to present—in contrast—some of the basic ideas I shall be employing. Kitcher's work is certainly commendable, and when I read it long time ago (probably in 1995) there was an immediate feeling of intellectual resonance. There are interesting similarities in the way Kitcher and I approach the study of mathematical knowledge: he is a pragmatist, his approach is avowedly historical, and he also pays attention to cognitive issues; although he focuses on rationality and the word "objectivity" does not feature prominently in his book, significant features of his account anticipate the one offered here. A detailed discussion would reveal subtle differences which, in my view, are highly relevant to the success (or failure) of the enterprise. To give an example, Kitcher's account of elementary mathematics arising from concrete interactions with the world has obvious similarities with mine; but his empiricist, Millian version of the story is quite different from the more cognitive one I propose (which perhaps is, roughly speaking, "Piagetian" rather than Millian—or so I thought back in 1995). As a result, his account seems to relegate the concrete interactions with the world to "perceptual experiences" in the distant past, "deep in prehistory" (1988, 321 note 10), while in my view the strong cognitive roots of basic arithmetic are manipulatory as much as perceptual, and the story is reenacted in each generation and each individual (see Chapters 3 and 7). To give another example, Kitcher's attention is focused on "inter-practice transitions" (1988, 299; 1984 passim), while my main focus in on the interactions between coexisting practices, on their *interplay*.

It would actually be of great interest to continue discussing the similarities and differences between Kitcher's work and the present one, but I remain content with inviting readers to engage in such comparisons themselves. My main goal being to explore in some depth how far a certain approach to mathematical knowledge can go, most of my references to other authors are simply meant for preliminary orientation only, so detailed discussion of their ideas would be a distraction from the narrow purposes of this volume. The same applies to Kuhn's views, which greatly interest me, and which I know I am perhaps oversimplifying in the telegraphic comments offered here. For the same reason, I shall not even mention other works that have significant points of contact with mine and which could profitably be discussed if time and space allowed.

2.3. WHAT IS MATHEMATICAL PRACTICE, THEN?

Mathematical practice is a wider topic than "mathematical practices" in Kitcher's technical sense. His quintuples <*L, M, S, R, Q*> are relevant for understanding the practice of mathematics in a given context and time, but they remain very far from a reasonably complete analysis of the factors that regulate such practice. Kitcher's "practices" could also have been termed mathematical *frameworks* (an expression that I prefer), or "research programmes" following Lakatos, or even "paradigms" to borrow Kuhn's term. We shall come back to this question in Chapter 3, where I shall propose two specific notions of framework. Meanwhile and provisorily, in the rest of this chapter, readers who are conversant with Kitcher's ideas may take a framework to mean a quadruple <*L, S, R, Q*>, leaving out the "metamathematical" *M* (see Chapter 3 for the rationale of this exclusion).

Let me approach a reasonably broad, albeit flexible, characterization of the notion of mathematical practice by successively proposing several constraints.

First, reference to the community of mathematicians is essential, for a variety of reasons that we shall keep analyzing in Chapter 3 and subsequent ones. But one can state briefly the main reason to move beyond Kitcher's rather abstract frameworks: there is no practice without practitioners. Any reasonable analysis of the factors that regulate mathematical practice, making mathematical knowledge possible, must include the resources and abilities of the single mathematician, and his or her interaction with others in the community. As mentioned already in Chapter 1, nothing is gained by trying to study epistemology without a subject—better, without an agent, or, in fact, a community of agents. In all that follows, the role of the human agent will be assumed to be at the center.

By a "practice" in general I mean a recognizable type of activity that is done—and can be *taught* and *learned*—by human agents. We always associate mastery of a practice with one or several particular ways of solving some problems, and we only acknowledge that somebody knows a practice when some *success conditions* are met. In many cases there will be explicit norms and rules associated with a practice, but in *all* cases there are *implicit* norms: think of the practice of riding a bike, where we have explicit rules in the form of traffic regulations, as we have rules for competitions; but, even at the most basic level, there are success conditions and associated norms (the ability to maintain control and balance, to brake at or negotiate a curve, and so on).[13] In relation to such implicit norms, other agents typically act as regulators—the community is most relevant to an individual's attainment of proficiency; if you want to learn how to dance, or how to sail a yacht, you become an apprentice to those who already know. This is the way mathematics was learned back in time, e.g., around 1400.

A second requirement has been implicit in all that has preceded. Let me formulate it again as a criticism: attempts such as those of Kuhn, Lakatos, or Kitcher to establish a historical philosophy of science have been accompanied by the tendency to propose an all-encompassing notion (respectively, paradigm, research programme, "practice") that subsumes all of the knowledge shared by a certain community. This was the main reason why Kuhn ruined (in his own view) the perceptive idea of a paradigm, turning it into a kind of Hegelian zeitgeist (see the postscriptum to Kuhn 1970). One can hardly overemphasize the importance of avoiding such temptations to erect all-encompassing frameworks that supposedly would reign supreme over the conduct of science.

Our mathematical practices will be at a lower level of generality, and they will be linked with a recognizable type of symbolic framework (formulas or diagrams of a certain kind) employed by the agents to solve problems in the pursuit of particular goals. In retrospect, it will always be possible to link such mathematical practices with particular mathematical theories (given that we, today, possess a large body of mathematics). But, in fact, the natural thing is to keep separate, at the level of analysis, the practice of mathematics and its products. Attention to *processes*, and not merely to more or less perfect results, is characteristic of studies of mathematical practice. As for the results, we shall of course

---

[13] The topic of implicit norms has recently been discussed in influential ways by R. Brandom. You may want to consider other practices that are close to your own experience, say the participating in some kind of sport, the playing of a musical instrument, or the practice of teaching (in all its complexity, from the first day of classes to the final exam and grading); or perhaps practices associated with work (at a hospital, for instance), or laboratory practice, and so on.

consider problem solutions, theorems, etc.; but in order to be more specific about this point, it is necessary to discuss the issue of *goals*.

We are trying to elaborate on a notion of practice that is helpful in understanding mathematical knowledge, i.e., in developing an epistemology of mathematics. This particular objective recommends a certain understanding of the nature of mathematical practices oriented to the knowledge they produce, namely a bias toward know-what rather than know-how at the level of goals. Consider again the idea that "mathematics" is to some extent present in a great diversity of cultural productions, including the sand drawings of the *malekula* and the geometric patterns of artisans in many cultures; a noteworthy, sophisticated example is the presence of (not all but) most possible planar symmetry groups in the tilings employed by Arabic artists to decorate the Alhambra in Granada. It may be likely that some of those artists possessed knowledge of plane (Euclidean) geometry, but they did not have knowledge of group theory and the seventeen types of periodic symmetry groups in the plane.[14] It is perfectly possible to explore empirically diverse possibilities of covering the plane with symmetric patterns that are repeated periodically, and to retain noteworthy examples of them. Perhaps the Arabic artists did that with the help of some theoretical knowledge of Euclidean geometry. But the end result is significant practical knowledge: knowledge of the know-how, which can be transmitted without any theoretical underpinnings.

Mathematics, as we understand it, is not just practical knowledge but has at its core theoretical knowledge (we elaborate on this in Chapter 5). Paradigmatic examples of this shift are the transition from knowledge of particular instances of the Pythagorean theorem to attempts to prove the general result, or the transition from particular approximations of the diameter/circumference ratio to the notion of a precise value of π (compare, for instance, the Old Testament with Archimedes). Notice that the particular examples of Pythagoras and the concrete approximations to π are certainly more useful and important, with know-how mainly in view, than their "contemplative"

---

[14] For a recent discussion, see B. Grünbaum (2006). Grünbaum's answer is that only thirteen groups are present, and beware the criteria you use while counting; most importantly, beware the conclusion that groups of symmetry were relevant to the artisans!

counterparts: nobody needs thirty-five decimal places in the expansion of π for practical purposes! Our third requirement shall be that mathematical practices incorporate a theoretical orientation, which is characteristically linked with attention to some epistemic values (such as accuracy or precision, consistency, simplicity) and with the goals of the practice.[15]

Though other approaches may be acceptable for purposes other than epistemology, here we shall consider problems and proofs and theories as central elements of mathematics. There have been mathematical traditions that articulated their methods for solving problems in ways that do not comply with the Western ideals of proof (ancient or modern). My aim is not at all to deny this fact, nor to diminish the importance of providing a historical and philosophical analysis of the forms of knowledge developed by such traditions.[16] But the task of understanding mathematical practices and mathematical knowledge in the Occidental tradition is already complicated enough for a philosopher. This is the simple, pragmatic reason behind my choice. Notice also that we shall be interested in the paidological niches of problems and proofs and theories, which will turn our attention to other elements of practice.

Traditionally, philosophers of science have focused on theories, and in our field the focus has been on axiomatized mathematical theories, organized internally into a proof structure based on explicit principles, the axioms. Traditionally, theories are conceived as static, ready-made abstract objects. From the standpoint of practice, however, theories are one (and only one) of the products created by the mathematical community; and any given theory is in a dynamic process of reconception and changes in presentation. Other highly relevant products put forward by mathematicians are problems, conjectures, and methods. Hilbert, more than a hundred years ago, had already emphasized the immense importance of *problems* in the life of any branch of mathematics and in

---

[15] Arguably, this theoretical orientation is not exclusive to "Western" mathematics. Liu Hui's work, despite its great differences with classical Greek math, can be argued to have the crucial epistemic values (see the numerous papers of Chemla devoted to this topic, and the French edition of the *Nine Chapters* and Liu's commentary by Karine Chemla and Guo Shuchun, 2005).

[16] I refer to the work of K. Chemla, J. Høyrup, and others.

the conformation of mathematical knowledge.[17] Much more recently, Mazur (1997) called attention to the central role of *conjectures* in guiding and structuring research, especially in the twentieth century. As regards *methods*, it is the case that, very often, a theorem is more important because of the method it presents us with (which can then be applied in other related situations, or generalized) than by the relevance of its statement.

An analysis of mathematical knowledge must consider such products, too, problems, methods, and conjectures, alongside proofs and theories. Indeed, the question of how problems are posed already belongs in the analysis of mathematical practice. (Cross-referencing this topic with a previous one, notice that the emergence of theoretical problems is an obscurer issue than the emergence of practical problems.) But, for our limited purposes, one can remain quite traditional and consider problem solving, theorem proving, and theory shaping as the most important goals of mathematical activity. Conjectures and methods can be made sense of as revolving around those key goals: once a problem is posed, conjectures may start popping up, and some of these have a great importance in articulating further work and research strategies; once a problem is solved or a theorem proven, certain methods are set in place that may become salient, be further applied in other cases, or be generalized; and once certain methods are adopted or regarded as the most adequate, theories are reshaped accordingly.

What I have just called *theory shaping* is a very important activity, which has become more and more so in the process of the emergence of modern mathematics. The goal here is to select those methods that are judged most appropriate and relevant for a given subject matter, and to rework systematically a theoretical corpus so as to develop it in accordance with that approach. The theory will be reconceived and the proofs of theorems will be fine-tuned accordingly; often the process will lead to decisions about the axiomatic basis on which the theory is to be erected, about the assumptions that are decisive for establishing

---

[17] Cantor stated on the occasion of his doctorate: "In re mathematica, ars proponendi quaestionem pluris facienda est quam solvendi," i.e., in mathematics, the art of posing questions is more relevant than that of solving them.

what one regards as the central results in the field.[18] Value judgments are of the utmost importance in such cases—judgments about the most adequate conceptual framework; about the "purity" of the methods employed with respect to the subject matter; about which methods or ways of proving theorems are more fruitful, more explanatory; and so on (see section 2 for references). Historical examples abound, because it is often the case that those mathematical texts that are regarded as canonical have been so judged for the way in which they reshaped a certain theory. This is what Gauss's *Disquisitiones arithmeticae* (1801) did for number theory, what Hilbert's *Grundlagen der Geometrie* (1899) did for pure geometry.[19] A list of great theory shapers in the last two centuries is likely to coincide to a large extent with the list of most admired mathematicians: Gauss, Cauchy, Riemann, Dedekind, Hilbert, Weyl, Kolmogorov, the Bourbaki group, Grothendieck, etcetera.

We have suggested constraints associated with the role of agents, with the middle level at which we shall locate practices (in connection with symbolic frameworks and theories), with our interest in know-what (not just practical knowledge), and with the main goals of practice that will be our concern. It is now time to summarize. For our limited purposes of an elementary epistemological analysis,

mathematical practice is what the community of mathematicians do when they employ resources such as frameworks (and other instruments) on the basis of their cognitive abilities to solve problems, prove theorems, shape theories, and (sometimes) *to elaborate new frameworks.*

---

[18] An example is the question concerning the right combination of algebraic and topological assumptions to be used in a proof of the "fundamental theorem of algebra"—which, by the way, may be judged to belong more properly to the theory of holomorphic functions in complex analysis (the traditional name reflects the eighteenth century understanding of "algebra" and is rather foreign to more modern views). It is generally believed that there can be no purely algebraic proof of the Fundamental Theorem of Algebra, essentially because the field $\mathbb{R}$ of real numbers is a topological construct (a feature inherited by its algebraic completion, the field $\mathbb{C}$ of complex numbers).

[19] Consider also van der Waerden's *Moderne Algebra* (1930, based on lectures by Artin and E. Noether) or Alexandroff and Hopf's *Topologie* (1935, based partly on Alexandroff's own ideas, partly on the views of Brouwer and Alexander). For details concerning these and many other key treatises in the period 1640 to 1940, see *Landmark Writings in Western Mathematics*, edited by Grattan-Guinness (2005).

Notice that we have not required a focus on mathematical innovation or research, the elaboration of new problems, new results. I will intentionally *not* adopt the position that only research mathematicians do mathematics properly, and we shall be interested in how individuals attain knowledge of mathematics. Doing research and developing new mathematics is certainly of enormous interest, but we are interested also in how aspiring members of the community learn to do math and mathematical research, how they obtain their professional know-how. Thus, we are interested in the educational level, which is centered not on novel production but rather on the reproduction of mathematics, in developing knowledge and understanding among students—and in the reproduction of the community of mathematicians. We should also be interested in mathematicians who will not become researchers, and students in other communities who must understand and use math, such as physicists, engineers, economists, etc.

Almost all of the present work will be devoted to studying basic features of mathematical knowledge, and for this purpose it has seemed advisable to avoid excessive complications. Hence I shall be less interested in questions that are central for historians, even if in my view they clearly have an epistemological import: detailed analysis of paidological niches (to use the expression from section 1), of the activities that give origin to math, or of the processes of reception and appropriation of texts. After all, despite my training as a historian, and even though my conception combines cognitive, historical, and pragmatist traits (see Chapter 1), this book is about epistemology.

2.4. THE MULTIPLICITY OF PRACTICES

Kitcher's work on patterns of change in the development of mathematics was based on two ideas: first, the history of mathematics shows periods in which work and research is dominated by a given framework; second, the transition from a given framework to another is a response to innovations that cause instabilities or tensions among the elements of the quintuple $<L, M, S, R, Q>$ (see section 2.2), and such "interpractice" transitions can be seen to be rational. As already remarked, this approach was influenced by work on models of scientific change (Kuhn,

Lakatos), and indeed there is a very Kuhnian element in it, namely the assumption that the growth of mathematics passes through "normal" periods (reminiscent of Kuhn's "normal science" within a scientific discipline) when a given framework dominates the work.[20] In my view, this Kuhnian element of Kitcher's analysis is misguided, and even makes it impossible for his approach to address the key epistemological issues.[21]

It is a key thesis of the approach defended here that several different levels of practice and knowledge are coexistent, and that their links and interplay are crucial to mathematical knowledge. This becomes especially obvious when one clears one's eyes from the foundationalist reductionism that was a frequent feature of twentieth-century philosophy of mathematics. The purpose of this whole book is to present and develop these ideas, but perhaps it will be good to introduce them here briefly.

Let me give an example in some detail. In the course of a discussion of foundational views about math, and from a greatly simplifying perspective, one may argue that the period 1750–1950 saw the shift from mathematics understood as a theory of magnitudes to the mathematics of set-theoretic structures. In 1800 it was usual to define mathematics as "nothing more than a *science of magnitudes*" (Euler 1771, 15), a definition that immediately points to the real world and the so-called applications. Functions and even numbers were conceived as *relations* between magnitudes; their "existence" was derivative from the obvious, real-world existence of magnitudes such as physical lengths, times, and volumes. The "foundations" of mathematics was not a topic for mathematics—rather

---

[20] See, e.g., Kitcher (1988), 299–300, where the presentation suggests that geometry, algebra, and the Newtonian calculus form "the practice" (a single one) of the British community around 1700. This text at least suggests that the community of Leibnizians had a different "practice" from that of Newtonians, but—surprisingly—I do not find such a suggestion in Kitcher's book (1984, Chapter 10), which thus promoted even more the idea of an all-embracing "practice." It's probably unnecessary to add that Kitcher did not adopt other traits of Kuhn's work, especially those having to do with his understanding of revolutionary changes: where Kuhn emphasized Gestalt shifts and "incommensurability" in natural science, Kitcher saw rational transitions in math.

[21] Also in the history of the natural sciences, "normality" is less common than Kuhn argued, as evidenced by the phenomenon of research schools. Kuhn himself was well aware of the history of electromagnetism in the nineteenth century, which was an obvious counterexample to his account due to the coexistence for several decades of the "paradigms" of Weber, Maxwell, and Helmholtz.

for philosophy or "metaphysics," as was commonly said. Mathematics in this period focused on forms or formulas, explicit analytic or algebraic expressions and methods for their resolution.

In contrast with this rather empiricistic conception of mathematics, the modern conception appears to be idealistic.[22] In 1950 it was usual to emphasize that "all mathematical theories can be considered as extensions of the general theory of sets" ("as every one knows," Bourbaki 1949, 7), a definition that does not point to the real world, but to an inner world of "pure mathematics." Functions and numbers were now set-theoretic constructs, and their "existence" derived from the basic existential assumptions embodied in the axioms of set theory—on the understanding that mathematical existence has nothing to do with real existence.[23] Foundations of mathematics had become an important branch of mathematics (encompassing logic, set theory, proof theory) and it was conceived to be strictly mathematical and fully independent from philosophy. Mathematics in this period focused on *structures*, complex relational systems that can be characterized by morphisms.

And yet, in spite of these profound changes which some have called a rebirth of the discipline, despite all the foundational changes and the reshaping of theories, many mathematical practices remained largely the same. Take as clear examples a whole series of rules for the resolution of algebraic equations (regardless of the epochal change in algebra from a theory of equations to the modern theory of structures), for the finding of roots by approximation, or the integration and differentiation of functions (notwithstanding the famous rigorization of the calculus). Further obvious examples are the elementary practices having to do with the natural and the rational numbers, which have not changed much since the Renaissance. In spite of deep conceptional changes that affected the rationals (from ratios between commensurable magnitudes, they became equivalence classes of pairs of integers),

---

[22] The terminology has been employed in this connection by Du Bois-Reymond (1882) and others around 1900; much more recently, one can find it again in W. Lawvere.

[23] The idea was central to Hilbert, but already Frege had stated that the whole question revolves around acknowledging "a domain of what is objective and non-actual" (*sic*; Frege 1893, xviii). J. Tappenden has called my attention to Frege's terminological choice of "actual" (*wirklich*) due to its relationship with "to act" (*wirken*), "non-actual" being what does not act on our senses.

the elementary practices of representation and calculation with those numbers remained quite the same.

Acquaintance with these and many other practices was—and continues to be—an essential part of the process of becoming a mathematician, of gaining mathematical knowledge. This implies that, in any historical period, many different mathematical practices coexist, and the same applies to different strata of mathematical knowledge. It seems that the "normal science" approach followed by Kitcher was still heavily influenced by the typical reductionism of foundational studies and traditional philosophers.

One can look at this matter from another angle. It is very different to work with geometrical theories, or in the fields of algebra, or in areas of analysis—the problems, proofs, and methods are very different, including from theory to theory, within each one of those broad subdisciplines; even methodological questions that may interest a philosopher will change in moving from one to another. A reductionistic approach that simplifies all this complexity, turning everything into developments of a single system or framework, will not only flatten the interesting complexity of mathematical methodology, but also make it impossible to understand the process of building up mathematical knowledge. Let me emphasize again that the approach I am presenting is intended to avoid reductionism and excessive drive toward systematicity—a Fregean, Quinean, and more broadly foundationalist drive that actually gets in the way of understanding mathematical knowledge.

Those examples should also suffice for making it clear that mathematical practices coexist not only historically in a given period, but also within each agent, the individual mathematician. We do possess working knowledge (better or worse) of elementary number theory, of calculation with fractions, of so-called Cartesian geometry with its algebraic representations of figures[24] (also, in some cases, of Euclidean geometry with its diagrammatic proofs), of the elementary calculus, of set theory, and so on.

In all cases I can think of, the analysis of a certain mathematical framework (of its concepts and symbols, statements, methods of reasoning, and

---

[24] I do not mean here the concrete historical practice of Descartes and his followers, but rather the more modern developments referenced by his name.

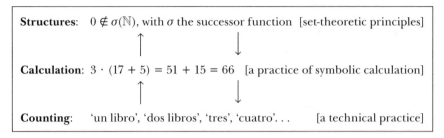

Figure 1 A schematic picture of interrelated practices. In each case, only a representative instance is given, out of indefinitely many.

problems) requires us, at the very least, to consider its connections with another level of practice (or more than one), its links with other frameworks and knowledge strata. This is particularly the case when the framework under analysis is not a fully developed formal system, because only formal systems have been designed to "stand alone," so to speak. Consider a very simplified representation of the practices that a modern mathematician associates with his or her understanding of the number concept:

In the pedagogical evolution of an individual, these different practices and ideas come to be known at very different times: counting is learned at home in childhood, calculation is studied in elementary school, and the structure of the number system is eventually studied at the university. Diachronically, counting has emerged time and again in different cultures, calculation has been approached employing very different practices (the abacus, Greek *logistikē*, the Hindu-Arabic numerals and their algorithms), and structural analysis emerged with so-called modern mathematics as one of several alternative ways of theoretically approaching the science of numbers. But the most important aspect is that knowledge of a new, higher practice—or of more advanced knowledge—does not in the least eliminate the previous ones, even when from a theoretical point of view one may consider the higher practice as "more fundamental" and regard the lower practices as derived from it.[25]

---

[25] How this applies to counting is a problematic issue, and in fact a celebrated problem that was central to Frege's thought. Arguably, from an epistemological point of view, acquaintance with the lower practice is a prerequisite for learning the higher ones (see the next section).

The thesis, then, is that we have *working knowledge of several different practices and strata of knowledge, together with their systematic interconnections.* This causes links that restrict the admissible—for instance, when a new framework is being developed—and that are responsible for much of the objectivity of mathematical results and developments. The interplay of practices acts as a constraint and a guide. As we shall see, it guides the formation of new concepts, the adoption of new principles, and the configuration of new frameworks. It forces constraints upon new mathematical developments, so that innovation can hardly ever be arbitrary.

2.5. THE INTERPLAY OF PRACTICES AND ITS BASIS

The scheme of a web of interrelated practices, with their systematic links acting as constraint and guide, can be applied in the analysis of very different levels of mathematical activity. Consider the following series of interrelated practices and frameworks:

A. Counting practices   / reckoning arithmetic / structure of $\mathbb{N}$[26]
B. Measuring practices / fraction arithmetic / proportion theory
C. Practical geometry / Euclidean geometry / Cartesian geometry[27]

The first one was mentioned in the previous section, and includes some crucial steps in the evolution of the concept of natural number (without any attempt to cover all the historical steps, and without bringing the story to the present day either). The series B has to do with the emergence of the sophisticated theory of proportions in antiquity (attributed to Eudoxos), and can be revisited as the initial evolution of the concept of real number in light of the seventeenth-century definition of numbers as proportions (offered by Newton and others). Series C includes some crucial steps in the evolution of

[26] An interesting variation of series A is suggested by the *Elements*: take the third practice to be the geometric study of number theory, as in Euclid's books VII to IX.

[27] Here, in view of what I say below, it is best to interpret that we are referring to Descartes's himself.

classical geometry, from antiquity to the seventeenth century (again with no attempt to be complete). My interest here is not so much in calling attention to important historical developments, as, rather, in choosing relevant examples taken from some of the core notions of mathematics. Present-day exposure to practices and frameworks has not changed so deeply from earlier forms of education—at least some of the mentioned practices are preserved.

According to the thesis of the multiplicity and the interplay of practices, in order to obtain knowledge of any of the frameworks that lie to the right-hand side, one needs (or is required) to be conversant with the practices that lie to the left. You may think that fraction arithmetic is an exception, since it is perfectly possible to present the rules for calculating with fractions in a formal way, without establishing its links with measuring. This is true, but it makes it all the more significant that, at all times, the connection between fractions and measures has been given not only as motivation, but as an aid to forming a representation of what fractions are (their meaning and use).

The whole history of mathematics can be presented as the gradual development of a network of links connecting different core notions that initially lie separate, i.e., as the creation of a delicate tapestry or spider web establishing bridges between the discrete and the continuum, number and geometry, algebra and analysis—and, later, topology—etcetera. The process has been aided by moves of generalization, by steps of identification of significant hybrid notions (*mixtes* as some have called them), by the introduction of new hypotheses, and so on. Arguably, the case of the real numbers can be regarded as the quintessential example of such linking activities, even if it may appear to be too elementary for the modern eye. In the present work, we shall consider several examples, at different levels of complexity. Let me now consider some of the simplest cases in order to make a couple of remarks.

2.5.1

In those simple cases, such as A, B, C, the connections with a technical practice (left-hand side) appear to be crucial. By a *technical* practice I mean a recognizable type of activity that is done—and can be taught and learned—by human agents, involving direct manipulation of objects in the world, through the use of human-made instruments. Consider, for instance, the practices of measuring times by means of different kinds of clocks. By "measuring" as a technical practice I mean *concrete* measuring with conventional units—such the measuring of distances in feet or cubits—and not the abstract structures that one may recognize behind that; and by the drawing of geometrical shapes I mean the kind of practical technique that was taught to masters and apprentices in guilds, employing ruler and set square and compass, a practice that was presented, e.g., by Dürer in the sixteenth century.[28]

Now, the cases of measuring with the ruler and other instruments (e.g., for measuring volumes or quantities of grain) and of drawing geometricals with the compass, the ruler, and the set square (in practical geometry) are so paradigmatically technical that they do not require further comment. However, the claim that counting is also a technical practice may require some explanation: what is the human-made instrument here?

Consider counting as we teach it to little kids in Western societies: there is a coordination of body movements and objects (more generally, perceived phenomena) through the vehicle of oral language, by employing the spoken numerals: the child recognizes (let's say) books, points at them one by one, and produces the numerals "uno, dos, tres . . . ," one by one, while simultaneously pointing. My claims are, first, that this kind of practice is too basic and mundane, too widespread and important anthropologically, to be considered a part of mathematics proper—it is a basic cultural technique; and, second, that the spoken numerals and their correct coordination with the remaining elements (the conditions for which are indeed complex and too lengthy to elucidate here, but guarantee that the outcome of the counting process will be unique) is here the human-made instrument.

The more ancient practices of employing tallies cut on a stick (or body parts, as in Papua,[29] or a bundle of pebbles, *calculi*) instead of words come in handy for establishing the link between the spoken

---

[28] A. Dürer, *Instruction in measuring with compass and straightedge* or the *Four books on measurement* (1525). This is not a book of deductive geometry, but a somewhat systematic exposition of methods of graphical solution for practical geometrical problems emerging from the arts and crafts.

[29] See Lancy's classic study, Chapter 5 of Lancy (1983).

numerals (properly used) and more typical kinds of instruments. In our culture, oral language is playing the role that sticks, cutters, and tallies (or different body parts) played in other counting practices. Counting is a most peculiar way of doing things with words! Linguists have emphasized that the numerals form a very special part of language, and this already suggests their peculiar nature as special instruments.[30]

The idea is that, at the basis of mathematical knowledge and understanding, there lie other practices that are not mathematical, properly speaking: the techniques (technical practices) of counting, measuring, and drawing geometrical forms. This would seem to imply that the cognitive abilities and resources that are necessary for such practices will be found necessary for doing mathematics.[31] This idea is, quite obviously, compatible with (indeed natural for) the pragmatist standpoint that I have adopted, as mentioned in Chapter 1. But if we read it conversely, it amounts to a thesis of *irreducibility of mathematical knowledge to the level of purely symbolic systems*. Here, I have to stress the word *knowledge*, since I am not saying that mathematical frameworks cannot be reduced to formal symbolic systems: my claim is about knowledge. The formal systems only come to life, so to speak, when they are interpreted in connection with a network of practices—only then can they be said to incorporate or codify knowledge. This standpoint may be considered radical, but as we shall see along the way, it makes it possible to understand a number of paradoxical phenomena.

2.5.2

The role of technical practices as poles of reference for the emergence and the transmission of mathematical practices was taken on, in the course of historical development, by scientific practices. Consider, e.g., the case of astronomical models of planetary motion, from Eudoxian

---

[30] Cf. Hurford 1987. For instance, the numerals are much more resistant to change than other words.

[31] In cases when some impairment, e.g., blindness, makes it difficult to participate in some of those practices, an individual's ability to develop mathematical knowledge will depend on her ability to compensate for the missing faculties. For instance, the blind are no less capable of counting objects they touch, or processing the sounds they hear; they are also able to develop a practical conception of measuring; and so on.

models (concentric rotating spheres that carry each other along) to Newtonian ones (conic trajectories due to inverse-square forces of attraction); undoubtedly, these scientific models and their theoretical study have stimulated the development of geometry and analysis. But this is not to mean that such practices play the same crucial, motivating, and cognitive roles that the above-mentioned technical practices do. However, it is still possible that some scientific practices do play a central role. Perhaps the most natural candidates are certain modeling practices in mathematical physics, which are related with the concept of a function; nevertheless, this is a topic that I shall not pursue in the present work. This becomes an empirical question that might be studied in future work in practice-based philosophy of mathematics.

Notice also that, by putting an emphasis on the interconnections between mathematical practices and other kinds of practice, the proposed approach ceases to pose the problem of the "applicability" of mathematics as external to mathematical knowledge itself, and it becomes internal to its analysis.[32] This should not be regarded as unnatural, for, after all, the very terminology of the "problem of the applicability" of mathematics is biased and should become suspect. This way of framing the problem assumes that mathematical knowledge, as some kind of "pure" form of knowledge, is totally independent from the world we live in and the acts of human agents. It seems likely, to me at least, that the problem will only be solvable when we accept that mathematics does not belong to another world, but is related to the concrete activities of the human species in this world.[33] Incidentally, the world I mean to talk about should not be called "the outside world," because this terminology is again biased in a similar way—as if the "inside" of our minds were somehow "outside" the world.

---

[32] For further details see Ferreirós 2013.

[33] Of course, there is another way of solving the problem of the applicability of math as it was originally posed. We can accept that mathematics belongs to another world, the realm of the mind-soul, and still explain its applicability; the most natural way is the traditional one, assuming that both the mind-soul and the world were created by God, that "mathematician of a very high order" (Dirac). Borges once wrote: "What god behind God begins the round / of dust and time and sleep and agonies?"

3

Agents and Frameworks

The approach to mathematical knowledge and practices that we are developing is—I have argued—pragmatist, historical, and cognitive or agent-based. In the previous chapter we elaborated on some of the reasons for this orientation; now we shall begin to implement it. I have argued that one should resist temptations to over-generalize and introduce an all-inclusive perspective of the kind of Kuhn's paradigms, Lakatos's research programs, or Kitcher's "practices." Also, the notion of practice cannot remain too abstract and disembodied: I emphasized the weak claim that there is no practice without practitioners—the actor or agent must be centrally placed in any perspective on practice worthy of this name. In this chapter I shall propose a "core scheme" to be applied whenever one analyzes mathematical practices that can be presented in a simplified way by means of a couple: Framework–Agent. The Framework–Agent pair is *not* to be identified with a mathematical practice, but is at the core of practice, and of the production and reproduction of knowledge.

Mathematical practice always involves communities of agents in interaction, and a diversity of theories or frameworks, linked and combined skillfully so as to produce results such as problem solutions, theorems, and proof-methods. This of course is fully consistent with the fact that a proof or a solution can be produced in isolation by one mathematician. But the theories, frameworks and practices she masters cannot! Nor will the proof or solution become knowledge without first being accepted, and often revised or modified, by the relevant community. These remarks may be sufficient for the reader to see that mathematics in practice will typically depend on the performance of several Framework–Agent couples, intertwined in several possible ways. It will be my task in this and the following chapters to explain this in detail and elaborate on it through some case studies.

Besides presenting the notions of framework and agent, Chapters 3 and 4 will sketch what I regard as an important theoretical possibility opened by adopting our pragmatist, historical, cognitive approach: namely, the idea that thought and meaning do play a central role in mathematical knowledge and mathematical practices. In this simple sense, mathematics is not just a formal science.[1] (As we shall see in Chapter 6, there is also a second sense in which math is more like theoretical work in the sciences than we have normally assumed.) This general topic will be introduced by presenting a principle of complementarity in math, according to which symbolic means and thought (or formulas and meanings) are complementary in mathematics: both must be present, neither being reducible or eliminable. Math is not cold and bloodless as the system presented by Gödel in section 1 of his book on the consistency results in set theory (1940)—for us, mathematics is primarily mathematical *practices* and *knowledge*.

3.1. FRAMEWORKS AND RELATED MATTERS

The Framework–Agent pair will play the role of a core scheme in the analysis of mathematical practices and knowledge. In order to approach the matter smoothly, it may be useful to start from the Kitcherian quadruples $<L, S, R, Q>$ of language, accepted statements, reasoning methods, and questions (Kitcher 1984, chapters 7 and 8). Actually, we shall consider two different kinds of frameworks, "theoretical" and "symbolic." Our main aim with these is to introduce tools that can be used quite naturally and flexibly in a direct analysis of the documents that typically are the concern of historians and philosophers.

In this chapter we shall relocate "metamathematical views" for reasons that will be explained in section 3.6. The "metamathematical" views are no doubt very important, but we have to proceed carefully (and, it seems to me, flexibly and gradually) in attempts to analyze them. My first move is to consider them not as part of the Framework

---

[1] It seems likely that many formalists have not tried to eliminate thought from our views of mathematics; unfortunately, however, the insistence on the formal seems to have affected deeply the public perception of mathematics. For an example, see Wigner (1960, 2–3).

but rather as belonging to the Agent, as part (but only part) of the agent's constituency. Once I flesh out what I mean by talking of agents (section 3.3), it is my hope that you will see this as a reasonable choice.

To begin with the Frameworks, another proviso. The Kitcherian quadruple $<L, S, R, Q>$ is only a bit more complicated than a theory, the basic unit of analysis in mid-twentieth-century philosophies of science and mathematics. Likewise, it is still focused on a linguistic analysis of the production of scientific knowledge. Given the agent-based, practice-oriented nature of our approach, it seems natural to pay much more attention to non-linguistic elements, including tacit forms of knowledge. However, with the simple move of emphasizing the couple Framework–Agent, one gains full leeway for analyzing those other cognitive elements (which have to do with the constitution of agents and their interaction with frameworks and with each other), while at the same time retaining the possibility of stressing linguistic and symbolic elements.

Thus our frameworks shall remain very close to a direct analysis of the documents that typically are the concern of a historian. But this does not mean that the approach developed here is a close follower of the linguistic orientations that were dominant in the philosophy of science before the 1980s—once more, remember that our analyses will involve several intertwined Framework–Agent pairs, not Frameworks alone. From the beginning, we renounce the intention of making these notions capture all relevant aspects of practice. With this renunciation we gain, I believe, both naturalness and flexibility. Instead of trying to turn the notion of a Framework into a very deep concept for analyzing mathematical knowledge, our main aim will be to present a flexible instrument with which to analyze some important elements of actual practices, namely the *linguistic and symbolic elements.*

3.1.1. On Symbolic Frameworks

We distinguish between two kinds of frameworks that are central to the analysis of mathematical knowledge and practices: the symbolic frameworks employed by mathematicians, and the conceptual or theoretical frameworks that they establish. By "*symbolic*" I mean the special kinds of characters that mathematicians employ in representing problems and solutions, in deploying methods to solve problems and establish results. The symbolic elements are ideograms or graphic symbols that represent certain notions and that jointly may constitute a representation system. In semiotics one learns that ideograms may be conventional (comprehensible only by familiarity) or "pictographic" in a broad sense (not necessarily visual), conveying their meaning through resemblance to a physical object or phenomenon.[2] Here we shall follow this viewpoint and exploit it by considering figures or diagrams (such as the diagrams of ancient Greek mathematics) as symbolic characters. Other more obvious examples are the ciphers, fractions, and other symbolic forms employed for representing numbers and operating on them (as in the medieval Abacus schools);[3] or, more advanced, the series representations of functions that were so central to Euler's practice in eighteenth-century analysis; or the commuting diagrams that are central to twentieth-century category theory.

This first kind of framework, then, is directly related to the language L considered by Kitcher in his tuples. Kitcher's languages typically mix a vernacular language and symbolic means; in the natural-language sentences produced by mathematicians, symbols of the peculiar mathematical kind may appear. Likewise, there normally are specialized *technical expressions* that constitute a peculiarity of language use in the "form of life" adopted by mathematicians. The technical words actually establish bridges between the vernacular language and the symbolic expressions, and learning to understand those technical words will typically involve learning to employ the symbolic expressions. Practically speaking, in a given context, these distinctions—especially that between vernacular and technical words and symbolic means—are relatively easy to make, but on a strictly theoretical level there may be hidden difficulties.

Let me offer two well-known examples. In the *Elements*, Book I, the Greek vernacular is supplemented (or regimented) by the strict use of

---

[2] Some readers may prefer the terminology of *sign* for the most inclusive category in semiotics (as in Firth 1973); if so, what I term "symbolic framework" they would prefer to call "*signific* framework". I prefer the opposite move, expanding the category *symbol*, since this conforms much better to the usual terminology in connection with math.

[3] See Heeffer 2013.

some terms, such as "straight line," "plane," "right angle," "diameter," "isosceles triangle," and "parallel"; and the Greek letters A, B, Γ, . . . (employed according to some conventions which we shall not discuss) facilitate the connection between sentences and diagrams, and especially the assignment of concrete referents to the technical words. The carefully controlled Greek diagrams we shall consider as part of the symbolic framework, thus stretching the meaning of "symbol" beyond its usual boundaries. (Normally, semioticists employ "symbol" in strong contrast to "sign," with the connotation that there is no previous relation, and certainly not an intrinsic relation, between the symbol and its referent. This is not the case for diagrams, as ◊ represents a rhomb intrinsically, and so does Δ for a triangle; however, in ancient Greek diagrams, Δ may represent any triangle—not necessarily an isosceles one—and this already shows that there is a marked symbolic component in the way diagrams are interpreted. I refer the reader to Chapter 5 for more on this topic.)

Among the many other examples that could be given, let us consider a closely related one: Descartes's *Géométrie* of 1637. In the work of Descartes, diagrams reminiscent of the Greek ones appear, and also many of the same terms as Euclid's, and the use of letters as links. (The diagrams do feature lines of a more complex kind, such as the conics, and this is related to a significant expansion of the domain of acceptable curves, and a "redefinition of geometrical exactness"; see Bos 2001, Serfati 2005, and Panza 2012.) But we also find a whole new layer of symbolic expressions, algebraic equations in real variables written according to syntactic conventions that are quite close to the present ones, and these expressions are systematically interconnected with the diagrams via coordinates (Descartes 1954, e.g., 12–15 and 96–99). Since this material was very novel at the time of publication, Descartes explained carefully the nature of his calculus, the way in which arithmetic expressions are used in geometry, "how multiplication, division, and the extraction of square root are performed geometrically," and how equations can be used to solve problems of different kinds (plane, solid, supersolid).

What we are calling a symbolic framework can thus be mainly identified (at least for publications since the seventeenth century that bear a strong imprint of Western techniques and ideas) with the notations and rules of formation and calculation employed. But at least three provisos are needed. Diagrams do count as symbols, and their role has to be determined and analyzed contextually—in particular, one cannot take for granted the post-1800 attitude to diagrams, paradigmatically expressed by Lagrange in his famous "you shall not find Figures in this Work." As Hilbert wrote,

> The arithmetical symbols are written diagrams and the geometrical figures are graphic formulae; and no mathematician could spare these graphic formulae, any more than in calculation the insertion and removal of parentheses or the use of other analytical signs. (1900, introduction; Ewald 1996, vol. 2, 1100)

Second, it seems natural to associate the idea of a symbolic framework not only with the ideograms (symbols and icons in special written characters) but also with the technical expressions of mathematical dialect which are central to establishing links between the symbolic framework and the theoretical framework.[4] As mentioned above, learning to use these expressions is at the same time learning to employ the symbols.

Thirdly, there is a very important difference between the usual kind of symbolic framework, often called "informal," and the very specific case of formal languages. This difference has been expressed by Italian philosopher Cellucci by saying that the latter are *closed systems*, while the former are *open systems*.[5] For a formal system to be well determined, it must be possible to check mechanically whether a sentence is well formed, and whether a proof is correct. In order to make this possible, the language is precisely determined from scratch, the inference rules must be fully specified, and no new expressions (new symbolic means, new concepts) may be introduced subsequently. Most often, however, the actual language and symbolic frameworks of mathematicians are not precisely specified from scratch—especially if we move from the

---

[4] The symbolic elements in mathematics may be either symbols or icons in the strict sense of Peirce. Thus the letters and operations of algebra are symbols (have convention-based relationships with their objects), as are the technical expressions; but, if we adopt Parsons's idea (2009) of the objects of geometry being quasi-concrete, geometric diagrams are icons (have specific properties in common with their objects). It seems that indexes (having a natural or causal relation with their objects) are absent from mathematical practice.

[5] See Cellucci 1998.

calm world of textbooks into the dynamic scenarios of research. Indeed, new expressions, new symbolic means, and new concepts may be introduced along the way by innovative practitioners (in particular cases, the new symbolic means may include a formal system of the kind invented by Frege). Let me say, in passing, that in considering this difference it should be obvious that the celebrated incompleteness theorems of Gödel do not have immediate application to the practice of mathematics—for they apply only to closed, formal systems (sufficiently rich to codify some recursive arithmetic).

Some basic forms of mathematical practice may present little by way of a conceptual or theoretical framework justifying the procedures that are employed in problem solving. In such cases, however, there is certainly more than just ideograms and technical expressions—most importantly, one finds certain methods and procedures set to work. This suggests the problem of where exactly to locate *methods* in an analysis of mathematical knowledge and practices; notice that in these cases it is crucial to count problems and problem types among the elements defining the practices at stake. In such practices, it often turns out that methods have to do with the manipulations and rules that an agent will apply while confronting a certain configuration of symbolic means, i.e., written ideograms. Those methods may be taught by example, e.g., through a given paradigmatic manipulation that is imitated or applied analogically to other cases (think about ancient Babylonian mathematics, for which see Høyrup 1994 and Robson 2008). It may be convenient in such cases to regard the methods as practical know-how that emerges at the interface of expert agent and symbolic framework. In other cases, corresponding methods may have been reflected upon and formulated explicitly, becoming part of what I shall call a theoretical framework.

These considerations suggest also that very elaborate symbolic frameworks may already be incorporating certain methods and procedures that, in other simpler cases, would be included in the theoretical framework. It seems to me that such fuzziness is not dangerous, for it is desirable that the analytic tools we are introducing remain flexible and adaptable to quite different situations. It is probably wiser to leave decisions on such matters to the historian or philosopher who is studying a particular constellation of knowledge and practices.

In fact, attempts to justify and systematize the procedures and methods in use are one of the main sources of the conceptual developments and theoretical justifications that we tend to associate most intimately with mathematics. May the above suffice for presenting the basic idea: we shall develop it by exploring particular cases in due course.

3.1.2. On Theoretical Frameworks

It seems reasonable to assume that conceptual ingredients are present in any mathematical or proto-mathematical practice, regardless of the level of explicit justification or theorizing that we may find in it. "Why" questions always emerge—simply because human agents are cognitive agents—but they can be dealt with in different ways. In some cases it may seem enough to unify and simplify the methods employed in problem solving, so that their fruitfulness and their apparent uniformity may satisfy the inquirer. But mathematicians went beyond that a long time ago. Consider Liu Hui's attempts to justify certain mathematical procedures contained in the *Nine Chapters* by deploying in a systematic way certain basic procedural algorithms to establish the cogency of the target procedures.[6] The mathematician or philosopher may doubt whether one should apply the word "theory" to this particular form of justified knowledge, because in our tradition "theory" is closely associated with some characteristic (proof-)procedures that can be found in ancient Greek and later European mathematics. However, from our present standpoint it seems clear that we are confronting a theoretical framework. As before, it seems best to keep our approach relatively flexible, leaving it to the experts to decide and argue about whether and how to use the terminology established here in certain cases and contexts.

Generally speaking, and for the purposes of the case studies investigated in this work, it is quite useful to speak of theoretical frameworks. Theoretical frameworks are a bit more complex than theories, mainly because we consider the heuristic element of distinguished problems, following Lakatos, Kitcher, and many others. According to Kitcher, we have to consider *statements* or propositions *S*, licensed *forms of reasoning*

---

[6] See on this topic the extensive work of Chemla.

R (including methods), and a series of *questions Q*, which may be basic or advanced, solved or unsolved. Many of the heuristics might be properly placed at the interface between questions *Q* and methods *R*, for instance by way of conjectures about the possibility of solving this or that problem employing a particular method. In the remainder of this section, I shall be content with making some general remarks, without entering more deeply into the important subject of heuristics.

Nobody will doubt that statements or propositions are important elements of theoretical frameworks; they will often be called theorems, lemmas, or corollaries. In principle this category would include *all* the accepted statements relevant to a theoretical domain that have been established by practitioners at a given time. Naturally, even this is an idealization, since no mathematician is able to know all the relevant literature; but she can have access to the statements as needed, at least in principle. It should be obvious to the reader that theories are being conceived here in the concrete, constructive sense of actual practice: the propositions in *S* are actually given and/or proven, typically in published literature. That is to say, in dealing with theories we shall be more interested in faithfulness to the complexity of actual cognitive practices than in neat, formal (i.e., mathematical!) definitions. We shall not be thinking of a theory in the idealized sense of logic—an infinite set of statements in a fully specified formal language, closed under logical consequence—but rather in the concrete, quite constructivist sense they have in actual practice.

That in no way implies a denial that some mathematical practices have actually dealt with theories in the ideal logical sense. But if we are going to analyze the practices of Euler and his contemporaries, or even the practice of a present-day professor teaching an undergraduate course in analysis, it is better to avoid passing through an idealized rational reconstruction of the logical kind. This does not reflect any bias toward constructivism as a philosophical or foundational position. The (finite!) set of statements *S* may or may not include axioms, but in many cases it will; for instance, the traditional theory of magnitudes was assumed to be based (among others) on the famous "common notion" of Euclid: "the whole is greater than the part."

The propositions or theorems are often proven, at least in the Western tradition of "high" mathematics. At least two quick remarks must

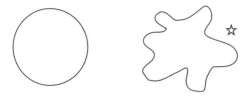

Figure 2 The circle may represent the idealized logical notion of a theory, while the rough section aims to depict a theory as concretely given in practice. The star might represent a conjecture and its ramifications.

be made concerning proofs. Not infrequently, a proof is more relevant because of the methods it employs—which can then be applied to many other instances, generalized, or transferred to related areas—than because of the proposition it establishes. This is of course a central aspect of mathematics, and the methods have a procedural character, which defines mathematics. (A noteworthy example might be the fourteenth-century polymath Gersonides, who employed mathematical induction to prove some combinatorial results, e.g., about permutations, in his *Maaseh Hoshev* of 1321; he talked about a process of "rising step-by-step without end."[7]) As one can see, analyses of licensed methods and forms of reasoning cannot be neatly distinguished from the proofs—nor from questions about the arrangement of the corpus of knowledge. The panoply of methods that a mathematician must be acquainted with is typically very broad, some methods associated with the process of learning to solve particular problems, some with the proofs of particular theorems.

Secondly, we—like mathematicians—are interested in multiple proofs for a given statement. This phenomenon stills comes as a surprise to some laymen and even some philosophers, but for those with a special interest in mathematics it should be more than familiar. Reasons why multiple proofs are interesting abound, including theoretical arrangement

[7] A very comprehensive paper in French is Chemla and Pahaut (1992). Victor Katz discusses Levi ben Gershon (Gersonides) in his well-known text (Katz 1993, p. 279): ". . . Levi introduces, somewhat more explicitly than his Islamic predecessors [al-Karaji, ibn al-Banna], the essentials of the method of mathematical induction, ... In general, . . . he first proves the inductive step [from k to $k+1$], then notes that the process begins at some small value of k, then finally states the complete result."

(to find out about the ordering of results and what is possible and what is convenient), method (can this result be obtained purely algebraically, or does it need topology?), axiomatic dependence (which parts of this theory depend on this axiom?), and so on.

It remains to consider the series of *questions* (basic or advanced, solved or unsolved) that are actually emphasized by agents in a community. Concerning the importance of this element, Hilbert's celebrated address, "*Sur les problèmes futurs des mathématiques*," i.e., his "Mathematical problems" (Hilbert 1900), is still very relevant. Problems may be very concrete—Can we provide a formula that solves with radicals a given equation of the fifth degree? Can we construct a 17-gon with ruler and compass?—but they may also be of great generality—Do the integers in any algebraic number field $F \subset \mathbb{C}$ obey a law of decomposition similar to the Fundamental Theorem of Arithmetic? What properties must a measure on a point set satisfy in order to serve well as a basis for integration theory? As Hilbert said (1900),

> The deep significance of certain problems for the advance of mathematical science in general and the important role which they play in the work of the individual investigator are not to be denied. As long as a branch of science offers an abundance of problems, so long is it alive; a lack of problems foreshadows extinction or the cessation of independent development. Just as every human undertaking pursues certain objects, so also mathematical research requires its problems. It is by the solution of problems that the investigator tests the temper of his steel; he finds new methods and new outlooks, and gains a wider and freer horizon.
>
> It is difficult and often impossible to judge the value of a problem correctly in advance; for the final award depends upon the gain which science obtains from the problem. Nevertheless we can ask whether there are general criteria which mark a good mathematical problem. (Ewald 1996, vol. 2, 1096–97)

Of particular importance for the development of mathematics are some problems called *conjectures*, which have become a characteristic element of modern mathematical practice (Mazur 1997), since they play a guiding role in the articulation of established frameworks and even the emergence of new ones. The way in which mathematicians confront conjectures has very interesting methodological characteristics.[8]

Notice that the Framework approach is applicable quite directly to the analysis of historically given practices (such as those of Descartes in his *Géométrie*, Euler in the *Introductio in Analysin infinitorum*, and Klein in the Erlangen Program), among other reasons because the language and the theory need not be formal, and we have avoided the idealized logical notion of a theory. But it is also possible precisely because the scheme we are considering builds around the couple Framework–Agent. Normal frameworks such as that of Euler (or those developed by university professors teaching algebra or analysis) simply cannot be made to stand alone: the linguistic and symbolic components alone, separated from the way the practitioners use them and gesture around them (so to speak), lack clear conditions of meaning and use.

To summarize, we have highlighted three elements in symbolic frameworks: ideograms, I, technical expressions, E, and symbolic methods, M_S—where ideograms can be diagrams too, and the (optional) symbolic methods have to do with their manipulation. In the case of theoretical frameworks, we have emphasized four elements: statements or propositions, S, proofs, P, theoretic or proof methods, M_P, and open questions, Q, including conjectures. The reader need not memorize those letters, for actually we shall not follow the logico-mathematical practice of writing $<I, E, M_s>$ and the like.

3.2. INTERLUDE ON EXEMPLARS

It is characteristic of Kuhn's paradigms to have represented a shift away from the dominant linguistic orientation of his time, for what he termed in 1969 "disciplinary matrices" included elements such as values and exemplars. It is also well known that Kuhn's conception of exemplars and their role in guiding research (by analogy) was the original idea that led him to use the word "paradigm," a word which (unfortunately) he subsequently abused making it play different roles, until he

---

[8] See Ferreirós and Gray (2006), Introduction, and Chapter 6 below.

felt the need to abandon it. I am not alone in thinking that Kuhn was here offering a deep, insightful contribution to the analysis of scientific knowledge. Unhappily, most of the people who read his work do not pay attention to this aspect of his contribution.

I believe that exemplars play a crucial role in mathematical knowledge too. Some years ago, while proposing a historical philosophy of modern logic, I employed the notion of an exemplar "as an extremely useful tool in understanding the historical evolution of this branch of modern mathematics" (Ferreirós 2001, 442). Indeed, it was my argument that some exemplars (the Aristotelian syllogistic figures) had been essential ingredients in the configuration of modern logic, in particular for the incorporation of the quantifiers on a par with the truth-functional connectives as basic logical notions. Commenting that I would refrain completely from using the broader and much vaguer notion of a paradigm or disciplinary matrix, I continued:

> I would go as far as saying that I consider this notion [of an exemplar] capable of shedding much light on the development of branches of mathematics in general. Obvious examples could be found in the case of modern abstract algebra, specifically in the exemplary role played by group theory. First-order logic itself has served as *the* paradigm for a modern logical system and its metatheoretical study (it can be argued that the treatment of other systems, e.g., modal logic, has mimicked that of first-order logic). In this connection, readers might consider the present paper as a case-study in the employment of that Kuhnian tool for the historiographical and philosophical analysis of mathematics. (Ferreirós 2001, 442)

The reference to group theory was meant in the sense that general theories of rings, fields, etc. were modeled upon the paradigm provided by the somewhat earlier development of an abstract theory of groups, which was taking shape around 1890/1900. But there is little doubt that exemplars started playing a role much earlier and at much lower levels.

Probably the process begins with the way in which particular problem solutions and the methods employed in them are paradigmatic for solving other problems; and it continues with the ways in which particular proofs of results become paradigmatic, so that other proofs are articulated by analogy. Any trained mathematician will, I believe, agree with me on this point. And the fact is that those examples are absolutely close to what Kuhn had in mind. Tim Gowers wrote in his blog about "general tricks that can be applied all over the place" in mathematics when you are trying to find a proof or a solution to some problem: "Try to draw analogies with other contexts," "look at similar problems where you *do* know what to do"; and he comments: "This is an incredibly important method used in mathematical research—I'm tempted to say that it is the most important method we have."[9] That is explicit recognition of the importance of exemplars in the practice of mathematics.

There are also beautiful historical examples of exemplars at work in the production of novel mathematics, at a level of complexity lower than that of group theory or logical metatheory. Consider the geometric representation of the real line (one dimension), which was the model upon which the geometric representation of the complex numbers, the complex plane (two dimensions), emerged. And this, in turn, gave rise to efforts to proceed analogously into the third dimension, efforts which famously confronted characteristic difficulties—as it was not enough to handle three units $(1, i, j)$—and led to Hamilton's discovery of the quaternions as the system needed for representing three-dimensional vectors (to use the modern term).[10] Below, I shall say a few more words about the complex and hypercomplex numbers.

This immediately raises the question: should we not include an additional element in our frameworks, the exemplars? For a long time I was inclined to answer yes. But this reaction was perhaps too influenced by the Kuhnian tradition of paradigms (disciplinary matrices).

Let me insist that frameworks are *not* meant to capture all the relevant aspects of practice, but only to build a flexible instrument with which to analyze some important elements of actual practices,

---

[9] See point 8 of http://gowers.wordpress.com/2011/11/18/proving-the-fundamental-theorem-of-arithmetic/. Accessed on July 17, 2013.

[10] Quaternions are a hypercomplex number system with three complex units, i, j, k, so its numbers are of the form $a + bi + cj + dk$; the units are such that $i^2 = j^2 = k^2 = ijk = -1$.

the linguistic and symbolic elements. One can acknowledge the great importance of exemplars and analogy in the configuration of knowledge, while not including them in the schemes I have just proposed. This is mainly because the phenomenon of paradigm-following has more to do with the ways in which agents tend to use frameworks, apparently in a rather natural form. It is not our business here to try to analyze where this phenomenon comes from, but it is tempting to find its roots in learning by imitation, and even in the basic workings of neural networks.

Eventually, then, it may be useful and important to emphasize that certain problems, methods, or proofs are playing the role of exemplars in mathematical practice. I recommend this idea, but I also recommend avoiding the inertia of the Kuhnian tradition: we do not need to include this as an element of the frameworks.

A related matter is worthy of mention. When applying the notion of theoretical framework to the analysis of a concrete historical case, it may often be useful to consider *distinguished statements* among the statements S: those propositions that are particularly salient, typically found in most textbooks, regarded as basic for attaining "sufficient knowledge" of the theoretical domain in question. A basic example is the Intermediate Value theorem in Cauchy-Weierstrass analysis, or the Bolzano-Weierstrass theorem that every bounded infinite set of real numbers has a limit point (or point of accumulation). Of course, many other propositions of greater "complexity" or "depth" may be deemed necessary before the experts acknowledge a novice as having "sufficient knowledge" of the theoretical domain.

Similarly, it may be important to lay emphasis on distinguished elements among the questions Q, salient problems such as the question about the zeros of the Riemann zeta function $\zeta(s)$ (in connection with the celebrated Riemann Hypothesis), and distinguished elements among the methods M, say the diagonal method employed by Cantor in 1891 and later turned into a widely used procedure in mathematical logic, or, in the case of analysis (and topology), the method employed for proving the Intermediate Value theorem.

Salient unsolved problems are important because of their guiding role in mathematical research, to the point that sometimes a whole program of research takes shape as a potentially productive way of finding a route to the solution of a relevant problem. The reasons why distinguished statements, problems, and methods are important include the frequent need to appeal to them while proving results or solving problems, and not least their role as exemplars to be imitated in the practice of mathematics.

3.3. ON AGENTS

The thesis that multiple practices coexist and are interrelated, in a way that is crucial for the conformation of mathematical knowledge, has one interesting consequence. It means that my analysis of mathematical knowledge has to be crucially centered on the agents. This is because the links between different frameworks and practices are not given abstractly, but are *established concretely* by an agent, thanks to his or her cognitive abilities and mastery of practices.

When I speak of *agents*, what is meant are human beings, since up to the present, all mathematical agents are humans.[11] It is not my intention to diminish the interest of developments such as automated proof and formal verification, nor more generally the impact of computers on mathematical practice (in calculation, simulation, and communication). Certainly these are topics that deserve close study.[12] Machines have become an important addition, to the point that during the last half century the pair framework–agent has evolved into a *triad*, framework–agent–machine, with very significant impact at least on some fields of mathematics. This is not unprecedented, since the tradition of mathematical instruments is old as history—the abacus, the quipu,

[11] Although there has been a significant amount of thought and speculation devoted to the possibility of other kinds of agents—extraterrestrial or machine agents—the fact remains that no example of the former is known, despite NASA efforts, and also no instance when the latter have developed *knowledge* of mathematics in a form sufficiently close to the performance of human agents. Hence I shall leave it to others, better prepared for the task than myself, to develop such lines of thought (see the interesting paper by Ruelle 2000).

[12] See, e.g., Avigad 2008. As an interesting etymological aside, it is well known that from the seventeenth century until about 1950, "computer" referred to humans (often women) who worked calculating in the context of astronomical observatories and the like.

the compass, etc.—and it includes instruments as complex and sophisticated as the differential engine.[13] But it has certainly scaled up to a point that it has changed the profession and the practice. However, the human agent remains central, mostly because our computers are implementations of formal systems and they have not been developed into systems that can implement technical practices and interact with symbols in such a way that they evolve human-like knowledge of mathematics. This may very well be possible, but as of today it remains mere possibility.

No special emphasis will be placed here on agents with somehow abnormal or supernormal capabilities, such as mental calculators (human computers) or mathematical geniuses. We shall be content with considering normal, *average* agents with "usual" cognitive abilities—abilities such as visual perception, memory, and language, about which there is general agreement that they are essential for mathematical cognition.[14] The idea is simply that those cognitive abilities (and the capacity to learn and develop basic practices) are, apparently, common to Euclid, Ruffini, Russell, and a typical undergraduate student of mathematics today.

As mentioned before, by speaking of agents we mean to emphasize human action, in particular concrete physical action—which in our view is always associated with knowledge; we mean to emphasize the continuous loop between knowing and acting. Considering agents is not the same as considering subjects, among other things because the notion of an agent has much stronger presuppositions: agents live in a world of things and other agents, they have bodies and social lives, and they interact constantly and in numerous ways. Such presuppositions are, of course, very unsatisfactory from the viewpoint of *first philosophy*, of traditional theory of knowledge developed according to the style of the mind's theater (with its main spectator, the I/Self). But we have abandoned *philosophia prima*; we are naturalists in the minimal sense that the project of a first philosophy has been abandoned as hopeless, and the account of knowledge we develop is non-foundationalist,

---

[13] Readers are encouraged to visit the collections of mathematical instruments at London's Science Museum, or at the Musée des Arts et Métiers in Paris.

[14] See the *Handbook of Mathematical Cognition*, ed. by Campbell (2005).

having no claims to possessing greater evidence or certainty than scientific knowledge. Reliance on scientific results becomes an option, and knowledge is conceived as produced by communities of living beings (the members of our species) that engage not only in interactions among themselves, mainly through language and images, but also in interactions with their physical environment.[15]

Much philosophy of mathematics, guided by the traditional assumption that math is a priori knowledge, has been developed on the assumption of a very minimally furnished world: those philosophers have imagined themselves living in a rather empty space where there is only natural language, formal languages, abstract objects, and perhaps space-time (or, lowering our standards of acceptance, the objects of the "fundamental" theories of physics, such as energy and elementary particles). There is nothing more, and one should not presuppose anything else. But our assumptions are much more exuberant: we assume given a world of physical and other natural objects (stars and proteins, as much as subatomic particles and energy-matter)[16] and agents endowed with practical and linguistic abilities. Practical abilities include our competence in handling measuring rods and clocks—and other tools, such as microscopes, X-ray generators, and spectrometers.[17] For our purposes we do not need to consider more that this, although it is important to keep in mind the practical ability of using pens to write on paper, or keyboards to "write" on a computer. Notice how this implies the ability to perceive differently shaped letters as tokens of the same type (otherwise one cannot continue writing, one cannot recognize the previously written).

Thus our agents are quite unlike a "transcendental" subject; they are not like Brouwer's *creating subject* (Brouwer 1948), which he conceived as an idealized mind in which all of mathematics takes place, abstracting away from inessential aspects of human reasoning such as limitations of space and time and the possibility of faulty arguments.[18] Our agents are not idealized; they do not have unbounded memory and immediate

---

[15] For a more detailed discussion of my stance regarding naturalism, see Ferreirós 2010.

[16] Reductionism never worked perfectly anyway, see, e.g., Dupré (1993).

[17] This sentence was written with Einstein and Bohr in mind.

[18] For a phenomenological analysis of the creating subject as a transcendental subject in the sense of Husserl, see van Atten 2007.

recall; cognitive restrictions are relevant to them. Yet these limitations are not directly relevant to most of my discussion, since we can debate the main epistemological issues without taking into consideration how much time and effort an agent may have taken to learn, to find out about something, and so on. But please be aware that we never engage in the fiction of imagining science (math here) to have been the product of a single mind: we shall be discussing the work and productions of many agents, fully immersed in socio-cultural interaction.

Unlike Brouwer's, our agents live *in* a world and a human society—I reject solipsism. They have bodies and their minds are not independent of their bodies—I also reject dualism. More to the point, Brouwer's subject only *knows* something when he experiences it in full consciousness: when the subject experiences something with a propositional content, then he knows a true proposition; when he experiences an entity, then *there is* an object fully present and (fully or partially) defined in the subject's mind. But, in the epistemology we are developing, *to know* is not understood in the Cartesian way of having something fully present before the mind's eye. We can know stuff on the basis of expertise and previous experience that is memorized, as when we know a proof because we keep in memory some of its salient steps and its main points (from key ideas to little tricks) and because we trust in our ability to reconstruct the argument.[19] More to the point, we can know something on the basis of hypothetical assumptions when we establish what follows (necessarily) from them; ergo, we can know something even if that thing might not be true in a metaphysical sense. This whole book is an attempt to make this thesis plausible, to flesh it out (see, in particular, the arguments about the continuum, which form a *leitfaden* throughout the last chapters of this book).

Already Peirce, in his writings on math, emphasized the idea that mathematics "is the study of what is true of hypothetical states of things. That is its essence and definition. Everything in it, therefore, beyond the first precepts for the construction of the hypotheses, has

---

[19] One may even talk of "knowledge" in relation to results for which we have solid evidence based on the expertise and authority of others (we know the Wiles-Fermat theorem has been proved, even if we don't fully understand the proof); but this is certainly not our topic in this book.

to be of the nature of apodictic inference."[20] Mathematics, according to Peirce, is the only one of the sciences "which does not concern itself to inquire what the actual facts are, but studies hypotheses exclusively"; and he adds, noteworthily, that this is "a principle perfectly established today"![21] Knowing what is "true of hypothetical states of things" without concern for "the actual facts" deviates strongly from the more metaphysical understanding of knowledge that was characteristic of Descartes or Kant, and that is upheld even today by many philosophers. I agree with Peirce and others on the idea that mathematical knowledge (as we find it in contemporary or even in "classical" mathematics) has to be reconceived along such lines, divorced from the assumption of necessity or even truth in "actual fact."

Insofar as this viewpoint was "perfectly established" around 1898, as Peirce claimed, perhaps exaggeratedly, it was due in large part to the influence of Riemann. The first Peircean quote I have given seems to be written with Euclid's geometry in mind—its axioms are precepts for constructions, e.g., of regular polygons or regular solids—and their hypothetical nature was underscored by Riemann (1854) within a context of non-Euclidean geometry. We tend to forget that such ideas spread rather widely in the late nineteenth century: for instance, Poincaré's well-known *Science and Hypothesis* (1902) should be a remainder.

Brouwer's subject is not allowed to entertain hypothetical thoughts, especially of the kind when an object that is not fully perceived, conceived, and defined (insofar as I can think of it, or picture it) is assumed to have permanent properties, partially unknown to us, and is investigated with regard to them on the basis of partial specification, say by means of axioms. Our agents are allowed to entertain such hypothetical thoughts, and come to conclusions on that basis. Our agents are even allowed to call the outcome of such thinking processes, usually brought about in conversation and debate with others—*knowledge*. This

[20] See Peirce 1974 (*CP* vol. IV), 4.233. He liked to quote his father Benjamin's definition of math (in 1870) as "the science which draws necessary conclusions," adding that "it is impossible to reason necessarily concerning anything else than a pure hypothesis" (*CP* vol. IV 4.232).

[21] From 'Reasoning and the Logic of Things' (1898); see also Peirce (1955). I'd like to add that modal characterizations of mathematical knowledge, such as that proposed by Putnam, seem to go essentially along the same lines as Peirce (at least in their initial conception).

difference in approach affects very seriously questions related to particular real numbers, say a transcendental number (such as π), and the question of the real-number continuum as a whole or structure.

In a nutshell, our approach is in the spirit of twentieth century fallibilism, unlike Brouwer's, and also in the spirit of the supersession of "subject philosophy." This creates an important distance not only with Brouwer's, but also with Husserl's thinking. It is hard for me to know whether I can agree with Husserl's principle that all genuine knowledge refers back, directly or indirectly, to intuitions (i.e., experiences in which objects are given as themselves to the subject). The idea is phrased in a subjectivistic way, referring to intuitions of a subject, while I prefer to avoid terms that suggest the notion of a mind representing a reality by means of "mental images."[22] If one assumes that Husserl must be understood this way, too, then perhaps I agree: genuine knowledge refers to concrete experiences. But I have concrete experience of the real numbers by studying a book on the topic, even if the book deals with the real-number continuum presented and studied in the "classical," postulational way. Knowing the reals does not require me to be able to experience each real number in isolation, so to speak, as it is enough to work with some of them ($\sqrt{5}$, π, e, etc.) and to have knowledge of the system of real numbers, be it through axioms or even through a definition like Newton's (see Chapter 8). This clearly deviates too much from Brouwer's perspective, but maybe not from Husserl's.

After all, our approach is aimed at making sense of the business of "classical"/modern mathematics, while at the same time being capable of making sense of constructivistic mathematics, or even math with dynamically changing objects (such as the choice sequences of Brouwer). In this book I do not try to adjudicate the matter between intuitionism, or at least constructivism, and classicism. According to Husserl, the mathematical universe is static: its objects are finished or complete, and mathematical truths and objects are omnitemporal (*all-zeitlich*, as he says; see van Atten 2007). Brouwer, in contrast, regards the universe as a construction of the mathematician; hence it is not

---

[22] We shall concentrate on public pictures, avoiding the idea of mental representation (if only for simplicity, as it can be done without loss), also when we discuss geometry in a later chapter.

omnitemporal; moreover it is dynamic in the sense that some objects (choice sequences) are open-ended and develop in time. For us, both standpoints are possible options in the development of mathematical knowledge and mathematical practices. They are different, of course, and it is important to understand the differences. One of them became mainstream, while the other remains an absolutely minority view (here I refer to strict intuitionism, not so much to constructivism); and we aim to understand why mathematical practices have evolved this way. This is the topic in what remains of this book.

3.4. COUNTING PRACTICES
AND COGNITIVE ABILITIES

Consider the practices of counting, which, according to our standpoint in Chapter 2, are basic grounding practices for math (alongside measuring and drawing geometrical forms). This means that such technical practices are important root elements in the web of practices related to mathematical knowledge. One obvious question is, what are the cognitive abilities involved in mastering counting practice? There is good evidence that counting (like spatial reasoning) invokes basic cognitive abilities that are hardwired in specific brain areas. But there is much more to counting.

Work during the last decade or two concerning the neuroscience and cognitive science of numbers has led to interesting results and achieved great visibility, having enormous success among scientists and the general public; the reader is likely to know something about it.[23] Such work has provided stronger evidence for the existence of specific neural circuitry involved in processing basic numerical knowledge, allowing mammals to *subitize* small numbers of things (i.e., directly grasp numbers up to about 4); there is also evidence for what Dehaene called an "accumulator," brain circuits for grasping approximatively and somewhat qualitatively larger numbers. All of this suggests that we possess something that might be described as an innate intuitive

---

[23] See, e.g., Dehaene 1997 and Butterworth 1999.

notion of numerosity, which in due course may serve to anchor a more sophisticated and advanced grasp of the notion of cardinality (in the sense that the cardinality of a collection of objects is uniquely determined, precise, and invariant). This is a highly relevant discovery.

Sometimes cognitive scientists speak more loosely, saying that the number concept is hardwired in our brains, but I must warn that this conclusion does not follow from the available evidence. It seems best, indeed necessary, to reserve the expression *number concept* for cases where we have the capacity to *precisely* conceive or (at least) determine in practice "how many," regardless of size. That is to say, size should not constitute an in-principle problem for the precise grasp of numbers (which is compatible with the obvious fact that no oral language possesses the means to determine very large numbers). I surmise that this condition is the same as requiring that numbers greater than a certain small threshold—say 10 or 15—still be precisely determinable and distinguishable from each other. Given the well-established results of cognitive scientists, in practice it will suffice to check whether a given individual or an average member of a community can discriminate precisely between 10 and 12, or between 15 and 17.[24]

For a precise grasp of cardinalities to be possible, it is not indispensable to have a system of numerals incorporating both the ordinal and cardinal aspects of numbers.[25] Some cultures are known to possess a limited numerical vocabulary, but their members are able to solve counting problems by means of concrete one-to-one correspondences. To ascertain how many coconuts there are, they may employ a bundle of sticks: the "Wedda" or Veddas, a culture originally of hunter-gatherers in Sri Lanka, "assigned one stick to each coconut, always saying 'this is one'. In this way they obtained just as many sticks as there were coconuts;

[24] Most known cultures would pass this test, although there are cultures and languages in Amazonia or Australia that do not possess linguistic or other cultural means to express quantities precisely, not even words to express (precisely) "one," "two," and "three." A famous recent example is the Pirahã living along the Maici river in the Amazonas; see Frank, Everett *et al.* 2008, and see also Gelman and Butterworth 2005; and for Australian cases see older linguistic studies such as Dixon 1980.

[25] *Ordinals* indicate position in a totally ordered collection: first, second, third, . . . *Cardinals* answer to the question "how many": one, two, three, . . . The *natural numbers* combine both in a single system.

nevertheless they had no number-words. But they were able to keep a record: if a coconut was stolen, one stick was left over when the assignment of sticks to coconuts was repeated."[26] A community that has reached this level of practical mastery of cardinalities can be said to have a number concept, even in the absence of number-words, and even if its members do not employ such an ability to enumerate beyond 20 or 40.

Notice that the origin of number signs in the most ancient written language, cuneiform, was actually related to such practices. Writing was created in Uruk around 3200 BCE for purposes of accounting, due to the needs of an emerging urban society (city-states), and "basic mathematics—numeration, metrologies, and fundamental calculation—was thus not only part but an essential constituent of the earliest Mesopotamian written tradition" (Høyrup 2011). Now, as is well known, the earliest numerical signs depicted little stones (*calculi*) enclosed in a bubble-shaped envelope, a calabash made of clay; the origin is a process basically like that of the Veddas mentioned above. According to Herrenschmidt (2007), the writing of numbers was the first form, from which the whole history of written signs in the West arose.[27]

Nevertheless, it is very common for human languages to have number-words beyond "four," indeed to have a system of numerals that in principle can always be expanded. (In practice, of course, the system of number-words in any given language—and also in computers—falls infinitely short of the number structure conceived as an infinitarian system.) For our limited purposes here, since we do not aim at full coverage, it will be sufficient to restrict attention to the number concept as associated with counting by means of numerals. Thus, when I

[26] Menninger (1969), p. 33. See also the case of the Kpelle in Liberia (who are farmers, not hunter-gatherers) studied by Cole, Gay, Glick (1974). It is reported that Glick, having asked the Kpelle to sort items into categories, found them sorted into functional groups (e.g., "eat" for apple); such functional grouping is something only very young children in Western cultures would usually do. They did not produce taxonomic categories (e.g., "fruit" for apple). Glick tried and failed to teach them to categorize items; thus he tended to think they didn't have the mental ability to categorize in this way. Then, as a last resort, he asked them how a stupid person would do this task. At this point, without any hesitation, they sorted the items into taxonomic categories: "They could do it, but in their culture, it was of no practical value. It was stupid."

[27] On this topic, see Høyrup (1994), Damerow (2007), Herrenschmidt (2007), and Robson (2008).

speak of *counting*, I do not mean mere enumeration as in practices of finding "how many" by concrete one-to-one correspondences, but as in practices that rely on ordering and on the use of number-words or number-symbols.

In order to count, it is practically necessary to *order* the things we are counting and to correlate them with representatives—often number-words, but possibly also body parts (famously the case in some cultures of Papua New Guinea) or tallies on a stick (the case of the Lebombo bone, reminiscent of calendar sticks still used by Bushmen in Namibia). Let us focus on the numerals. The indefinitely extensible system of words or symbols is correlated with the things on the basis of an ordering, and this introduces the ordinality of numbers, which is essential in practice. Number is commonly associated with recursive characteristics of oral language, which are exploited in the production of the number-words, but that ability is a more general trait, as one can see from considering the way tallies or written marks can do the work required for counting. The crucial point is simply to follow a rule and go on, with no finite limit in principle, either orally or practically in some other way. And this of course is learned: the incipient, vague notion of cardinality (Dehaene's *number sense*) may have a genetic basis, but ordinality certainly has a cultural basis too, and it is necessary for a definite conception of cardinals to emerge.

This ability and the number concept can apparently be developed by humans of any cultural background, but in many cases it is just latent.[28] To the best of my knowledge, it cannot be developed by our animal cousins. The number concept, in that sense, is *a cultural product* based on the specific cognitive abilities of the human species; it has been developed only with the help of cultural tools; it is not innate but culture-dependent. (Oral language suffices for obtaining it, although it might have been the case that written symbols were necessary, in the

---

[28] This refers to cases of cultures that do not have a full system of number-words: it is well known that many societies do not count precisely beyond three, four, or five, and I infer that such cultures have not developed or "fixated" a concept of number (although, of course, particular individuals in the society may be able to come to that point).

course of human cognitive evolution, for the number concept to be first grasped.)

The above observation does not have any implications of relativity or a lack of clear referents for numbers:[29] to put it paradoxically, the number concept is a *perfectly natural* cultural product; and knowledge of numbers—I shall argue—is characterized by *certainty* of a kind that has little counterpart even in other areas of math (see Chapter 7). As I wrote some time ago:

> To state that mathematics is a part of culture, and mathematical knowledge is a social product, does not imply the thesis that mathematical theories are mere cultural products, nor that the only relevant factors to explain their genesis be sociological factors. That is the distance between a soft and almost trivial version of the idea of social construction, and the strong versions defended by Bloor and others, which are based on an unproved and implausible hypothesis. I would rather suggest that the case of mathematics and its associated cultural invariants seems to show that cultures *are not* ultimate factors, self-sufficient and autonomous, without links with nature. (Ferreirós 2005, 64, Spanish in the original)

In fact, the nature/culture dichotomy is only valid on a first approximation; conversely, when studying cognitive factors that underlie mathematics, cultural elements must be taken into account. My standpoint is better described as a form of pragmatism rather than naturalism, if by the latter we understand—as tends to be usual—a reductionistic, scientistic approach (see Ferreirós 2010).

Notice that, since we are adopting precision and the attainment of a minimal size as defining traits of the number concept, practically speaking, from the anthropological point of view, this means that we

---

[29] One may want to say that numbers are abstract objects, and this can be done with the necessary provisos (Chapter 6 section 4); but the view that is more consistent with our approach is that numbers refer to *relational situations* of a perfectly objective (intersubjective) kind that involve the ingredients I) to IV) discussed in the sequel. (Already Gauss, around 1831, had proposed that mathematics has to do with relations and relations among relations, not *prima facie* with objects.)

have a simple criterion based on oral language.[30] To judge whether a given culture has or incorporates the number concept, it often suffices to inspect the number-words available and consider whether they allow the formation of larger and larger numbers. As a matter of fact, it seems that once the language has means for expressing numbers above a small threshold, say 10,[31] the system of number-words already has the required expressive power. At that point, already some recursive clauses are being exploited for the formation of number-words, and such recursivity has the potential for unlimited development. (Limitations of size are always present, of course, as any notational system faces practical constraints that make it usable only for finitely many numbers; but this is less relevant than the potentiality of such systems to surpass any given limit.)

Much can, and has, been done about differences in the conceptualization of numbers—e.g., the fact that 0 and even 1 do not fall under the concept "*arithmos*" of the ancient Greeks—but from a pragmatist, practice-oriented standpoint the real question is whether there have existed divergent counting practices in different world cultures. The answer seems to be no; there is perfect compatibility and translatability, in spite of all the cultural nuances of numbers and their symbolism in different cultures.[32] (Perhaps I should warn some readers that they should make an effort to concentrate merely on *oral* numbers, number-words. Our education makes us so used to identifying number-words and ciphers that we often do not realize how different the underlying systems are.[33])

---

[30] To be supplemented, if needed, by criteria having to do with material practices such as that of cutting tallies on a stick (consider the Lebombo bone, from roughly 35,000 years ago) or of employing body parts to count (Lancy 1983). We have remarked that, in principle, nothing prevents members of a human group from developing a number concept without corresponding words. Suffice it to say that most cultures that have arrived at that stage have developed number-words.

[31] I must leave it to experts, e.g., linguists such as Hurford (1987), to determine precisely where this threshold can be located.

[32] See Menninger 1969, Hurford 1987, Frank and Everett *et al.* 2008.

[33] To give an example, number-words in languages such as English or Spanish are formed according to principles closely similar to those of the Latin numbers (compare seventeen and XVII, twenty-one and XXI, and one hundred fifty-three and CLIII) and they are very far from the simplicity of our Indo-Arabic numbers (17, 21, 151).

The number-words are used for counting, and counting is a very peculiar (and highly stable) way of doing things with words. With counting, the imprecise innate grasp of numerosity is complemented by a notion of order (ordinality), and the notion of cardinality is refined into a precise concept (with the aid of specific words),[34] so that they combine to generate the number concept. That is to say, the idea of a natural number combines cardinality, ordinality, and notation. Now, let us consider again the question with which we started: what are the cognitive abilities involved in mastering counting practices?

Already at this level of technical practices we find interaction between even more elementary practices and skills. In the case of counting, beyond the core cognitive systems involved,[35] it seems necessary to emphasize at least four elements: perception, deeds, oral language, and the symbolic (semiotic). In the process of counting, as it is done daily, the following elements can be found:

I.   we deal with a certain collection of objects or phenomena that we categorize "conceptually," i.e., that we are somehow able to isolate or specify;

I follow Frege (1884) in associating the class or collection with a concept, but I put this word in brackets because there is a long cognitive story to be told, underlying our use of words as "concepts."[36] This however is not the place to enter into that topic. Also, I use the word "collection" to *avoid* "set" and its connotations: by a "collection" I mean those we engage with in everyday life; a collection is *not* a single object, and the operation "collect" is not iterated.[37]

II.   we survey the objects in a certain (often arbitrarily imposed) order, while orally producing a number sequence

---

[34] We might say that they are "technical words" but now in a broad sense. "Eleven" or "twelve" are highly technical expressions, due to their role in the technique of counting, which explains why they resist change so much more than other words of English (notice the contrast with "thirteen" and "fourteen").

[35] Feigenson, Dehaene, and Spelke (2004).

[36] Note the significantly different stories for concepts of different kinds; consider three examples, such as the concepts of "dog," "chair," and "prime number."

[37] A set is a single object, different from its elements and from the collection of the elements; the operation "set of" is iterated, even transfinitely. Cf. Chapter 9 and Ferreirós 2011.

(via number-words or number-symbols, or even strokes like a primitive shepherd);

III.   by so doing we establish a concrete one-to-one corre-spondence (an effective pairing up) between the number sequence and the counted objects; and

IV.   the number that we finally reach in the counting process is employed to determine the cardinality of the collection (invariant for alternative ways of counting, i.e., for different orderings of the objects).

I will be pleased if all readers consider this analysis a triviality, because my analysis aims to be naïve.

Counting so considered is nothing but a technical practice of man-kind, one of the oldest, and one of the earliest we consistently teach our children, generation after generation. Due to the restrictions of constructivity and effectiveness that I have indicated, it is obvious that counting is only possible for finite collections of objects. Counting typically requires visual perception, language-related abilities of cat-egorization, memory of the spoken numerals in their standard order, practical knowledge of their (recursive) formation rules, the ability to order a collection according to certain rules (no duplications, no missing object), and command of the practice of coordinating body movements—normally pointing a finger toward one object of the col-lection at a time—with the production of numerals. Say, e.g., that a child is counting toys or beans: beyond having sufficient control of the practice of producing numerals according to a combination of memory and grammatical rules (". . . ten, eleven, twelve, thirteen, . . . , twenty, twenty-one, . . . , thirty") and the practice of establishing *concrete* one-to-one correspondences, he or she must be able to correctly perceive and categorize the objects. In the typical case of counting a collection of visualizable things, the concrete correspondence is actually between a gesture simultaneous with an utterance, on the one hand, and a thing or an event, on the other; that is, from finger+word to one object or phenomenon.

Similar remarks apply to the case when a native of Papua New Guinea is using body parts to count. Even if no oral language were being used, perception, deeds, the conceptual skills that are typically mastered only with language, and the semiotic element of representation (with body parts standing for numerals) are nonetheless present.

Let me insist that an explanation of (I) will depend on a very complex theory of perception and categorization; (II) involves command of an ordering practice and often command of recursive grammar; and (III), simple as we find it in normal life, involves a very complex coordination of abilities in several different modes. The cognitive complexity of counting is certainly very high, incomparably higher than that of subitizing, but the most noteworthy aspects of this practice are its *stability, reliability, learnability, and intersubjectivity.* I dare to suggest the idea that mathematics, in its most elementary strata, may be the best expression of that which we humans have in common, merely by virtue of being human. To quote myself again:

> The diversity of 'natural' ways of employing the fingers and toes to count, as the origin of the bases of numeration employed by different cultures (base 5, 10, 20, 60) offers a good image of how the naturalistic roots of knowledge restrict the variability of historical products. It is tempting to think that mathematics, in its most elementary strata, may be the best known expression of that which humans have in common, by virtue of being humans. Its roots must be sought in the factors we have enumerated, and above all in the intersection of perception, deed, oral language and symbolic language. (Ferreirós 2005, 65, Spanish original)

Mathematical practices, properly speaking, such as reckoning arithmetic and other more sophisticated forms of number theory, are of course linked with the concrete practice of counting. But they go far beyond it. It is possible to reinterpret counting practices from an advanced mathematical standpoint, but this can only be done from the vantage point of modern math with its characteristic assumptions. The advantages are bought at the price of assuming the hypotheses on which advanced mathematics is founded (see Chapter 6); the conceptual clarity and precision of the analysis is paid for at the price of foundational difficulties. We shall come back to this matter in section 7.2.

3.5. FURTHER REMARKS ON
MATHEMATICS AND COGNITION

In Chapter 1, I characterized the approach we are discussing as cognitive, historical, and pragmatist. It should now be clear why it is cognitive at the core: we aim to understand mathematical knowledge as it is developed by human agents—physical beings in the world, with a characteristic biological constitution, cognitive abilities, and mastery of certain shared practices. Obviously the study of cognitive abilities such as visual perception, memory, and language is most relevant to our task.

One cognitive ingredient that is central in the basic technical practices mentioned above, and which continues to play an important role in key mathematical practices, is visual thinking.[38] In fact, there is abundant evidence that visual-geometric thinking is omnipresent in non-elementary math; important examples are the real line, the complex plane, topological structures such as Riemann surfaces, and even the cone representation of the set-theoretic universe (the usual picture of Zermelo's cumulative hierarchy; see Chapter 9). From the standpoint that I am presenting, this is understandable: the widespread role of visual and diagrammatic thinking in mathematics can be regarded as part and parcel of the general phenomenon of the interplay of practices.

Notice that those same abilities are also crucial for the command of written language and systems of written symbols. One must distinguish between the contributions to cognition made by oral and written language; at the same time, it is clear that mathematical practices typically involve the latter:[39] complex semiotic systems of written symbols. Examples range from the symbols that enter into reckoning practices to those present in advanced theories—such as Dedekind's ideal theory, where

$$\mathfrak{a} = \mathfrak{p}_1^{e_1}\mathfrak{p}_2^{e_2}\mathfrak{p}_3^{e_3}\ldots$$

---

[38] There would be much to clarify concerning visual thinking, including the idea that it is not "visual" strictly speaking, but rather perceptual. Thus, the blind can develop "visual"-geometric thinking despite their visual impairment, in some cases to extraordinary degrees. See Giaquinto 2007.

[39] Devlin (2000) claims that the capacity to do mathematics is closely related to the evolutionary changes in the human brain that made language possible. A difference is that here we emphasize cultural practices involving signs, symbols, and diagrams.

represents the unique decomposition of an ideal into a product of prime ideals.[40] Notice, too, that other practices, such as those of working with diagrams, enter into the most important forms of mathematical knowledge, and here too visual perception is playing a central role—as is manipulation.

Human agents (like members of any species) diverge quite widely in their abilities and skills. The notion of a "normal" or "average agent" is a theoretical construct or a statistical product, but it is nonetheless true that most of our fellow humans share with us minimal command of a good number of abilities (maybe more). Thus, such a notion is important for the elaboration of a theoretical model of what a typical producer or student of mathematics can do. Just as most of us are able to learn to drive a car or play simple tunes on a flute, most of us manage to learn to count and reckon and solve simple equations and understand basic Cartesian geometry. I wholeheartedly agree on the importance of understanding and emphasizing *differences*, which has been a characteristic element of biology since Darwin and of social science and psychology since Quetelet and Galton. But surely the reader will concede that simplified models can and must play an important role at the level of theory.

In the next couple of chapters we shall discuss a bit more the issues mentioned above. Here, I would like to say a few words about perception, with the aim of raising awareness about the fact that perception is a more complex ability than imagined by traditional philosophers (from Locke onward), or even by psychologists in the early twentieth century. What I am about to say resonates with themes discussed in Chapter 1: the philosophical tradition was based on a "spectator theory" of knowledge, with the subject (the "mind's eye") sitting in a theater, passive and expectant, and with sense-data, in flux, being displayed on the stage. Perception was thought to be the main form of relation between the

---

[40] An *ideal* is a certain kind of set of algebraic integers (*closed* for sum and certain products, i.e., such that the sum of two elements of the ideal also belongs in it, and the product of an element of the ideal and any integer is also in the ideal); a *multiplication* of ideals is then defined ($I \cdot J$ is the additive closure of the products $i \cdot j$, for all $i \in I$ and $j \in J$), and then a natural notion of prime ideal emerges. Dedekind was able to prove that, here, we obtain the analogue of the Fundamental Theorem of Arithmetic (that each natural number can be expressed uniquely as a product of prime numbers).

mind and the "outer world,"[41] and the key epistemological problem was that of the reliability of perceptual information (hence an interest in illusions and hallucinations). But human action, our acts, were absent from the picture.

Consider what a cognitive scientist has to say today: "In perception, the world appears before us as available to our thoughts and susceptible to our deeds. What we perceive shapes our thinking and guides our action" (O'Callaghan 2007). A crucial shift has been made, one that occurred somewhere in the middle of the twentieth century: physiologists, psychologists, and philosophers started insisting on the view that perception is not an immediate or primordial form of acquaintance with "the given," but rather a complex cognitive ability that emerges at the interface of sensory inputs and motor outputs.[42] One cannot sufficiently insist on how important the change is for eradicating the image of a passive spectator, the subject with its sense-data, and putting in its place the notion of a cognitive system that is only partly hardwired in our brains. Perceptual abilities progress and mature from birth, governed by stabilities in the interplay between the different senses and motor actions. Imagine the progress from the rather hallucinatory display of sights, sounds, smells, tastes, touches, and (mostly involuntary) motions experienced by a newborn,[43] to the increasingly ordered and coherent world of the child. For a concrete picture, imagine this: she hears the noise of a ball somewhere out of sight, turns around the corner, observes his or her schoolmates playing soccer, and runs to meet them joyfully for a game.

Cognitive science aims to explain the processes and "mechanisms" by which organisms understood as biological and cognitive systems perceive. It aims "to describe and explain how a creature accomplishes the feat of perceiving given constraints imposed by its physiology, environment, and goals" (O'Callaghan 2007). How do our bodies—our sensory

[41] The word comes from the Latin *perceptio*, verb *percipere*, and means "receiving, collecting, action of taking possession, apprehension with the mind or senses." Notice the many metaphors involved.

[42] In this connection one may mention the work of Gibson and of D. N. Lee, who coined the motto "perception and movement are two sides of the same coin, the coin is action."

[43] The literary work of William Faulkner is a wonderful attempt to capture (artfully!) such arrays of rough "sense data" in flux.

apparatus and our brains—ground our awareness of things around us based on sights and sounds?

In the midst of its oblivion of action, the Western philosophical tradition went on to insist that one's perceptions are a result of interplays between past experiences and the sense-data, by way of all important *interpretation*. This is clearly correct, but too easily understood in biased ways. While motor action and deeds were ignored, and kept out of the picture, the role of culture in shaping memory and guiding interpretation was insistently emphasized. I do not want to reverse the situation completely, ignoring culture for the sake of action, but clearly we need some balance. There are very good examples of experiences for which cultural background is quite essential; to take just one example, some people are horrified when they see the picture of a person's head (they seem to think it has been cut!). But when it comes to analyzing the famous Gestalt experiments showing what that school called the principles of "reification" or "invariance," the constitutive role of action in perception becomes central. To express it briefly and metaphorically, culture will not be able to turn our visual space upside down, the way that wearing inverted glasses for more than a day does.[44]

In what follows, we shall follow the usual tendency to employ the words "cognition" and "cognitive" to refer to basic abilities like perception, memory, and language, which we may classify as psychological or neurophysiological. But the idea of cognition may be somewhat ambivalent, as with the concept of *distributed* cognition developed by Edwin Hutchins.[45] It seems natural to say that our systems of written symbols are cognitive, and human cognition depends on the command of certain practices that involve objects we interact with, and other individuals, so that cultural components clearly enter into its constitution. It is precisely in this broader sense that I have stated that my approach is meant to be cognitive—not in the narrow sense of the more usual cognitive science.

---

[44] For the celebrated Stratton experiment, see Wade (2000), or the classic Merleau-Ponty (1962).

[45] See Hutchins (1995). According to this view, human knowledge and cognition are not confined to the individual; memories and knowledge are distributed on the objects, the tools, and the other individuals in our environment.

All of the above makes it necessary to ask a question of obvious methodological import: What is the proper way of framing the interaction between cognitive science and studies of mathematical practice? The results of cognitive scientists are of central interest to us; ideally, a research program such as the one proposed here should be developed by multidisciplinary teams. But let me be clear: we are interested in understanding mathematical knowledge as it is developed by human agents, and this task cannot be accomplished by biologists or psychologists alone.

It is my firm belief that it would be premature to frame the study of mathematical practice as a case for the application of currently accepted theories in cognitive science. This is too often the way cognitive philosophy of science is understood. I have granted that cognitive science can say a lot about the neurological bases of proto-numerical and proto-spatial representations. But even a basic number concept (in contradistinction to the innate, imprecise notion of numerosity) involves complex combinations of basic cognitive abilities and culturally transmitted cognitive practices: perhaps the skilled use of objects such as the abacus or the quipu; writing, for instance, special systems of number writing and reckoning; or, at the very least, a sophisticated, recursively produced system of number-words. Current cognitive science does not seem to have reached the point of being able to tackle satisfactorily complex combinations of cognitive abilities (say a delicate combination of perception, memory, and oral language). Cases when those abilities are supplemented by sophisticated cultural practices—like reckoning based on the Indo-Arabic numerals—are even more complicated.

My answer to the question about the proper way of framing the interaction between cognitive science and studies of mathematical practice is that the relation ought not to be one of simple application, but rather a two-way street of multiple interactions. Cognitive science and biology must play a dual role: (i) as a broad constraint for our models of mathematical practices, so that the elements we presuppose are not ruled out by biological or cognitive considerations; (ii) as a field of convergence, in the sense that we should expect our models and considerations to converge (at least at some future time) with developments in cognitive science. According to (ii), the hope is that the procedures we hypothesize will eventually come to be grounded in cognitive processes. On the other hand, the constraint (i) could be used, e.g., to rule out a faculty of intellectual intuition, such as the one Gödel apparently was proposing in 1964.

A simple one-way relation of application does not make room for surprise, but the situation we are facing is just the opposite; accordingly studies of mathematical practice can make substantial contributions to the development of cognitive science. Because the relation is and should be two-way, our work can help bring to light crucial relationships between the elements entering into complex cognitive developments and establish targets for cognitive science. (Ultimately the issue we are dealing with is related to questions about reductionism in science: the link I am proposing for cognitive science and studies of mathematical practice can be compared to the connections between physics and chemistry.)

The above remarks in no way preclude cognitive scientists designing and conducting careful empirical studies of such complex combinations of cognitive skills as we are dealing with, and offering important inputs into the study of mathematical practices. That is not what I have done, but I would welcome wholeheartedly any such attempt, provided it is well designed and carefully analyzed.

3.6. AGENTS AND "METAMATHEMATICAL" VIEWS

As we have seen, the most basic approach is to analyze the agent as a "normal" human, whose cognitive abilities lie each toward the middle of the Gaussian curve—and similarly for her command of the basic practices involved. Those of you who admire the great minds in the history of mathematics could immediately complain, "but Gauss was not like that; nor was Ramanujan certainly!" All right; but these pages aim to show how far one can go in understanding mathematical knowledge without having to deal with the idea of "genius" (which anyhow is a problematic idea: we know nothing about Gauss's cognitive abilities; we know nothing about his IQ; and although his brain is kept somewhere in formaldehyde, in all likelihood that will not help).[46]

But it is possible (in fact, easy) to think of ways in which the notion of the Agent can be enriched by putting more information into it, so to speak. One possibility would be to respond to the complaint above, considering "super-normal" agents, or geniuses. But there are other options.

One interesting possibility that helps bring our approach in line with actual historical work is to consider *historical* actors, with their specific metamathematical views and research agendas. This is one of the ways in which I would recommend recovering Kitcher's element of "metamathematics" from within the core scheme I am proposing. Let me remind the reader that this use of the word "metamathematics" deviates strongly from the usual twentieth-century meaning of foundations analyzed from the standpoint of mathematical logic (especially metatheory, and, in particular, proof theory).

What Kitcher (1984, 189ff) called "metamathematics" can be quite complex and variegated: think of basic ideas concerning the map of the mathematical disciplines, of changing criteria and values concerning rigor, of views regarding the proper way of articulating mathematical theories (e.g., the contrast between geometry-centered math and the nineteenth-century ideal of arithmetization), and so on. In actual history these elements admit many different configurations, to the point that two contemporary agents will, typically, show agreements and differences in a complex pattern.[47] As a result, I find it confusing to put such an element together with the other four, $<L, S, R, Q>$, without further ado. It seems obvious that, here too, Kitcher was influenced by the tradition of Kuhn's paradigms and related overarching, all-inclusive models of scientific change.

---

[46] One can suspect that Gauss must have ranked high in basically all cognitive abilities, since he appears to have been an impressive calculator with high visual-geometric skills, and also a great conceptual thinker; concerning his IQ, I see little need to think that it must have been above some figure, since probably there are other (subtler) explanations for his accomplishments.

[47] Famous examples are Riemann and Weierstrass, Dedekind and Kronecker, Hilbert and Poincaré, but one can multiply them indefinitely.

For instance, Carl F. Gauss revitalized the Euclidean ideal of a strictly deductive mathematics, searching for a "logical edifice" of propositions linked "by an iron chain of thoughts" to basic truths, which ought to be "generally acknowledged by the human mind." At the same time, he altered the traditional picture of the hierarchy of mathematical disciplines, with pure arithmetic sitting at the top and replacing the geometry of the ancients and early moderns.[48] Both motives combined in his great youthful contribution, the path-breaking *Disquisitiones arithmeticae*, and in his early proposal of an arithmetization of analysis and algebra. For Gauss, *pure* mathematics reached from number theory to algebra and real and complex analysis, but geometry was already "impure" insofar as (conceived to be the theory of physical space) it inevitably has an empirical component: according to him, it was not a priori like pure mathematics. Gauss also was a stern defender of mathematics as a clue to the secrets of Nature, on the simple basis that Nature is God's creation. And even in those cases when the complexity of a phenomenon defeats us, "God arithmetizes," and "the logic that goes through the whole cosmos" would be understood if we enjoyed a deeper insight into the number relationships.

A short remark is in order here: it is often the case that the views of a mathematician evolve in time, sometimes quite radically—think of Frege before and after the discovery of the paradoxes, shifting from the conviction that arithmetic is pure logic to the view that its foundation is geometrical. Thus, if we aim to produce a historical analysis including a characterization of an agent's metamathematical views, we must either be very specific about the time frame, or make room for an analysis of the dynamics of such views.

Such metamathematical views allow for innumerable combinations, which is one of the reasons why it seems wise to associate them with concrete agents. For instance, Weierstrass shared with Gauss all of the views I have just listed, but they differed in their particular conceptions

---

[48] See Ferreirós (2006). The quoted words, here and below, come from S. von Waltershausen, a Göttingen professor and friend of Gauss, in his obituary of 1856 (*op. cit.*, 236); they are highly likely to be faithful to Gauss's ways of expression. See also original quotes from Gauss in the article just mentioned.

of arithmetization and (relatedly) of how to build up "pure arithmetic." The core of the disagreement has to do with an *ideal of reduction* of all arithmetic (the real and complex number systems) to the theory of the natural numbers, which is characteristic of Weierstrass and also Dedekind, but foreign to Gauss's views.[49] To give one last example, although that ideal was normally linked with the exclusion of geometry from pure mathematics, in 1899 Hilbert was able to change this conception by relinquishing the notion that geometry is the study of physical space, and thinking of it essentially as the study of spaces, i.e., spatial structures.

The examples we have been reviewing show that a group of mathematicians can often share a good number of metamathematical views. One can describe the constellation of motives that Gauss, Weierstrass, Dedekind, Cantor and Hilbert shared, insofar as all of them were guided by the ideal of arithmetization—but also those other ideas that nuanced their conceptions and made them differ. This suggests a third way of employing the notion of an Agent, namely, at the collective level of research schools or mathematical traditions.[50] In this case, one would describe the metamathematical views and the mathematical practices that were shared by typical agents of a given community. Members of the famous Berlin school of mathematics, led by Weierstrass in the period circa 1865–1890, typically shared his views discussed above; had studied very carefully the elements of arithmetic, the algebraic basis of analysis, elementary analysis in the celebrated ε-δ style, the theory of analytic functions, and so on; and had fully absorbed "the need for painstaking care in proofs, in contrast to the mathematical 'romantics'."[51] Examples of the mathematical "romantics" recalled here by Schwartz, who dream about results but do not fully prove them, would be Felix Klein and notoriously Bernhard Riemann (later on, a famous case would be Poincaré).

The importance of considering not only metamathematical views, but also the network of practices actually commanded by the agent, is

---

[49] Petri and Schappacher (2006) offer a careful analysis of the many variants of arithmetization in the nineteenth century.

[50] See Hawkins (1981), the introduction to Ferreirós (1999), and Rowe (2003).

[51] Schwartz commenting on his "pride" in being a member of "the Berlin mathematical school" in a letter to Cantor, 1870. Quoted in Ferreirós (1999), 158.

indisputable also in the case of studying a single historical agent. But in this case it is perhaps less necessary to insist on the point, which is naturally taken into account.

3.7. ON SYSTEMATIC LINKS

What is meant by "systematic links"? The expression may be somewhat ambiguous, and needs to be clarified case by case. We have already mentioned some examples, the most paradigmatic being the interconnections between number systems, and the links between those systems and spatial configurations.

Consider one of our previous examples, set-theoretic arithmetic as interconnected with reckoning arithmetic and counting. Set-theoretic arithmetic identifies numbers with the elements of an infinite set; to concretize ideas, let us consider the von Neumann ordinals. They have clear ordinal and cardinal properties, so that the set '3' representing number 3 has indeed three elements and is readily seen to be the successor of '2' and predecessor of '4'. Induction is guaranteed to hold by definition of $\mathbb{N}$ as minimal closure of $\{0\}$ under the successor function, and this allows for the introduction of sum and product and the proof that they have the well-known algebraic properties. All of this creates very clear links; set-theoretic arithmetic produces an abstract, idealized representation of common arithmetic that can offer deductive justification for its well-known features.

Consider also the systematic links between counting numbers, natural-number arithmetic, and fractions. In this case we can already recognize the typical situation of a domain extension with immersion of the previous theoretical elements into the new ones.[52] The reckoning numbers are identified with certain fractions in such a way that the old laws of reckoning are validated; they are also extended to obtain a better-rounded theory. (In modern terms, we obtain the structure of an ordered field $\mathbb{Q}$, with dense order, instead of the commutative monoids that form the ordered, discrete set $\mathbb{N}$.) Obviously, this interconnection

---

[52] See Manders 1989.

was well understood long before there existed the conceptual means to formulate it as an immersion in the modern sense.

Very often, the links that may interest us while studying practices can indeed be analyzed explicitly with modern mathematical theories. This is because of the fact that mathematics often progresses by *thematizing*, developing through explicit theory what had previously been well-established practice, but was implicit. An important example is the correspondences, bijective or not, between elements of different mathematical theories, a notion that came to be formulated and theorized explicitly in the context of set theory through the idea of mapping.[53] Subsequently, the general idea of a morphism dominated twentieth-century mathematics, in both set-theoretic and category-theoretic guises. Thus modern mathematics is strongly recursive, recurrent, or self-referential, in the sense that it possesses powerful theoretical means that are applicable to mathematical theories themselves. But we must remind ourselves that this was not the case through all classical and modern history of mathematics, up to 1850 at least.

In speaking of systematic links between mathematical practices, we open up a wider field of study than what is normally treated in explicit mathematical or logical theories—even if, for the reasons given, one will easily find counterparts inside today's mathematics. Certainly it is not the same to attempt to clarify the grounds on which a systematic link was understood, presented, or taught in an old mathematical practice, and to make explicit, in crisp mathematical terms, what that link consists of under the lens of twentieth-century math. Notice that, analogously, the means by which a student of mathematics comes to understand a certain link may in practice deviate strongly from the sophisticated theoretical analysis of that link that her professor may prefer to consider.

3.7.1. Geometric links

There are also systematic links at the level of technical practices: consider the link between counting and measuring, in the sense that measuring procedures always depend on establishing a recursive procedure

---

[53] Interestingly, Dedekind was the first to do this, while Cantor never came to possess a general theory of mappings. See Ferreirós 1999, 225 (cf., e.g., 275, 288) and 228*ff.*

of adding basic units of measure, which is the basis for the correlation of numbers with entities that are not numbers. If we think of classical forms of measurement (without presupposing the real numbers) by means of units and subunits, what this produced—seen through modern eyes—was a correlation of fractions with quantities.[54]

Another privileged case is that of the geometry of Euclidean plane and space. First, one finds the systematic links established between Greek geometric practices and Cartesian geometry, which is an extension both conceptually and methodologically, enlarging also the range of problems considered. But the case of geometry is very rich historically and theoretically, Euclidean geometry was expanded but also relaxed (or weakened) in many different ways, and as a result it was immersed in more complex, richer geometries—projective geometry à la Klein, abstract geometry à la Hilbert, and, more radically, differential geometry à la Riemann. The Euclidean system is obtainable again by specification, for instance Klein showed how to introduce a metric with purely projective means, and then how to specify this metric to get not only non-Euclidean planes but also the Euclidean one.[55]

Once again, attention to systematic links in mathematical practice underscores other aspects beyond what stands out from a modern logical or mathematical perspective. Descartes's way of doing geometry is quite different from later presentations of analytic geometry, and in particular the role of diagrams in his proofs deserves attention, as does the way he establishes an interplay between algebraic methods and geometric ones.[56] These peculiar methods direct our attention to characteristic links between the new geometric practice and the ancient Greek one.

Particularly interesting and complex are cases when a new theoretical body is being developed with the aim of obtaining correspondences at certain levels with established notions or a theory of a different kind. The paradigmatic instance of this was the introduction of real quantities to obtain an isomorphism with the order relations (right-left) of points on

---

[54] A purely mathematical counterpart of those ancient links is the action of scalars on vector fields.

[55] See Gray 1989 and 2008, Bartocci 2012, and Hartshorne 2000.

[56] See Bos 2001, and forthcoming work of Manders.

the Euclidean line (more generally, with lines in a manifold). On the one hand, the introduction of irrational numbers, such as quadratic irrationals $\sqrt{3}$, $\sqrt{5}$, etc., was a forced move (assuming of course the "strange" idea that there should be a number corresponding to each possible measure of length—the basic hypothesis here), since existing geometric theory could prove their existence; the same, of course, goes for π. On the other hand, Greek geometry, and even Cartesian geometry, did not completely specify the range of irrationalities to be introduced, so eventually the arithmetic or set-theoretic definitions of the real numbers came to supplement geometric conceptions, introducing the idea of a *complete* line.[57]

This is a topic that will figure prominently in the rest of the book; but let us briefly consider another line of development. With the real number system, even in the informal and not fully developed form that it took around 1800, arithmetic operations and properties could represent the phenomena on the line. Then, in the first third of the nineteenth century, the new theories of the complex numbers turned out to specify geometric operations on the plane (e.g., with vectors), establishing an isomorphism with the arithmetic operations on complex numbers. With all this in place, the idea emerged of transferring such correspondences to the geometry of space, finding a suitable number system; this was the origin of the celebrated quaternions of Hamilton (the earliest example of a non-commutative field), and of the elaboration of vector calculus.[58]

3.7.2. Links, extensions, and revision

The interesting cases may typically require theory *extension* in order to obtain new methods that may solve questions posed in the context and with the means of previous theory and practice. Think of Galois theory: the initial question is to explain why there is no general solution to the fifth-degree equation by radicals (Abel, 1825) and in what cases higher-degree equations may admit solving by radicals (Galois in 1831).[59] This involved wholly new elaborations in the direction of both field theory

---

[57] This was extensively discussed in the late nineteenth century under the heading of the "Cantor-Dedekind" axiom: that a point on the line corresponds to every real number.

[58] See Crowe 1967 and Hamilton 1837.

[59] As is well known, his work was only published fifteen years later by Liouville.

(different interconnected "domains of rationality") and group theory—and particularly the structural interrelations between group-theoretic and field-theoretic formations (to describe it metaphorically), where the simpler group-theoretic elements calibrate and clarify the links between field extensions.[60] Eventually this led to mathematical work in which the initial motivation of studying equations and their solutions receded more and more into the background, and the modern conceptions of abstract algebra (field automorphisms, for instance) occupied the foreground. (Several important twentieth-century developments have been modelled on the example of the classical Galois theory; an example is Grothendieck's Galois theory, which links with algebraic topology and algebraic geometry. We find again an exemplar, this time a high-level one.[61])

It is well known that extensions may not be conservative. To go back to our first example, set-theoretic arithmetic can prove results that are not attainable in the base theory; classic examples are the Gödelian one of proving the sentence Con(PA), which asserts the consistency of Peano arithmetic; Goodstein's theorem; and the miniature version of Kruskal's tree theorem given by Friedman.[62] And it is important also to notice that an immersion may involve a strong reconception, as is the case with the *processes* of reckoning arithmetic when they are reconceived as *static* set-theoretic structures.

In other interesting cases, the systematic links between practices involve *revision* of the previous practice. This is apparently the most complicated situation, for in such cases there is no conservation of all previous results, nor of the methods, but a delicate balancing of conservation of some methods and results (or adequate reformulations of them) together with the rejection of other results. A clear example of this would be the transition from the infinitesimal calculus, with its infinitesimals and differentials, to Weierstrassian analysis. Notice that in this

---

[60] Whether a polynomial was solvable or not turned out to be equivalent to whether or not the permutation group of its roots—the Galois group of the polynomial—had a certain structure, whether or not it was a solvable group, i.e., a group that can be constructed from abelian groups by means of extensions.

[61] See Dubuc and Sanchez de la Vega (2000). http://arxiv.org/abs/math/0009145v1.

[62] For details about these, see Stillwell 2012; we return to this topic in more detail later on, in Chapter 7.

case we have (at the very least) a doubly and inversely oriented interplay of constraints: on the one hand, the pull of geometry and basic theory of quantities of the Stevin-Newton kind, tending toward restriction; on the other, the tendency to amplification in the direction of infinitesimal methods and results, which were to be preserved as far as possible.[63]

In the case of set theory, the systematic links are many, in many different directions, always acting as merely partial constraints, but (except for the case of the real numbers and infinitesimals, at the origins of set theory in 1872) with the aim of conservativeness. One may mention the links with the number systems, which set theory systematized; with the known geometries (Hilbert 1903 is clearly framed in naïve set theory); with algebra in its modernizing trends (starting with Dedekind 1871); and centrally with the theory of functions (Dirichlet, Weierstrass, Cantor, Baire, Lebesgue, among others).[64] We find the extension of elements and methods in all of these domains, aiming at the development of new results that may advance the theory of arbitrary functions (discontinuous functions, more general notions of integration, classification of functions, and so on); develop new and deeper insights into algebraic situations of different kinds (structures and morphisms in Galois theory, in advanced number theory, in algebraic geometry); provide methods to investigate new topological questions (limit points, neighborhoods, point set topology, general spaces); and—last but not least—also serve the elaboration of self-motivated or endogenous developments (cardinal and ordinal arithmetic, CH, etc.).

It is precisely the richness and intensity of this network of interconnections that makes set theory a very peculiar instance in the history of mathematical knowledge. It would seem that even category theory has not integrated so many diverse elements and orientations simultaneously.

---

[63] The ideas of Stevin and Newton will be discussed in Chapter 8. Historical studies of this famous transition can be found in Grattan-Guinness 1980, Bottazzini 1986, and Jahnke 2003; it is also discussed in detail by Kitcher 1984; but perhaps there is a need for new analyses from the kind of standpoint I am advocating.

[64] I offered a historical analysis of these links in Ferreirós 1999.

4

Complementarity in Mathematics

The approach I am sketching has the advantage of suggesting ways in which considerations about thought and meaning, about the conceptual side of mathematics, can be elaborated. There is of course a more or less traditional way of doing this, which consists in introducing a plane of semantic entities—concepts, relations, propositions—and considering simply the interplay between syntactic expressions and semantic entities. The approach I wish to propose is to explore a rather natural way in which semantic considerations *emerge* from the interplay between agents and frameworks, in the context of mathematical knowledge and practices. This way of proceeding avoids the difficulties involved in postulating abstract semantic entities.

In order to introduce this broad and difficult topic, let me begin with considerations about form and content in mathematics itself.

4.1. FORMULA AND MEANING

In the development of mathematics, we find noteworthy instances of two contrary tendencies. One has been the drift toward reduction of mathematics to a purely symbolic system; notable instances can be found in twentieth-century strict formalism, but also in Lagrange around 1800, Peano around 1900, and many others. A colorful quotation, expressing views akin to logical positivism and strict formalism, was provided by Wittgenstein in 1939:

> The only meaning they have in mathematics is what the calculation gives them . . . if you think you're seeing into unknown depths—that comes from a wrong imagery. (Wittgenstein 1976, 254).

This seems to presuppose a reduction of mathematics to merely the symbolic frameworks discussed above.[1] Kronecker once wrote to Cantor that he found real scientific value "only in mathematical formulas" and remarked that Lagrange's theories about the foundations of the calculus are forgotten, "but Lagrange's resolvent is here to stay!" (Meschkowski 1983, 238–239).

The other tendency has been the attempt to reduce mathematics to a purely conceptual system; noteworthy examples can be found in the nineteenth-century trend labeled "the conceptual approach," e.g., in Riemann and Dedekind, but also in some twentieth-century proponents of category theory.[2] In his doctoral dissertation, Riemann had already explained that his theoretical approach was aimed at determining the functions under study "independently of any form of determination by means of numerical operations," by starting from the "general concept of a function" and adding to it only the "attributes necessary for" specifying the laws of dependence; only after this, as a result, would one proceed to the study of "expressions" or formulas (Riemann 1851, § 20, 38–39). Dirichlet summarized this innovative idea with his famous phrase "to put thoughts in the place of calculations," and Dedekind would be a staunch defender of this "deep scientific idea"—he promoted structural notions and methods in mathematics. But this has never led to an elimination of symbolic frameworks, only to the introduction of new frameworks.

When Hilbert, continuing in that tradition, first encountered Frege in 1895, he reacted negatively to Frege's logical work, saying that his effort in mathematics "was oriented toward diminishing the repertoire of formulas [*Formelwesen*], rather than augmenting it." Frege had to explain that his aim with the new logical symbolism was just to capture "in symbols"—completely, briefly, precisely, and revisably—the new, deeper methods that had been developed by men like Riemann, so that he saw no incompatibility. "One can also think in symbols," he

---

[1] Whether this reflects Wittgenstein's mature opinion (probably it does not) is a question I must leave to experts in the topic.

[2] I have discussed the nineteenth century conceptual approach myself in Ferreirós 1999. For category theory, see, e.g., Mac Lane (1985), Lawvere and Schanuel (2005), and the historico-philosophical discussion in Krömer (2007) and Marquis (2009).

said, and this expresses his intent as a logician and mathematician very well.[3]

The impression one gets from studying these twin developments is that neither of them has been successful. For obvious (practical!) reasons, it is more difficult to deny the role and importance of the symbolic component in mathematics, but substantial arguments can be given for a similar conclusion concerning the conceptual component. Normal symbolic systems and theories, the so-called *informal* ones, cannot be made to stand alone outside of practice; they depend on the agents' ways of using them and the meaning they associate with them. On the other hand, when systems and theories are formalized and made to stand alone, the phenomena of nonstandard interpretations and incompleteness arise in a natural way (see below).

In my view, the failure of both the formalistic and the conceptualistic tendency is of the essence.[4] The standpoint I adopt emphasizes the need to consider the meaning or thought that accompanies formulas and calculations. This is no doubt shared by many other philosophers, but the question is *how* to do it. My proposal goes along the following lines: mathematical symbolism cannot be mastered without immersion in a practice, and by learning the practice—simultaneously absorbing knowledge—we agents learn to associate representations and meaning to the formulas. Moreover, even if such processes may suffer variations from agent to agent, they are not subjective in the usual sense.

Thus, I propose considering a *principle of complementarity of symbolic means and thought* in mathematics—each one complemented by the other, neither reducible to the other. For the present purposes, we

---

[3] See Frege's *Briefwechsel* (1976), 58. Letter of Oct. 1, 1895, after their encounter in Lübeck during the *Naturforscherversammlung*. See Tappenden (2006).

[4] I should mention that there has long existed an intermediate position, based on the hope that both of these attempts could fully converge. This was the idea that a symbolic system of formulas could be developed, expressing the true relations between thoughts or conceptual processes; see, for instance, Grassmann (1844), inspired by Leibniz's ideal of a *mathesis universalis*. In the Leibnizian revival of the mid-nineteenth century, many authors were influenced by this vision, including Boole, Frege, and Hilbert. Obviously, such goals have been of great consequence for the development of mathematical logic and foundational studies, and obviously, too, Gödel's incompleteness theorems established crucial limitations they face (which, of course, do not preclude limited partial realizations from achieving success, as witnessed by computer methods).

can be content with the modest claim that, in light of developments in mathematics and its foundations during the twentieth century, such a standpoint deserves to be seriously considered as an option.

The complementarity of symbolic means and thought explains the distance between formulas, including formal axioms, and conceptual understanding. A trivial example is the following:

$$2 + 2 = 5 \text{ is obviously correct,}$$

if we permute and take "5" as the cipher associated with the number four. Here we exploit the conventionality of the link between ciphers and numbers. One can offer a real example considering formal systems and the phenomenon of nonstandard interpretations.

Take for instance the real number structure $<\mathbb{R}, 0, 1, <, +, \cdot>$, characterized by some version of Hilbert's axiom system. Intuitively, by $\mathbb{R}$ we mean the set whose elements are all the real numbers, and intuitively we grasp very clear and systematic links between different ways of approaching this matter (in my jargon, different practices and knowledge strata). So we think that $\mathbb{R}$ will (i) have elements that correspond to any possible combination of the ciphers forming a number in decimal representation, (ii) correspond to equivalence classes of Cauchy sequences of rational numbers, defined along the lines of Cantor's work, and (iii) also admit a one-to-one correspondence between its elements and points on a geometric line (cf. Chapter 8). But all such links disappear from the picture once we decide to concentrate on a purely formal system. The formal axiom system in a strict first-order formulation leaves ample room for Skolem's paradox, that is, it allows for nonstandard, "non-intended" models: in particular, the axioms admit interpretation in a countable domain! This conflicts with set-theoretic considerations, with Cantor's basic result that no countable set of real numbers can contain all the real numbers. Despite the fact that $\mathbb{R}$ is the prototype of a nondenumerable set, its purely formal, first-order theory does have countable interpretations.

The above-mentioned paradox was established by Thoralf Skolem in a beautiful paper (1923) on set theory as a foundation for mathematics. It is a consequence of the (downward) Löwenheim-Skolem Theorem: any given system formalized in first-order logic that has an infinite model also has a countable model.[5] As the reader probably knows, there is no contradiction here, just a paradox in the strict sense of the word: something that goes against expectation. There is no contradiction, among other things, because the means of the presupposed axiomatic theory of $\mathbb{R}$ do not allow us to show that the domain of interpretation Δ can be put in one-to-one correspondence with $\mathbb{N}$. Such a correspondence can be established "from the outside," inside a broader theory, e.g., in axiomatic set theory and considering a wider domain E that includes Δ.[6]

Such models are called "nonstandard," but also often "non-intended," models. This customary way of speaking deserves some attention. The "intention" here is given by what we mean or think, just as with the "contrary expectations" mentioned above. From the standpoint delineated in the Wittgenstein quote of 1939 (see above), talk of "intended models" is sheer nonsense: we would only have what the formulas and calculations give us. And yet our expectations are not at all arbitrary; and I surmise that the approach adopted here (with its emphasis on the *interplay* of knowledge strata and associated practices) offers means to explain how and why they are not.

I should make clear that I am not taking for granted that formal systems must be first-order; on the contrary, I have nothing against higher-order systems, properly interpreted (see below, and Ferreirós, forthcoming). On the other hand, I am assuming that sentences in a formal system must be capable of concrete exhibition, in principle at least—in other words, they must admit recursive generation. This is a position in the tradition of Frege, Gödel, Quine, and others that takes for granted that a constructive restriction belongs in the very notion of formal system.[7]

The paradoxical in Skolem's paradox comes from the distance between the formal axioms and our conceptual understanding. We may

<hr />

[5] Later, Tarski established the Upward Löwenheim-Skolem Theorem; in its light, the formal theory of $< \mathbb{R}, 0, 1, <, +, \cdot >$ admits models of any cardinality—$\aleph_0$, c (the continuum), $\aleph_2$, $\aleph_\omega$, or any aleph you wish.

[6] On Skolem's paradox, see Jané 2001.

[7] See, e.g., Gödel (1944, 132), which speaks of "the fiction that one can form propositions of infinite (and even nondenumerable) length."

be inclined to say that, in a non-intended Skolem model, the formally given set $\mathbb{R}$ "cannot really have" as elements all real numbers. By doing so we manifest our impression (our thought) that the formal system does not capture what we meant. The complementarity phenomenon that I described above is thus emphasized.

4.2. FORMAL SYSTEMS AND INTENDED MODELS

Similar considerations apply to almost all other formal systems, e.g., to the Dedekind-Peano axioms for arithmetic. The topic of nonstandard models of first-order Peano Arithmetic (PA) constitutes a rich field of mathematical work. How is this possible? One can conclude the existence of nonstandard models as a consequence of both the completeness and the compactness of FOL, or from the celebrated incompleteness of PA. Every model of PA contains an initial segment isomorphic to the "intended" structure $\mathbb{N}$, i.e., with all the standard naturals; but in general, models contain further, nonstandard elements. In fact, there are not only countable models of PA, but also uncountable ones (by the upward Löwenheim-Skolem theorem).[8] Thus we have that neither the paradigm of countability, $\mathbb{N}$, nor the paradigm of uncountability, $\mathbb{R}$, can be appropriately captured by formal languages and formal systems.

Many logicians see a way out from that through the adoption of so-called "full" second-order logic, also called "standard" second-order logic (SOL). What is meant here is the *intended* set-theoretic interpretation of the second-order logical calculus—and with this hint, the reader should understand that we are moving in circles. As a matter of fact, the standard semantics of SOL cannot be based on the assumption that the full powerset of a given domain is uniquely and correctly determined—except in case we want to consider SOL as applied informal mathematics, applied (naïve) set theory in this case. For the phenomenon of nonstandard models applies just as much to axiomatic set theory as it does to Peano arithmetic.

---

[8] A standard textbook is Kaye (1991).

The question affects PA through its crucial Axiom of Induction: in FOL, this is an axiom scheme, which concludes that all numbers fulfill $\varphi(x)$, symbolically $\forall x \, \varphi(x)$, from

$$\varphi(0) \text{ and } \varphi(n) \to \varphi(n'),$$

where we write n' for the successor of n. There are infinitely many particular axioms that instantiate this scheme, one for each well-formed formula $\varphi(x)$ in the formal language: for in first-order PA, rather than considering all number-theoretic properties, attention is restricted to well-formed formulas $\varphi(x)$ in the formal language $\mathcal{L}_{PA}$. The expression "all number-theoretic properties" is, of course, ambiguous. Is there a well-defined totality of all number-theoretic properties? We know (by proof) that the language $\mathcal{L}_{PA}$ does not capture all number-theoretic properties; and in expanded languages, more properties become expressible. The natural broad interpretation of the Induction Axiom is as an *open-ended scheme* or even an open compromise: whenever a property of numbers becomes acknowledged, in any given language, the axiom will be applicable to it.[9]

Of course, this sounds too vague to a committed logician or mathematician, and so the desire arises to make precise in some way the idea of a totality of number predicates. One option is to go Fregean: one can understand properties as Frege did, as independently given abstract things; indeed there is no reason why formulas in a first-order formal language should correspond to "all" number properties thus conceived, and we have very precise evidence of the contrary. Yet the idea of properties as abstract entities is not only philosophically contentious, it is also understood differently by different people. To make it precise, we would need to axiomatize "property theory," and even those who tried hard to solve this problem (Gödel is an example) have not been able to offer a well-developed system. Another typical approach, which should be carefully distinguished from the one just mentioned, is to interpret "all number-theoretic properties" set-theoretically, to mean all arbitrary subsets of $\mathbb{N}$; with this interpretation, one can also prove that first-order formulas will never exhaust "all" number-theoretic properties. But new

---

[9] See Feferman 1991. The notion became relatively well known in the English-speaking world through a paper due to McGee, 1997.

worries arise here: if the identification of number-properties with open formulas in $\mathcal{L}_{PA}$ undershoots, its identification with subsets of $\mathbb{N}$ most likely overshoots. (Do you really mean that there is a property for each arbitrary set of natural numbers? To Frege, this would have sounded quite absurd.) Yet this mathematical idealization, like any other, could be quite satisfactory even if it overshoots. The real problem lies elsewhere, in the fact that nobody has been able to characterize precisely the set of all subsets of $\mathbb{N}$; we think we understand what the "real" power-set $\wp(\mathbb{N})$ is, and so we have an "intended" model here too. But fixing one intended model by relying on another is certainly not a good example of introducing mathematical precision! Meanwhile, the intended model of $\mathbb{N}$ is much better understood than the intended model of $\wp(\mathbb{N})$.[10]

Despite all that, we understand fully well what is meant by the standard, "intended" model of Peano Arithmetic, and in discussions of this topic (e.g., while lecturing) mathematicians remain content with simple statements such as: The standard model of arithmetic consists of the set of usual natural numbers $\{0, 1, 2, \ldots\}$; in nonstandard models there are elements x greater than any of the usual numbers; $x > 2$, $x > 3$, and in general $x > n$ for any counting number.[11] (In more sophisticated contexts, to explain what the standard model is, reference may be made to ω, the set of all finite ordinals, which in set theory plays the role of $\mathbb{N}$. But since the Skolem paradox applies just as well to ZFC, one is again trying to explain the clearer from the obscurer.) Why is the simple reference to $\{0, 1, 2, \ldots\}$ enlightening? Because a link is being established with most solid practices that preexist and are antecedently known by the mathematician—the technical practice of counting and the knowledge and practices associated with basic arithmetic and even with number theory.

Let me approach the matter sideways. Careful logicians such as Frege and Dedekind were very unhappy with the dots "..." in the

---

[10] For more on this topic, see Ferreirós 2011 and Chapter 9; also my (forthcoming).

[11] For instance, Wikipedia (accessed Oct. 6, 2010) states, "In mathematical logic, a nonstandard model of arithmetic is a model of (first-order) Peano arithmetic that contains nonstandard numbers. The standard model of arithmetic consists of the set of standard natural numbers $\{0, 1, 2, \ldots\}$. The elements of any model of Peano arithmetic are linearly ordered and possess an initial segment isomorphic to the standard natural numbers. A nonstandard model is one that has additional elements outside this initial segment."

expression $\{0, 1, 2, \ldots\}$. They felt that rigor was missing because of this simple reliance on a practical expedient that had not been logically grounded, and they set out to eliminate the dots and provide a fully precise logical definition of the naturals. From our multilayered standpoint, which insists on the interplay of different practices and strata of knowledge, there is a natural way of understanding the epistemic role that the dots ". . ." play. I suggest understanding them as indicators of a systematic link, of an interplay within the web of practices. They indicate a systematic connection with perfectly well-known, antecedent practices: we know how to count since preschool, and we know very well how to produce the successor of a number once this is given in any notation (be it positional, as in Indo-Arabic, or additive, as in Roman numerals; be it decimal or digital). The receiver of that information may also have good knowledge of the systematic role that the successor function plays in a deductive presentation of arithmetic.[12]

When Skolem presented his paradox in the 1923 paper, he explained its import for axiomatic set theory: first-order Zermelo-Fraenkel set theory has denumerable models. In such models of ZFC, we have both $\mathbb{N}$ and a powerset $\wp(\mathbb{N})$, but $\wp(\mathbb{N})$, being part of a countable model, is itself countable. The paradox was quite impressive, since set theory is meant to be mainly a theory of the nondenumerable; but when one adopts a formal system that complies with the exacting demands of foundational studies—namely a first-order system, possibly many-sorted[13]—the axioms have no power to enforce models into being nondenumerable. Once again there was no contradiction, because *in* the domain Δ there is no one-to-one correspondence between $\wp(\mathbb{N})$ and $\mathbb{N}$, but "from the outside" (in a wider model) one may establish the existence of such a correspondence. In this case, the paradox underlines the discordance

---

[12] Needless to say, no criticism of Dedekind and Frege is implied: for their project of a reconstruction of arithmetic *ab ovo*, purely from logic (and set theory), it was necessary to try what they did.

[13] See Ferreirós 2001 and Väänänen 2001 and 2012. We can adopt second-order logic, but for the reasons already given, not the set-theoretic "full" SOL (which oversteps the limits of logic in an ungrounded way). General semantics may be regarded as the standard semantics of SOL, and then SOL has the same power as a many-sorted FOL; this is why the Skolem paradox remains unaffected. See Ferreirós forthcoming.

between the principle of Powersets and its formal implementation in first-order ZFC (Ferreirós 2011).

The Axiom of Powersets stipulates that to any given set S there corresponds another set $\wp(S)$, called its powerset, whose members are the sets R such that $R \subseteq S$. Intuitively, by $\wp(S)$ we mean the set whose elements are all—*really all*, we might say in a useless effort to emphasize the point—subsets of C. But the formal axiom in the axiomatic system ZFC only ensures that, in a domain Δ that models the formal system, there is an object that bears the ε-relation to all objects in Δ that "*act*" as subsets of C in the domain Δ.

Notice that recourse to second-order logic is of no help here. A second-order version of ZFC can only exclude non-intended models provided that the second-order quantifiers are given a special meaning. The difficult problem involved here is usually trivialized by saying that this special meaning is "the standard semantics" of second-order logic. Good rhetoric, but no theory behind it. Notice that, if we make a second-order system stand alone, general semantics is just as natural as the preferred one—actually more natural in light of the common-sense notion of property. The "full" semantics is preferred because it agrees with our preferred set-theoretic standpoint; but the formal system per se is unable to do the job. Belief in the primitive nature and the "logical transparency" of the idea of all subsets of any given domain may be widely held, but it is just that—belief, a thought in the minds of some logicians or mathematicians. As widely held as this belief may have been—especially among logicians of an earlier generation—it does not go beyond the general thought of combinatorial maximality (a.k.a. quasi-combinatorialism) as a special understanding of set theory.[14] It is the task of set theory to spell out the quasi-combinatorial viewpoint in a detailed way.

It is the business of set theory to make explicit the principles of set existence, especially those principles having to do with existence of infinite and arbitrary sets; to relegate some of these to an underlying logic and its "intended" (a.k.a. "standard") semantics is merely to obscure that key goal. Strictly speaking, this way of resorting to second-order

---

[14] On this topic, see, e.g., Maddy 1997, the classic Bernays 1935, and Ferreirós 2011.

logic does not respect the rules of the game of formalization; it is certainly admissible in normal mathematical practice, but not in the context of a critical study of set theory. To put it differently: if our solution is to have recourse to meaning (through the special reading of the quantifiers), we might as well read the intended meanings into the first-order axiom system. When one makes the move of presupposing the "intended" interpretation of powersets, by adopting a second-order version of ZFC as if this were magically to solve the problem, formal systems and thoughts become entangled in an uncritical, unclear way.[15]

4.3. MEANING IN MATHEMATICS: A TENTATIVE APPROACH

In fact, a novice mathematician will more easily comprehend how the axioms in a formal system represent certain aspects of the intended concepts than the ways in which the formal system per se fails to capture the intended thoughts. The question is: How is this phenomenon possible?

The phenomenon suggests that, in the (developing) practice of an individual mathematician, working with a formal system ties up with certain preexisting practices and meanings. The situation is not much different from that of any other new framework (and new practice) that may be in the process of production or learning. When one studies set theory, the teacher will introduce examples such as sets of people and sets of numbers ($\mathbb{N}, \mathbb{Z}, \mathbb{Q}, \mathbb{R}$), and the latter shall immediately trigger the process of linking practices. Axiomatic approaches to the real number system are always introduced after one has learned a lot about the real numbers as decimal expansions, as correlates of points on a straight line, and as the possible outcomes of (ideally precise) measurements. The phenomenon, of course, is reproduced at other levels: in order to study category theory, sophisticated knowledge of several branches of mathematics (modern algebra, topology, algebraic geometry) is a

---

[15] For further arguments concerning this controversial matter, see Shapiro 1991, Ferreirós 2001, Väänänen 2001, and Jané 2005b (as well as the paper by Stewart Shapiro to which it responds).

prerequisite, for only then will the new concepts of object and arrow and commutative diagram and functor obtain their significance and show their potential.

All of this falls quite naturally within the scope of our approach, since we are emphasizing the coexistence of a multiplicity of practices and the centrality of their interplay. This is to be analyzed by reference to the intertwining of several frameworks handled by one (or perhaps several) agent(s). And the hope is to capture the conceptual element in mathematics, the origin of the meanings and thoughts which accompany our symbols, by considering what the agent can do—and understand—through proper use of her cognitive abilities and proper command of the relevant knowledge and practices.

Taking seriously the idea of strict complementarity between formula and meaning, the first implication is that no strictly formal system can capture the conceptual element in mathematics; nor will a hybrid combination of several formal systems. This is the same as saying that concepts cannot emerge on the basis of symbolic frameworks alone, but can only come forward when agents (with certain normal abilities) learn to employ frameworks in the way that characterizes a certain practice. To put it differently, meaning dawns upon agents when they have proper command of a certain web of practices and knowledge. And perhaps it is unnecessary to provide a philosophical explanation of what "meaning" or "concept" is, beyond laying out carefully the structural conditions for the emergence of meaning. It should be clear, anyhow, that our approach to meaning insists on use and pragmatic conditions.

This is how the agent-based, practice-oriented approach I am proposing offers the promise of making possible an analysis of the conceptual side of mathematics. The view that proper command of a certain web of practices is required for meaning or concept formation implies that the process depends both on the agent's cognitive abilities and the practices she has learned. Since we simplify our viewpoint by assuming we are dealing with (uniformly described) normal agents, some readers may have the impression that the weight of the analysis falls on the web of practices. However, this would be an incorrect inference: the fact that agents are regarded, so to speak, as interchangeable does not in the least imply that the agent's role is negligible.

In the previous chapters, we have emphasized the role of proto-mathematical practices—counting, measuring, drawing geometrical figures—at the basis of mathematical knowledge. I propose the idea that such technical practices lie at the roots of meaning. We shall flesh out this view in the next few chapters (especially Chapters 5 and 7), but I will say a few more words in this and the next section. But the proposal that technical practices lie at the roots of meaning should not be understood as reductionistic: just as the attempt to reduce "meaning" to "use" faces enormous difficulties, it would be wrongheaded to make the corresponding move in our field. Therefore, we shall be committed to the view that conceptual understanding depends on the web of practices and knowledge strata that an agent commands. That is to say, adding a new layer to the preexisting practices, in the form of new symbolic and theoretical frameworks, will in general have the effect of modifying or modulating the notions that are affected by the new systematic links. As I shall explain, this position has interesting implications both for understanding communication between mathematicians and for math education.

Consider for instance the connections between number theory, elliptic curves, and modular forms involved in Wiles's famous work on the so-called Last Theorem of Fermat. Connections such as those are one of the most fascinating phenomena of mathematics, and they arouse extraordinary interest among researchers. It is often the case that, having found new such links, researchers will start talking about having discovered "the essence" of a certain problem or subject area, and they may even come to believe that "only now" they understand what lies at the core of a certain mathematical idea. This may happen even with such basic concepts as that of natural number, for instance, in light of recent advanced work on number theory of the Wiles type.

That phenomenon shows that new interpractice connections effect an alteration in the mathematician's conceptual understanding. This can be seen with the very simple example given in Figure 1 (page 38), for arguably a child's understanding of "number" changes with his or her acquaintance with reckoning arithmetic—and an adult's understanding of the number concept is different before and after his or her having been exposed to an analysis of the structure of $\mathbb{N}$. The more advanced cases—for instance, when the natural numbers are conceived from perspectives that may be topos-theoretical, or in relation to the study of algebraic curves—would simply be another instance of the same general phenomenon.

This suggests the following thesis about concepts and conceptual understanding in mathematics, which we shall be exploring: *conceptual understanding arises from interaction of the different practices*, from the interplay of the different frameworks and cognitive resources involved. We claim that the agent's grasp of meanings cannot ultimately be analyzed without recourse to the connection between mathematical frameworks and practices such as counting or measuring—a connection that may be quite elaborate and remote. In some cases, the roots of mathematical concepts can be followed (it seems) all the way into the constitution of the neural system; we have seen how there exists an intuitive notion of *numerosity*, however imprecise it may be (for which reason we had better avoid calling it "cardinality"), that is neurally rooted. The same is likely to happen with at least some aspects of spatial conceptions, but this is a topic that needs to be studied very carefully in the future. In most cases, however, it seems more plausible to place the roots of meaning at the level of concrete techniques, learned by the agent and culturally transmitted.[16] The notion of fraction, for example, has roots both in problems of metrology and in questions of calculation, particularly when division leaves a remainder; hence the contexts of measuring practices and calculational practices are both relevant.[17]

A rough consideration of the history of mathematics shows a tendency to move from the concrete to the abstract, by which I mean from more direct engagement with those technical practices to more remote or disengaged notions. The mathematics of "number" was once based on the idea that numbers are ratios (proportions) between homogeneous magnitudes, and this makes possible a rather direct connection with the measurement of lengths, volumes, or weights. Modern structural mathematics, in either its set-theoretic or its category-theoretic form, clearly deemphasizes notions of magnitude, proportion, and

---

[16] Since anything we do has neural correlates, this viewpoint is by no means incompatible with research in neural imagery.

[17] For a sophisticated revision of the history of fractions avoiding presentism, see Benoit, Chemla, and Ritter 1992.

measurement, making more remote and elaborate the link between the basic mathematical notions and the proto-mathematical practices. The same is found when one considers the evolution of geometric notions. The main gain of such abstract conceptions seems to be that one can establish significant contact with a wide variety of properly mathematical (and other scientific) practices. The price we pay is that they become harder and harder for an agent to grasp, requiring more elaborate processes of education and the command of an increasingly complex network of practices.

The symbolic and theoretical ingredients that conform to the different frameworks employed by mathematicians do make crucial contributions to concept formation (when properly employed and understood). To give a simple example, this means that the number concept is modulated or subtly modified when the web of frameworks in which it is embedded becomes more complex. In particular (see Chapter 2, sections 4 and 5), one should accept that the introduction of a set-theoretic structural analysis of $\mathbb{N} = <N, \sigma, 0>$ as a "simply infinite" set (well-ordered by the successor function, characterized by an induction condition) did modify the mathematicians' concept of natural number. It would be naïve to interpret what happened as an instance of conceptual analysis that merely clarified an already-given concept. My thesis is in principle open to empirical scrutiny, because learners of mathematics do not encounter this type of structural analysis before a certain stage of university education.

Such processes can, at least in principle, always go on; perhaps analogous tendencies are present today due to the impact of recent developments in geometric number theory. Another good example is analytic geometry in the seventeenth century. As is well known, the new framework, exploiting a coordination between algebra and Euclidian geometry, was more than just a new method of proving and developing results in the same old geometry. The new development had implications on, among other things, the range of geometric configurations under study, leading to a new conception of geometric and mechanical curves. Once again, it would be naïve to interpret this as merely a methodological change within an already-given, unchanging field. To change the web of practices is to make a conceptual shift, or at least to introduce relevant nuances and new possibilities for further development. In this case, certainly, enormous new possibilities for further development opened up, as witnessed by the emergence of the calculus. (Parallel considerations could be applied to the calculus itself, understood in a context of theory of magnitudes vs. a context of arithmetization vs. an abstract set-theoretic context of arbitrary functions and measure theory.)

As conceptual understanding depends on the web of practices that an agent commands and links with the concept in question, agents who command different webs of practices may differ in their conceptions, and in their concepts. This, I believe, suggests a very interesting perspective for understanding the cooperation between mathematicians that has become a pervasive aspect of mathematical research practice since the early twentieth century. The interesting corollary is that a different web of practices is a different conceptualization, which applies to mathematicians who have different areas of expertise. The interesting case being when those areas may be interrelated in innovative ways, and mathematicians with different expertise cooperate.

But also, as we shall see in Chapter 6, the process of building up mathematics presents us with characteristic complexities, especially what I shall discuss as the introduction of "constitutive hypotheses." What this means is that the configuration of mathematical knowledge is not simply inherited from basic cognition, nor can it be understood somehow as a bottom-up process. There are top-down elements, constitutive hypotheses that determine (partially) some of the new theoretical frameworks and associated practices. Thus, our picture has consequences already for the assumptions of some specialists of math education, since the process of learning cannot be conceived merely as a bottom-up process: the top-down element is no less important, which means that learning agents ought to be immersed in properly mathematical practices at some relatively early point.

4.4. THE CASE OF COMPLEX NUMBERS

In order to illustrate the idea of the complementarity of symbols and thought, perhaps it will be useful to deal with an elementary example, albeit historically most significant: the imaginary unit $\sqrt{-1}$ and the complex numbers. It is very well known that the notion of a root of $x^2 =$

–1 appeared for the first time in the sixteenth century with Cardano (1545) and Bombelli (1570), but the imaginary numbers only obtained full citizenship in mathematics in the mid-nineteenth century. Euler said that they are "numbers which from their nature are impossible," and Leibniz had waxed poetical when he said they are a "miracle of analysis, the *monstrum* of the ideal world, almost an amphibian between being and non-being."[18] Even Cauchy, in his celebrated *Cours d'analyse*, wrote that imaginary numbers are "symbolic expressions which cannot be interpreted according to generally established conventions, and do not represent anything real" (translated in Kline 1972, vol. II, 635). To facilitate their acceptance, and eliminate the reproach of their supposed unreality vis-à-vis the so-called *real* numbers, Gauss felt the need to baptize them anew as the *complex* numbers.

The imaginary unit $i = \sqrt{-1}$ is an example of a phenomenon that is not so common in mathematics, namely, the introduction of a fundamental innovation through manipulations at the purely symbolic level. It was discovered by merely plugging certain coefficients into the formula for the solution of cubic equations published by Cardano. However, for 250 years this was not considered enough to deem them numbers proper— not even after witnessing the wonders that Euler could do handling them like a virtuoso.[19] There was a symbolic framework for reckoning with imaginary numbers, and all the procedures and formulas were exactly like those for other numbers. The algebra of imaginary numbers was just normal (a commutative field, as we say), unlike the situation that would later be found with the quaternions and other hypercomplex numbers. They even presented mathematicians with such a harmonious law as the existence of n solutions to any given polynomial of nth degree—the Fundamental Theorem of Algebra. Yet they were "impossible," not acceptable as numbers, and this because the existing conceptual definition of a number could not be applied to them. This explains why, when Gauss proved the Fundamental Theorem of Algebra in his 1799 dissertation, he revised the formulation of the theorem and

---

[18] See Euler 1771, part I, § 143; quoted in Bottazzini 1986, 40. See Leibniz (1702), 357. For further interesting quotes, see Remmert's chapter in Ebbinghaus et al. 1991, 57*ff.*, and the book Flament 2003.

[19] For example, in deriving formulas for cos x and sin x, in the famous *Introductio* (1748), § 138.

his proof methods so that they would only involve real numbers (Gauss spoke of decomposing into linear and quadratic factors).

For mathematicians in 1800, and a long time before, numbers had three main characteristics: (i) they could be operated according to rules or laws that are essentially coincident in all cases; (ii) they could be obtained as ratios or proportions between homogeneous magnitudes (e.g., the ratio between two volumes, or between two segments), where it was implicitly assumed that domains of magnitudes are continuous; (iii) they could themselves be regarded as magnitudes, so that a principle of trichotomy applied—numbers were linearly ordered, and two given numbers a, b were either equal, or $a < b$, or $a > b$. The *positive* real numbers were the paradigmatic example of "number," as they complied with all three requisites; they were typically associated with points P on a straight line, or, better, with segments OP from the origin, measured relative to the (arbitrary) choice of a unit segment. On that understanding, already the negative numbers presented a problem, because however you choose two homogeneous magnitudes, their proportion is always positive; they violated point (ii). Yet, since it was so easy to extend to them characteristics (i) and (iii), and their meaning was de facto clear as inverse segments $P'O$, in practice there was no particular problem using them.[20]

The imaginary numbers, although algebraically normal (point (i) above), presented great difficulties since they were not linearly ordered (one cannot define a total order, $<$, among them) and because they are not ratios or proportions. Euler discussed them in his *Algebra*, but he was unable to fully justify them:

> All such expressions as $\sqrt{-1}$, $\sqrt{-2}$, $\sqrt{-3}$, $\sqrt{-4}$, &c. are consequently impossible, or imaginary numbers, since they represent roots of negative quantities; and of such numbers we may truly assert that they are neither nothing, nor greater than nothing, nor less than nothing; which necessarily constitutes them imaginary, or impossible. But notwithstanding this, these numbers present themselves to the mind; they exist in our imagination, and we still have a sufficient

---

[20] The foundational question was intensely discussed; see Nagel 1935. For the history of the complex numbers, see Ebbinghaus et al 1991 and Flament 2003.

idea of them; since we know that by $\sqrt{-4}$ is meant a number which, multiplied by itself, produces -4; for this reason also, nothing prevents us from making use of these imaginary numbers, and employing them in calculation. (Euler 1771, part I, § 143 and 144)

This kind of pragmatic approach, which apparently satisfied Euler, would not be enough for Gauss. Notice that the salient feature, according to Euler's exposition, is that the imaginary numbers are not greater than 0, nor less than 0, nor equal to 0;[21] consequently they are "impossible," yet useful, fictions. Probably Euler meant that one can employ them in calculation *as long as* they lead to acceptable results, but for this he had no justification in general; Gauss, meanwhile, introduced a higher standard of rigor in mathematics, and the same applies to Cauchy, who kept rethinking the basic concept of imaginary numbers throughout his life, in spite of having developed basic ideas of complex analysis.

Particularly important was the Fundamental Theorem of Algebra, that every polynomial of degree n has precisely n roots if we allow complex numbers; conjectured by Girard in 1629, d'Alembert made a famous attempt to prove it in 1746, and Gauss proved it (though still with some gaps) in his 1799 dissertation, and again (now two full proofs) in papers of 1816. By that time, Gauss was in the possession of a general conceptual justification for the complex numbers, which he would defend as "full citizens" of mathematics, enjoying "equal rights" to the reals, as he emphasized in a letter to Bessel in 1811.[22] It is of relevance here to quote his later words on the topic:

but these imaginary numbers, as opposed to real quantities . . . have been merely tolerated rather than given full citizenship and appear therefore more like a game played with symbols devoid of

[21] Hamilton (1837) also emphasizes this before offering his theory employing ordered pairs of reals.

[22] Quoted in Ferreirós 2006, 227: "First of all I would ask anyone who wants to introduce a new function into analysis to clarify whether he intends to confine it to real magnitudes (real values of its argument) and regard the imaginary values as just vestigial [*ein Überbein*]—or whether he subscribes to my fundamental proposition [*meinem Grundsatz*] that in the realm of magnitudes the imaginary ones $a + \sqrt{-1} = a + bi$ have to be regarded as enjoying equal rights with the real ones. We are not talking about practical utility here; rather analysis is, to my mind, a self-sufficient science, which would lose immeasurably in beauty and symmetry from the rejection of those fictive magnitudes."

content in itself, to which one refrains absolutely from ascribing any visualizable substratum. (Gauss 1831, 170)

Clearly, a well-rounded and highly useful symbolic framework is not enough, because mathematics is not a game played according to simple rules with meaningless signs.[23] The situation began to change when Wessel and Argand around 1800, and then Gauss (1831), proposed a *geometric representation* of the imaginary numbers. The geometric representation of the real line served here as an exemplar (see Chapter 3 section 2); it was possible to interpret all the algebraic operations on $a + bi$ as correlates of geometric operations on directed segments in the plane, so in place of (iii) an analogue of the correspondence between the line and the reals (now with the plane instead) was available. This allowed one to ascribe a "visualizable substratum" to the expressions $a + bi$.

Yet, Gauss had more stringent requirements in mind. Previous doubts had begun to fade away, but there remained the question about the "numberness" of complex numbers, which forced abandonment of (ii); that is to say, there was a need to revise the number concept itself. (Simultaneous with this, there was also the problem of how to conceive of functions of a complex variable, $f: \mathbb{C} \to \mathbb{C}$, and how to define the operations of the calculus in this realm. But this is a long story and we cannot enter into it here.[24]) The famous geometric representation is, for Gauss, only an "intuitive, spatial exemplification" of broader considerations about general arithmetic. This is not often emphasized, but it should be obvious to careful readers of Gauss's 1831 paper, the *Selbst-Anzeige* or review of his own "Theory of biquadratic residues, II" (note that these are two different papers, one technical, the other a review adding foundational remarks).[25] The geometric representation

---

[23] Hilbert of course agreed, as you can read very explicitly, for example, in the lectures of 1919/20 edited by Rowe; see Hilbert 1992, 14: "Mathematics is in no sense like a game, . . ."

[24] Gauss presented some ideas in the 1811 letter to Bessel, and then came the important contributions of Cauchy, and the systematizations of the theory of analytic (holomorphic) functions in the hands of Riemann and Weierstrass. See Bottazzini 1986, and the recent book by Bottazzini and Gray (2013).

[25] There is an English translation in Ewald 1996, vol. I, p. 306. For comment and analysis, see Ferreirós 2006. I also recommend the work of Hamilton (1837) as it throws much light on the issue. Another influential nineteenth-century development that was linked with this same issue is the British school of symbolical algebra.

of complex numbers was offered merely to show those (Kantians or neo-Kantians) who defined mathematics to be a construction of concepts *in visual intuition (Anschauung)* that these numbers were just as visualizable (*anschaulich*) as the reals.

Gauss, however, was not a Kantian in that sense, and he insisted that mathematics is an abstract study of *relations*. Thus he abandons the traditional, sanctioned idea that numbers are proportions. To show that some type of numbers are admissible, it will suffice to show that there are indeed situations (call them relational configurations) that require that kind of number for their description. Gauss did not develop the idea in sufficient detail, but he indicated that one-dimensional "manifolds" — such as straight lines or smooth curves—justify the introduction of real and negative numbers (Gauss 1831, 175; 311), while two-dimensional "manifolds" are the justification for complex numbers:

> ... suppose the objects are of such a sort that they cannot be ordered in a sequence (even if it is unbounded) but only in sequences of sequences; or what comes to the same thing, suppose that they form a manifold of two dimensions. Then if the relations of one sequence to another or the transitions from one sequence to another behave in a similar manner to the transitions just mentioned [in a one-dimensional sequence, the real number case] then clearly . . . we need besides the previous units +1 and −1 two others, also inverse to each other, $+i$ and $−i$. . . .
>
> The mathematician abstracts entirely from the nature of the objects and the content of their relations; he is concerned solely with the counting and comparison of the relations among themselves. . . .[26] (Gauss 1831, 176; 312)

This attempt remained sketchy and cannot be said to have fully succeeded, yet it must have seemed to move on the right track; Weierstrass, for one, gave him credit in this regard. Also, Hamilton's theory of complex numbers as "ordered couples" of real numbers came

---

[26] In other texts, Gauss made it clear that the "comparison" of relations envisaged here is not necessarily quantitative, but can be qualitative as with topological properties. See Gauss 1849 and Ferreirós 2006.

to complement this new line of thinking, and authors such as Hankel or Dedekind completed the transition, ultimately in the context of set theory. Gauss's view of mathematics as an abstract "theory of relations" intriguingly pointed the way toward later conceptions, and in retrospect can be regarded as an antecedent of twentieth-century structural thinking.

By 1860, the view that complex numbers are a natural extension of the number concept was becoming generally accepted, promoted by developments in number theory (Gaussian integers, algebraic integers), in algebra (study of complex and hypercomplex algebras), and in analysis (the theory of analytic functions, elliptic functions). There were also fascinating combinations, such as Riemann's use of complex functions in analytic number theory,[27] linked with the problem of the distribution of the prime numbers, which led him to the famous Riemann Hypothesis. Simultaneously, results about the properties of hypercomplex number systems obtained by Weierstrass, Dedekind, and others increasingly led to the conclusion that complex numbers were *the* natural closure for the number system, a generally accepted viewpoint in the last decades of the nineteenth century.

The example we have shortly reviewed is certainly complex, and I make no attempt to analyze it fully, but it shows that *considerations of meaning* have typically been central to the development of mathematics. Perhaps a strict formalist would consider all this as examples of the murky facts having to do with the context of discovery of mathematical notations. Descartes's distrust of $i = \sqrt{-1}$ and Gauss's need to find some conceptual rationale for its admission would be psychological facts about these two people—probably understandable given the poor historical conditions in which they lived, and the low level of progress achieved by humankind before 1850. The only meaning behind all this would be what calculations with i offer: e.g., you can always factorize your polynomial formulas into linear factors $(x-a)$, and $i \log i = -\frac{1}{2}\pi$.

From our standpoint, by contrast, considerations of meaning are a typical trait of mathematical practice and mathematical knowledge, indeed an indispensable trait given the complementarity between

---

[27] This novel field was opened by Dirichlet. We return to the topic in Chapter 7.

formulas and thoughts. I hope the foregoing will suffice as an introduction to what seems to be a promising viewpoint deserving further development.

There would be many interesting, more advanced, questions to consider; I will offer two examples. In recent years, Rav (1999) and Dawson (2006) have raised the question of how to distinguish between different proofs for a given theorem, which is quite hard to treat from a formal standpoint. It seems to me that considerations of meaning along the previous lines may shed important light on this topic, and on the way real mathematicians distinguish proofs (quite independently of possible formal treatments of the question). Notice also that it is perfectly natural to think of a framework being transformed, as happened with elementary analysis from Bernoulli to Euler, or with basic set theory from Cantor to Zermelo to Bernays. There emerges the question why we should be entitled to consider those theories—obviously different from a formal point of view—as different versions of "the same" theory. Once again, it would seem that considerations about meaning, and also about position in a web of practices and knowledge, are crucial for the answer.

5

Ancient Greek Mathematics

A Role for Diagrams

There are strong reasons to distinguish between elementary and advanced mathematics in relation to their epistemological status and the kinds of presuppositions they assume. The arithmetic of the naturals, of fractions, and—despite cognitive obstacles—of negative numbers remains within the bounds of the elementary. But with the real numbers we are already on the other, "sublime" side,[1] to which the complex numbers, functions, sets, and most other important mathematical ideas belong. Elementary mathematics is strongly rooted in our cognitive systems, in the normal agent and her everyday practices. Advanced mathematics, on the other hand, builds on constitutive hypotheses. I propose the view that the borderline of elementary mathematics is strictly linked with the adoption of hypotheses in the sense that will be explained in Chapter 6.

The question we shall discuss in the present chapter is the status of a most celebrated and enduring mathematical theory, Euclidean geometry, and of how to apply our analytic tools to this central case. Needless to say, this topic has been studied intensively by a large number of authors, from Aristotle to Kant and on, and includes very recent work on the role of diagrams in proof practices by Manders and others. To some extent this will alleviate the need to go into details, but on the other hand it makes a case for incorporating at least some of this work into my exposition. This chapter is not meant as an original

---

[1] Advanced math was called *sublime* in the eighteenth century. Later, we shall give an example from Gauss; among the many others that could be cited, Leibniz often spoke of "sublime geometry" (see Grattan-Guinness, ed. 2005, 56); the respected German mathematician W. J. G. Karsten wrote *Mathesis theoretica elementaris atque sublimior* [*Theoretical mathematics, elementary and sublime*, i.e., *advanced*] (1760).

contribution to that large body of literature, but it is important in the present work for two reasons: it offers a wonderful example of what it means to analyze mathematical practices, and geometry plays a crucial role in the developments that we shall analyze in later chapters.

Three of the theses I want to defend are (1) that Euclidean geometry can be understood as a theoretical, idealized analysis (and further development) of practical geometry; (2) that by way of the idealizations introduced, the *Elements* builds on hypotheses that turn them into advanced mathematics; and (3) that the axioms or "postulates" (*aitê-mata*) of Book I mainly regiment diagrammatic constructions, while the "common notions" (*koinai ennoiai*) are general principles of a theory of quantities.

5.1. FROM THE TECHNICAL TO THE MATHEMATICAL

At the basis of mathematical knowledge lie a number of technical practices that form a substantial part of civilized human life. In 1887, German polymath Hermann von Helmholtz proposed an epistemological analysis of counting and measuring, regarded as the basic "procedures" that are at the basis of mathematics and form "the foundation of the most fruitful, sure, and exact scientific methods known to us at all" (Helmholtz 1887, 728). We have added a third elementary technical practice, that of drawing geometrical figures.[2] These three techniques will eventually lead to the core basic structures of mathematics, the number systems $\mathbb{N}$ and $\mathbb{Q}$ and $\mathbb{R}$, and the structure of Euclidean space. But there is a long way in between.

It should not be necessary to emphasize again that I do not consider the procedures of counting, practical measuring, or practical geometrical drawing as mathematics. It seems adequate—in the present context—to preserve the word "mathematics" for certain kinds of

[2] The list of basic practices could be expanded, but here we aim more at pregnancy than all-inclusiveness. Feferman (2009) lists counting, ordering, matching, combining, separating, and locating in space and time. But matching and ordering are implied in counting as we define it, and locating in space and time (like other practical problems) lead to measuring. Mac Lane (1985, 35) offers a very long list.

theoretical inquiries and the related practices, while counting, measuring, and drawing are technical practices, *techné* in the Greek sense, direct intervention in the world at hand. One difference between those technical practices and mathematics is that the latter adds a new layer of symbolic frameworks, a mediating system of representations that creates distance from immediate action on the world of objects and at the same time opens new possibilities for exploration. But, is the presence of (written) symbolic representations enough to define mathematical practices and distinguish them from other practices? Certainly not.

Two further elements seem necessary to distinguish mathematical practices, one of them having to do with topic, the other with approach. As regards topic, the old idea seems valid that mathematical knowledge is ultimately rooted in the pair Number/Figure (*form* in the geometric sense), from which arithmetic and geometry were constituted, leading eventually to hybrid products like the real numbers and Cartesian geometry. Notice that the natural number concept only requires the mastery of technical practices and their underlying culture, which typically includes oral language, and the same seems to apply to concepts of plane or solid figures: rectangular triangles and pyramids long antedate geometrical theorems, and so does the grasp of an underlying general concept.

Something else is needed for mathematics proper, and it has to do with approach. The standpoint adopted here is that mathematics begins when a step is taken toward theoretical knowledge, characterized by the exactness of results, and the peculiar nature of the goals and values that guide it. Theory is often also characterized by idealization.

Let me attempt to clarify this important idea of the shift in goals and values associated with transition from a technical practice to a mathematical one. It is an old idea that technical activity is different from scientific activity in virtue of the ultimate goals and, derivatively, of the central values that regulate the practice. Technique is aimed at building artifacts, science is aimed at producing data and representations, and as a result of their different aims the activity is differently evaluated. Electron beams employed in an electronic device should produce a desired effect reliably under a host of different circumstances, and it is not so important whether some fine effects of electron behavior are present or absent. By contrast, in a scientific experiment performed in a research context, everything is typically about those fine effects, and in fact about effects finer than before, while reliability and replicability, however important they are, are relegated to a secondary role. Thus, the values associated with scientific results include precision and (internal and external) consistency, those linked with techniques include reliability and robustness. Some values, like fruitfulness and simplicity, may be associated with both science and technology, but typically with distinguishing nuances.

In the world of mathematics and its relation with techniques one finds a similar contrast. To encounter it, one does not need to go into advanced or "sublime" mathematics; the case of arithmetic and number theory already offers obvious examples. For practical purposes, reliable counting, robust storing of the data, and simple, reliable procedures of calculation (by pen or machine, say the abacus) are essential in the world of numbers. But number theory is a different issue, guided by the search for precise results. When the Pythagoreans, around 500 BCE, became fascinated with the fact that all numbers fall under the odd/even dichotomy,[3] they were already in the business of "contemplation" as the Greeks said, of theoretical thinking. (Of course they also knew subtler results, such as the fact that consecutive square numbers are linked by an odd gnomon, so that $9 = 4 + 5$, $16 = 9 + 7$, and in general the nth square is equal to the $(n-1)$th square plus the nth odd number.) Questions about twin primes or the infinity of primes or the distribution of primes are theoretical, and in principle have no practical value. Mathematicians such as Gauss, Jacobi, and Hilbert have actually emphasized this fact—in the spirit of Platonic or Neoplatonic humanism—as

---

[3] As the reader probably knows, they regarded Odd (which they associated with Limit, *peras*) and Even (associated with Unlimited, *apeiron*) as the principles of number, and hence of all things terrestrial or celestial. For the famous list of ten pairs of principles, beginning with the two just mentioned, see Aristotle's *Metaphysics* 985b23. From a later standpoint, the dichotomy they had found is rather shallow—why not classify numbers, also exhaustively and dichotomically, into triple/non-triple (based on whether $n = 3 \cdot m$ or not)? But once we see this possibility, infinitely many other options become visible. The deep classification is into *prime* and *composite* numbers (with the Fundamental Theorem of Arithmetic as the central result), but it was only in modern times that number theory became focused on prime numbers and divisibility properties.

a distinguishing trait and glory of mathematics. Hilbert said in a radio speech in 1930:

> . . . all mathematicians have refused to let applications serve as the standard of value for mathematics. Gauss spoke of the magical attraction that made number theory the favourite science for the first mathematicians, not to speak of its inexhaustible richness, in which it surpasses all other parts of mathematics. Kronecker compared number theorists with the lotus eaters, who, when they had sampled that delicacy, could never do without it. The great mathematician Poincaré once attacked Tolstoy with astonishing sharpness, who had suggested that pursuing "science for science's sake" is foolish. The achievements of industry, for example, would never have occurred had the practical minded existed alone and had these advances not been pursued by uninterested fools. The glory of the human spirit, so said the famous Königsberg mathematician Jacobi, is the single purpose of all science.[4]

An even clearer example is provided by fractions employed for measuring purposes in contrast with the theoretical study of those fractions in the context of measuring geometric figures. Much ink has been wasted on the contrast between practically driven *approximate* results (where a fraction is always sufficient as solution) and theory-driven *exact* results. A relevant case is the value of π and the many approximations offered for it throughout history. The Bible (I Kings 7, 23) shows a practical concern with that value for the purpose of architectural construction, being satisfied with 3 as an approximation; meanwhile, the Egyptians had already employed a much better approximation, 25/8 = 3.125. But we find the crucial shift with Archimedes, in whose work there is a clear distinction between the *exact value* of π and its approximations (for which he calculates values corresponding to $3\frac{10}{71}$ and $3\frac{1}{7}$).[5]

---

[4] Translation by James T. Smith, taken from http://math.sfsu.edu/smith/Documents/ Documents.htm (accessed October 21, 2010). It is well known that number theory has become a practical branch of mathematics in the age of internet and communications security, and it is obvious that the image of math cherished by Gauss and his followers is no longer vibrant in our culture. But we cannot enter into this interesting topic; see Bottazzini, and Dahan-Dalmedico (2001).

[5] *Measurement of the Circle, c.* 250 BCE. Liu Hui in the third century CE calculated a value corresponding to 3.14159 as an approximation, although he did not establish an upper

The point is important, as it is generally agreed that the conception of perfectly precise results established deductively is a key novelty of the mathematical practice of Greek geometers.

Almost nothing has been done concerning the cognitive science of the practices of measuring and geometric drawing, but fortunately the kind of philosophical analysis I am after—although aimed at converging with that future cognitive theory—does not need to wait for it to emerge (see Chapter 3 section 5). As we have seen, at this level of the technical practices we already find interactions between even more elementary practices and skills. In all cases, it seems necessary to emphasize at least four elements: perception, deeds, oral language, and the symbolic. Practical action, or deeds, are at the basis of any of these techniques; in fact the scenario becomes much more complex—a true technology in the broad sense, based on employing sophisticated tools—with measuring and geometric drawing. Notice, for instance, that measuring practices involve the choosing of a relevant unit of measure, to which laying out appropriately and counting are applied.

In this connection, some authors have speculated that well-developed counting systems may have had their origins in the solution of measurement problems. The idea is that life in relatively simple cultures, even after the Neolithic revolution, would not have required precise counting (cf. the case of the Kpelle farming culture in Liberia, studied by Cole, Gay, and Glick 1974). Only relatively complex societies such as those in Mesopotamia, ca. 3000 BCE, facing complex problems of measuring and accounting due to the collection and distribution of goods on a large scale, would have developed an advanced system of numerals—and particularly number symbols.

5.2. THE *ELEMENTS*: GETTING STARTED

The name "Euclid" says almost nothing more than "the author of the *Elements*," so we shall employ the man's name as a synonym for the style of work. (Even the dating of the *Elements* is controversial, since the evidence that led to a date around 300 BCE is thin; some experts

---

bound by using a circumscribed circle; the Islamic mathematician al-Khwarizmi (*circa* 800) used 3.1416, already written in positional Indo-Arabic notation.

suggest that it may even stem from around 180 BCE, after Archimedes.) The plane geometry expounded in the *Elements*, Books I–IV, can be regarded as a theoretical study of practical geometry.[6] It is a study of the figures that can be drawn practically using a straightedge and a compass, and also—especially—of theoretical problems emerging therefrom. Its theoretical nature comes not only from the new goals and values guiding this practice (see below), but also from the idealizations introduced. We shall emphasize *idealization* and *exactness* of results as two of the crucial—and intertwined—traits of the "divine geometry" that interested Plato and his followers, the geometry that occupied Euclid and Archimedes.

This is important, for (to repeat the idea) it is generally agreed that the conception of perfectly precise results—such as incommensurability theorems—established deductively by a series of reasonings and constructions (on the basis of previously agreed notions and postulates) is a key novelty of the mathematical practice of Greek geometers summarized in the *Elements*. There certainly was mathematics in other cultures, before and after the ancient Greeks, but usually it was satisfied with offering solutions to problems that in practice did not need to be precise solutions. It has been emphasized time and again that, from a point of view less idealized and theoretical than Euclid's, a good approximation is not just sufficient, but often preferable to an absolutely precise analysis.

5.2.1

Notice that the idea of regarding Greek geometry as a theoretical study of practical geometry, linking its analysis to elementary mathematics, goes against the tendency to reflect abstractly on the subject, usual since Pasch and Hilbert. Generally speaking, reflections on the *Elements* since the seventeenth century have tended to presuppose the contemporary understanding of mathematics, offering a reading of the

---

[6] Here we shall disregard the rest of the *Elements*, which of course consists of highly important books. Books XI–XIII deal with the geometry of spatial figures; Book X, "la croix des mathématiciens" (Stevin), offers an (incomplete) classification of incommensurables. Of great foundational value, and a cornerstone for all of Greek geometry, were Books V–VI, dealing with proportion theory, and offering a generalization of the method of "application of areas." Finally, Books VII–IX discuss arithmetic and number theory.

Greek procedures from a more modern standpoint. Our tendency is, of course, different, and we make an effort to incorporate the faithful reading of texts that is characteristic of historians.

Mathematicians have an understandable tendency to emphasize continuity in the historical development of their discipline, claiming that modern axiomatics is a perfection of Euclid's, and saying often that Hilbert "perfected" Euclid's axiomatization. This is far from being true, and I would venture to say that there is no less truth in the statement that Hilbert's geometry represents a total change of topic from Euclid's. One has to understand that the distance between both is hard to bridge: the practice of proving and the very nature of the postulates or axioms (see section 5.3) changed most significantly.

5.2.2

Perhaps it is in order to emphasize a bit more the difference between practical and theoretical geometry. We have drawn a boundary between technical and mathematical practices in view of the theoretical goals and values that guide scientific activity, and we exemplified it with the contrast between practically driven approximate results and theory-driven exact results in the context of fractions and measuring. Geometry offers wonderful examples of the same feature. In practical geometry, the so-called Pythagorean theorem is needless in the great generality given to its statement and proof in Book I, Prop. 47.[7] Much more useful practically are concrete Pythagorean triples, concrete triads of positive integers that serve to construct rectangular triangles, in particular the simplest one 3-4-5, the simple fact that a rope with twelve or thirteen equally spaced knots suffices for drawing a very good right angle on the floor.[8]

This, of course, is likely to have been discovered independently several times, as it has been known in many different places and cultural contexts. The Pythagorean triple appears on the Babylonian tablet Plimpton 322, dated to the early second millennium BCE; such triples can also be found in the Shulba Sutras from India, assumed

---

[7] Not to mention the generalization contained in Book VI prop. 31, that similar figures constructed on the sides of a rectangular triangle add up to the similar figure constructed on the hypotenuse (say, rectangles or other polygons, circular portions, etc.).

[8] The triangle of sides measuring 3, 4, 5 is rectangular, and $5^2 = 3^2 + 4^2$. (There are 16 primitive Pythagorean triples with integers ≤ 100.)

to stem from before 200 BCE.[9] But we shall call this result the Gougu-Pythagorean theorem; *gou gu* means "base and height" or "hook and leg" in Chinese. It was the expression employed for the study of right-angled triangles, e.g., in Chapter 9 of the famous classic *Jiu zhang suan shu* (*Nine chapters of mathematical procedures*), a text that took its present form between 200 BCE and 100 CE. The diagram in Figure 3 is taken from a thirteenth century edition of the *Zhou Bi* (*Gnomon of the Zhou*), an even earlier text, dealing with the same triangle (at least in the commentary of Zhao Shuang, third century CE). This was another classic, an astronomico-mathematical treatise whose title is sometimes rendered as *Classic of the gnomon and the circular paths of heaven*, written between 500 and 200 BCE. It begins with a short dialogue between the Duke of Zhou (who ruled as regent near the start of the Zhou dynasty) and a sage named Shang Gao, in which the triangle of sides 3-4-5 is mentioned. The *Zhou bi* is one of the principal surviving documents of early Chinese science, but dating its preface is controversial: Needham (1959, vol. 3, 22) takes it to be of "rude antiquity," probably thinking of the optimistic dating around 1000 BCE, but the most pessimistic locate it about 100 or even 200 CE.[10] Be that as it may, the figure is employed by Zhao Shuang in his commentary, written around 200 CE, and it is intimately linked with the contents of the famous third-century commentary of the *Nine chapters* by Liu Hui (see the French edition by Chemla and Shuchun, 2005).

The interpretation of this diagram is quite interesting, since it not only proves the Gougu-Pythagorean theorem in the particular case 3-4-5, but also—and perhaps more prominently—presents a result that is specific to this case: namely, that the tilted square is $5^2 = 4\Delta + 1$, where Δ is the area of the 3-4-5 triangle, which the diagram also shows to be $\Delta =$ six small squares. The figure accompanies a text that presents the 3-4-5

[9] Also relevant in this connection is the famous Babylonian "hand table" YBC 7289, from before 1,500 BCE, carrying a very good approximation of the value of $\sqrt{2}$, which presupposes knowledge of the theorem in the simplest case.

[10] Cullen (1996, 153–156) argued that this seemingly old text is likely to be a later addition, and the figure is known to have been added by Zhao Shuang around 200 C.E. Also the translation of the preface given by Needham (1959, vol. 3, 22) is incorrect, mistakenly suggesting a proof of the Gougu theorem; for details and a sober literal translation, see Cullen (2002).

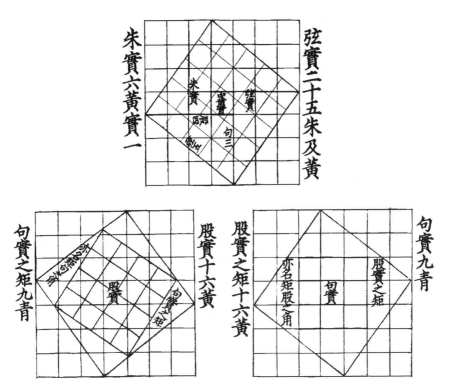

Figure 3 Diagram from a late edition of the *Zhou Bi* (an early text in astronomy), related to third-century commentaries of this work and of the *Jiu Zhang* or *Nine Chapters*. In China, the theorem "of Pythagoras" was known as the "Gougu rule" (勾股定理).

triangle as a folded rectangle, and the bigger square as the result of having placed four such rectangles "round together in a ring." Contrary to what Needham suggested (1959, vol. 3, 22), the preface contains no evidence of a very ancient proof of the validity of the Gougu-Pythagoras result in the particular configuration discussed. By contrast, the result is known to the third-century commentators Zhao Shuang and Liu Hui, who extracted from the figure the generalization we would modernly write as $c^2 = 2ab + (a - b)^2$ (with $a - b = 1$ in the particular case depicted above; ab is the rectangle = 2Δ, which is "folded" into the triangle).[11]

[11] See the Cullen edn, 1996, 174.

The difference between the practical versions of Gougu-Pythagoras—which one can find in Mesopotamia before 1000 BCE or in China and India before 200 BCE—and the perfectly general and accurate result found in the *Elements*, I.47, is no doubt of great importance. Somebody asked and somebody answered the question: Is this relation between squares on the sides valid also for other kinds of right triangles? How generally does it apply? The emergence of such theoretical problems marks quite clearly the transition from technical practices to the realm of theory, in our case to mathematical practices. As already emphasized, it is mostly the goals and values that differentiate the realm of technique from that of science—*generality* in the present case, going far beyond practical reliability.

But notice that this transition can happen at the most elementary levels too, as we remarked in connection with number theory. So that difference does not imply, by itself, that the boundary between elementary and advanced math (as we are laying it out) has been crossed. However, in the case of Greek geometry some hypotheses enter the arena (involved in its idealizations) that, I shall argue, imply the advanced nature of this very old practice and the knowledge it produced. We thematize this question in section 5.3 of this chapter and especially in Chapter 6.

5.2.3

Diagrams such as those above are not significantly different from the diagrams that can be found in manuscript copies of Euclid's work. Both the problems and the proof methods elaborated in this tradition of geometry were based on diagrams, i.e., on figures drawn by simple methods. The precise way in which this works will be analyzed in section 4, but the basic idea is that diagrams feature essentially in the conduct of proofs—and yet the results are perfectly controllable and reliable. As Newton said:

> The learner should first be taught to describe [lines and circles] accurately, before he enters upon geometry, which then shows how by these operations problems may be solved . . . and it is the glory of geometry that from those few principles, brought from without, it is able to produce so many things. . . . the description of

right lines and circles, upon which geometry is founded, belongs to mechanics.[12] (Newton's *Principia*, Preface)

The tendency to regimentation, so important in the context of axiomatic work, shows already in the fact that Euclid restricts drawing techniques to ruler and compass; this helps reduce the assumptions made in the postulates to a minimum. Geometric investigations in different cultures have employed cut-and-paste methods, translations and rotations, or even origami methods; Euclid goes a long way to avoid any of these.

But one must immediately add that geometrical theory and practice, in the ancient Greek style, is based on assumptions that transcend practical geometry. This causes great difference in the way agents *read* the diagrams and reason from them; that is the issue with the "Euclidean" idealizations (see especially section 5.3), and it is linked with the vexing question of the nature of the objects of ancient Greek geometry.

The scheme proposed in Chapter 3 considers a symbolic framework and a theoretical framework, plus agents handling them in a certain controlled way. In the case of the *Elements*, the symbolic framework is formed by the diagram symbols, the letters that link text and diagram, and a good number of concepts. It seems perfectly fit to regard the lines that form the diagrams as *symbols*, even though their nature is not quite the same as that of other mathematical symbols. For it is in the nature of symbols to present differences of kind and degree. All mathematical symbols are representations, and all of them present a conventional ingredient; but the conventions and the representing do not work in the same way.[13] Consider the well-known proof of the Gougu-Pythagorean theorem in Book I of the *Elements*: the drawn triangles and squares in the diagram simultaneously play the role of objects of study and representations of the objects of study. Stated differently, in the terminology

---

[12] This he took, quite interestingly, as an opportunity to illustrate the relations between geometry and mechanics, writing: "Therefore geometry is founded in mechanical practice, and is nothing but that part of universal mechanics which accurately proposes and demonstrates the art of measuring." (Newton's *Principia*, Preface).

[13] This should be obvious; compare the diagrams that we reproduce below with, say, the way equation $\frac{x^2}{a^2} + \frac{y^2}{b^2} = 1$ represents an ellipse centered on the origin; or, more extremely, with the purely conventional way in which $\aleph_0$ gets its meaning.

of Parsons, they are *quasi-concrete* objects: one can cogently defend the thesis that the geometer is studying the properties of these objects, not anything beyond them (such as Plato's ideal triangle), but she is not studying their properties as empirically given—the geometer handles the lines, triangles, and squares according to a certain way of *conceiving* them, not as empirically given.

What we have just said would be paradoxical, even impossible, if we had not regarded the agents as central to our scheme of analyzing mathematical work. What matters to geometrical work, to the practices of drawing diagrams and inferring from them, and thus to geometrical knowledge, is not the symbolic framework per se—but *as manipulated* by the agents. The diagram lines per se cannot be dually objects of study and representations of the objects of study; but the lines as conceived by the agent (according to certain stipulations) are the objects, and at the same time those lines, as empirically given, represent the objects. (When applying geometry, e.g., to study geometrical optics or mechanics, new layers of representation enter the picture.) More on the topic of the mathematician's notion of an object will be found in Chapter 6.

We said that the symbolic framework is formed by the diagram symbols, the letters that link text and diagram, and a good number of concepts. The concepts are explicitly dealt with in *definitions*, and they are visually realized in the diagrams; they correspond to what we often call "technical terms," for example, surface, right angle, figure, circle, quadrilateral, parallel, tangent, inscribed, and circumscribed. In Book I, more than twenty-three concepts are defined; in Book III we find eleven, and some five (or seven) in Book IV. Everybody who reads this book knows how to draw diagrams that exemplify the concepts that we have just named. We shall have a bit more to say on definitions in section 5.4.

5.2.4

There is a popular tradition among philosophers, mathematicians and logicians of drawing a parallel between the organization of the *Elements*, Book I, and that of twentieth-century treatises in foundations of mathematics. I suppose no expert in the Greek treatise has ever followed this line, but still it is probably worthwhile criticizing it explicitly. According to that parallel, the "common notions" (*koinai ennoiai*) would play the role of a general logical framework, and the postulates (*aitêmata*) would be like modern axioms for a specific theoretical domain. Indeed, Proclus commented that the postulates are peculiar to geometry, while the other axioms are "common to all sciences that deal with quantity and magnitude." If we forget the last six words, we could read Proclus as supporting the free logical interpretation of modern times. The problem is that this would misrepresent the ideas of Proclus in the fifth century, of Greek geometers in 300 BCE, and of almost all mathematicians influenced by Euclid from the Middle Ages to, say, 1850.

It should be clear that the axioms called "common notions" are in fact basic general principles of the science of quantity—except perhaps for one that might not make sense outside the domain of plane figures ("things which coincide with one another are equal to one another"), and is hence not general but specific to planar quantities or plane surfaces. The common notions are not meant as general logical principles, but as properly mathematical ones (using the word "mathematical" as Plato and Aristotle did).[14] Throughout history, the basic idea about quantities has been that, whatever their kind, they have to be susceptible of augmentation or diminution; common notions 2 and 3 are linked with these two ideas. Another famous principle is the "the whole is greater than the part," which is behind many proofs in the *Elements*, and was invoked oftentimes in relation with the (im)possibility of infinite "quantities"—most famously by Galileo.

As for the postulates, I shall argue in the next section that they are very far from modern axioms. In the course of this chapter it should become clear that the mathematical practice of the *Elements* is quite far from the later algebraico-symbolic traditions that led to the conception of modern logic.

Kant insisted that mathematics is not a form of knowledge based on conceptual apparatus and logic, but involves necessarily the "construction of concepts in intuition."[15] Whatever the merits of the modern notion of proof and associated methods, I contend that Kant's viewpoint

---

[14] The fact that the "common notions" include principles for *equality*, while in modern logic *identity* is treated as a logical notion, does not count at all. To put it syllogistically: "If you identify equality and identity, then you are a bad logician."

[15] See, e.g., *KrV* A713–A714, or *Prolegomena* § 7.

yields a much more adequate analysis of the practice of classical geometry. His philosophy of mathematics (and knowledge, more generally) rested on the theses that there exist forms of pure a priori intuition, and pure mathematics is based on pure intuition. Even without following him, one may understand Kant's most basic views concerning mathematical knowledge as a very good analysis of the situation with classical geometry, one of the best ever produced. Consider the following heterodox reading: rather than accept the notion of a pure intuition independent of anything empirical, take "intuition" to be perception of real or imagined forms, including also forms produced or constructed by drawing. Classical Greek geometry indeed involves by necessity the construction of concepts (and relations) in "intuition" so conceived, and this is done in practice by means of diagrams. The original geometry of Euclid lacks the means to derive its theorems by pure logic, but it presents us with a most interesting and fruitful way of proving results by diagrams.

In the wake of the algebraization of mathematics in the seventeenth and eighteenth centuries and its "arithmetization" in the nineteenth (Weierstrass, Dedekind, etc.), geometric intuition and reliance on diagrams fell increasingly under suspicion. It became more and more of a commonplace that diagrammatic reasoning is faulty, and that essential reliance on constructions or diagrams is an obvious defect in mathematics. Consider Lagrange's *Méchanique analytique*, where he proudly stated:

> No figures will be found in this work. The methods that I have presented do not require constructions, nor geometric or mechanic reasoning, but only algebraic operations subject to a regular and uniform procedure. (Lagrange 1788, *Avertissement*)

Even before, in a manuscript written in 1766 (published in 1786) on the question of whether Euclid's parallel axiom can be deduced from the Euclidean postulates and common notions, J. H. Lambert stated:

> For . . . this question one can abstract from all that I have previously called representation of the matter [*Vorstellung der Sache*]. And since Euclid's postulates and remaining axioms are already expressed in words, it can and must be required that in the proof one never leans on the matter itself, but carries forward the proof

in an absolutely symbolic way—if it is possible. In this respect Euclid's postulates are as so many algebraic equations, that one already has as given, and that must be solved for *x, y, z,* &c., without looking back to the matter itself. (Lambert 1786, 162)[16]

It has to be understood that this is a complete transformation of geometry vis-à-vis its tradition, in the wake of the algebraization of mathematics in the modern period (seventeenth and eighteenth centuries). In fact, Lambert himself was not in a position to carry out the project in full, but in due course this program led to very novel work in geometry. Famous later statements of a similar program, now satisfactorily developed, can be found in Pasch's *Vorlesungen über neuere Geometrie* and in Hilbert's *Grundlagen*, more than a century after Lambert.[17]

Another outcome of this process, and of the simultaneous reorganization of the calculus within an "arithmetized" analysis,[18] was the emergence of a precise logical notion of proof—one that is strictly linguistic, logocentric, and above all formal—that has been codified in modern logic. In the light of its rise, late in the nineteenth century, the *Elements* was subject to criticism and logico-axiomatic analysis, and the conclusion was arrived at that many of the required axioms had not been made explicit by the Greek author. This of course was a by-product of the highly different perspective of modern mathematics, which can be characterized as being based on formal inferences. Once again, in order to understand ancient Greek geometrical practice it is essential to distance oneself from such conceptions.

5.3. ON THE EUCLIDEAN POSTULATES: RULING DIAGRAMS (AND THEIR READING)

Proclus proposed an idea that I find thought-provoking: he said that the difference between postulates and *koinai ennoiai* is like the difference between problems and theorems. Readers of the *Elements* are familiar with the fact that some propositions end by saying "(being) what it was

---

[16] Engel and Stäckel (1895), 162.
[17] See, among others, Hilbert 1903, Nagel 1953, and Torretti 1978.
[18] See Nagel 1935, Ferreirós 1999, and Corry 2008.

required to do" (quod erat faciendum—q.e.f.), while some finish by saying "(being) what it was required to prove" (quod erat demonstrandum—q.e.d.). Propositions 1–3 of Book I are of the first kind; they are "problems" and offer constructions for solving certain problems (e.g., for constructing an equilateral triangle on a given segment), while Props. 4–6 are "theorems" and establish certain results (e.g., the angles at the base of an isosceles triangle are equal).

There is some difficulty in maintaining the parallelism suggested by Proclus, but in my view it does point to a very important trait of the postulates, one that unfortunately is sometimes lost in bad translations of the text. *Aitemata* means literally "things that are requested" or "things granted." The first three postulates begin with the same active or constructive phrasing that one finds in problems:

1. To draw a straight line from any point to any point.
2. To produce a finite straight line continuously in a straight line.
3. To describe a circle with any center and radius.

Proclus is calling our attention to the fact that the postulates have a proactive and pragmatic nature: they license the geometer to do certain things, to operate with diagrams in certain ways, and they force a possible opponent to accept certain diagrammatic moves. This should be clear and obvious for the first three postulates, which thus appear as procedural rules that regulate diagram construction.

On the other hand, postulates 4 and 5 look somewhat different, and have traditionally puzzled commentators:

4. That all right angles are equal to one another.
5. That, if a straight line falling on two straight lines makes the interior angles on the same side less than two right angles, the two straight lines, if produced indefinitely, meet on the side on which the angles are less than two right angles.

In fact, there is a whole tradition of texts of the *Elements*, apparently coming down from the master edition by Theon, in which the postulates are only three (the first three) and the other two appear among the "axioms" or common notions. (The first printed Greek edition, and all those editions based on it up to 1800, are of this kind; the parallel axiom became "axiom 11" in these editions.) This can be regarded as historical evidence of how old the puzzle of the status of postulates 4 and 5 is. However, it is possible to offer arguments to explain why they were catalogued as postulates by Euclid.

Postulate 4 is a particular assumption about the equality of certain quantities; thus it resembles the common notions. Of course there is the difference that the common notions are general principles of the science of quantity, while postulate 4 is particular to certain plane figures, namely angles of a specific kind. One is easily led to imagine Euclid himself doubting whether to put this one requirement in one group or the other. Below, I will say more about postulate 4, offering a suggestion as to why it is included among the postulates; but let us now move on to the notorious postulate 5.

Postulate 5 is of course famous because of its relative complexity, compared to the other postulates, and because of the long history of attempts to prove it on the basis of simpler principles, culminating in the conviction that this is not possible (inside Euclidean geometry) and in the discovery of non-Euclidean geometries.[19] Also puzzling is the fact that the author of the *Elements* would choose this particular formulation, when there are alternatives such as the well-known "That only one line runs parallel to a given straight line through a point which does not lie on the straight line." This is often called Playfair's axiom, although it was known from the time of Proclus, who credited Ptolemy around 150, and it was also employed by ibn al-Haytham around the year 1000.

An interesting reading of the original version of the postulate emerges from following Proclus and regarding it as akin to a problem, i.e., from emphasizing its proactive, operational nature. This, coupled with the ancient Greek attitude toward the infinite, provides a rationale for choosing the particular form given by Euclid and for placing it among the postulates. Notice that postulate 5 has the following schematic form: "That, under certain circumstances, one can produce two lines and determine a point." Despite the fact that the postulate begins with a "that" clause, its content is undeniably operational, and hence

---

[19] See Bonola 1906, Gray 1989, and Rosenfeld 1988.

similar to that of the first three postulates (and not to the problematic postulate 4). Euclid's form of the postulate, unlike the Ptolemy-Haytham-Playfair version, specifies a procedure that provides the construction of a point within a finite plane region.

It is well known that the Greeks preferred to avoid assumptions and formulations that somehow implied actual infinity, in particular the infiniteness of space. Whatever they may have thought about the actual infinite, they had a preference for formulations that were compatible with the notion that space is finite. In Greek geometry, lines are always finite, although in accordance with postulate 2 they are always extendable; there is never (given) an infinite straight line. This and the above interpretation of the postulates helps explain the choice in the *Elements*. If it is true, as we are arguing with Proclus, that the *aitemata* are meant to be construction procedures, operational rules that govern the production of diagrams, then it is natural that Euclid would prefer a postulate offering a construction that terminates in a point; and given the problem of infinity, it is certainly to be saluted that the construction involves only a finite region.

My emphasis on geometric constructions regarded as operations that terminate will obviously be seen by the reader as conflicting with the nature of postulate 4. In my opinion, there is no way we can deny that the fourth postulate is of a different nature from the rest. What is then the rationale for its inclusion among the postulates? Two things can be suggested. Postulate 4 could be proved by relying on common notion 4 ("things which coincide with one another are equal to one another"), as Proclus explained in his *Commentary*, but the proof requires translation of one right angle onto the other. This would imply a major revision of the regimentation of diagrammatic constructions established by Euclid, and hence was undesirable. To include postulate 4 among the postulates was a small price to pay for the preservation of that strict regimentation: all is done by drawing lines, never by moving figures around.[20]

On the other hand, I believe Manders's distinction between exact and co-exact information in Greek geometry offers the key to another part of the explanation. The distinction is explained below, but the main

---

[20] There is one exception in the proof of *Elements* I.4, where Euclid assumes that a triangle is moved and "applied" or superposed on another.

idea is that co-exact information is read directly from the diagrams, while exact information depends on what is asserted in the accompanying text. A proposition saying that two angles are equal provides *exact* information—and hence that cannot be simply read from a drawn diagram. The author of the *Elements*, Book I, feels the need to specify one and only one postulate concerning exact information, namely that all right angles are equal. This cannot be proven, but everything else can be proven on the basis of this axiom of plane geometry (together with the common notions and the diagrams).

This is why postulate 4 can be found among the *aitemata*. The *aitemata* provide the necessary rules for geometric constructions, which will be the basis for co-exact inferences, plus one postulate having to do with exact content. This stipulation, which is not constructional but propositional, will be needed to read required information from the diagrams. From this point of view, its difference with respect to the other postulates is that the remaining four regiment the *construction* of diagrams, while this one regulates the *reading* of diagrams.

Let me come back to the often-drawn parallelisms with modern foundational studies of mathematics. To emphasize the differences between Greek and modern geometry, one could follow the modern tendency to use the word "axiom" only for *linguistic information* that can serve as the basis for purely linguistic or symbolic inferences, independent of any visual representation or drawing. Then I can say that, among the postulates in the *Elements*, there is only one axiom—postulate 4. All the rest are rules for geometric construction that do not deserve to be called axioms. (By contrast, all the common notions—which are, as we have seen, general principles of the science of quantities—can be called axioms.)

5.4. DIAGRAM-BASED MATHEMATICS AND PROOFS

Learning to do classical geometry and contrasting its methods with those of modern mathematics is a most formative experience, one that—pace Dieudonné—should be routinely practiced by those who prepare to become math teachers. Kant insisted that mathematics is not a form of knowledge based solely on conceptual apparatus and logic, but involves the "construction of concepts in intuition."[21] One need not follow his apriorist philosophy of mathematics to appreciate his very good analysis of the situation with classical geometry. I propose the following heterodox reading: rather than accept the representationalist notion of pure intuition, take "intuition" to be perception of physical forms such as those found on a paper drawing, or, in due course, also imagined ones. Classical geometry indeed depends on the construction of concepts and relations in "intuition" so conceived, and this is actually done in its practice by means of diagrams constructed according to highly regimented procedures. The original geometry of Euclid lacks the means to derive its theorems by pure logic, but it presents us with a most interesting and fruitful way of proving results by diagrams.

A number of recent authors, including Knorr (1975), Fowler (1999), Netz (1999), and, especially, Manders (2008), have developed an up-to-date analysis of Greek geometric practice.[22] As they have shown, Euclidean proofs rely on diagram-based inferences, and the diagrams are not eliminable; however, their use is strictly controlled and allows perfectly reliable inference, to the point that Euclid, Apollonius, and Archimedes are virtually faultless. Manders has offered a careful "logical" analysis of Euclidean reasoning, and later work by other philosophers and logicians has for the most part developed his insights further.[23]

The experts often underscore that Greek diagrams are not reproducible from the text alone,[24] and they even insist that the diagram and constructions are more central to the proof than the text. In this connection, some philological evidence is very revealing: as Netz underscored, starting with Aristotle, the word *diagrammata* (plural of *diagramma*) can mean "mathematics," while *diagramma* itself certainly means "a mathematical proposition." Moreover:

---

[21] See, e.g., *KrV* A713–A714, or *Prolegomena* § 7.

[22] Other important names are those of Ian Mueller and Sabetai Unguru, who led a heated debate with figures such as B. L. van der Waerden and André Weil.

[23] See Mumma 2010, Avigad, Dean and Mumma 2009.

[24] Although some logicians have tried to throw doubt on this. It seems to me that isolated lost figures can be reconstructed, to be sure, but only once people know the diagram-based practice from many other instances.

The word *diagramma* is never used by Greek mathematicians in the sense of "diagram." When they want to emphasise that a proposition relies upon a diagram, they characterise it as done *dia grammōn*—"through lines," in various contexts opposed to the only other option, *di' arithmōn*—"through numbers." (Netz 1999, p. 36)

Here we shall not go so far as to defend Greek geometry as purely diagrammatic,[25] but we regard it as proven that it is not "logical" in the modern sense—i.e., limited to linguistic or symbolic inference (now I use "symbolic" in the restricted sense of mid-twentieth century logic). Once we adopt this standpoint, following Manders, it becomes clear that the meanings of "axiom" (postulate, *aitema*) and "axiomatization" are very different in Euclid and, say, in Pasch and Hilbert. If one understands the scope and limits of diagram-based inference as practiced by the Greek, there is no need to say that the *Elements* lacks any of the required axioms—it suffices to underscore that Greek "axiomatization' is quite unlike modern axiomatics.

Fowler said that ancient Greek mathematics is about drawing a figure and telling a story about it, but a philosopher must qualify and make this more precise. Deductive mathematics grew out of the Greeks' drawing lettered diagrams and offering arguments about them or by means of them. In order to make this satisfactory, in the face of intense skeptical scrutiny, they went very far in the direction of *regimentation* both of diagram-construction and of textual argument. The extent to which Euclidean geometry limited means of construction (to the drawing of circles and lines for Books I to X) is an old topic and has been discussed already. One should note, however, that not all Greek geometry remained within these strict limits; actually both Archimedes and Apollonius use geometric constructions which go beyond compass and straightedge; but not too much beyond.[26] As regards the text, Netz (1999, p. 120) has called attention to the fact that the vocabulary employed is also "dramatically small—not only in specifically mathematical words, but in any words, including the most common Greek

---

[25] For an attempt, see Macbeth 2010.

[26] For instance, there are the arguments by *neusis*, which take advantage of marking a certain distance on a ruler. Euclid's straightedge, not a metric instrument, is not marked.

grammatical words. It is strongly repetitive within authors and between authors."

To get a grasp of ancient geometric methodology, it suffices to understand the methods laid out in the *Elements*, Book I (although it is true that the essential set of geometric tools employed by Greek geometers includes Books II, V, and VI). I urge readers to grab a copy and go through this first section of the *Elements*, from the equilateral triangle to the celebrated theorem "of Pythagoras" (and his Chinese and other counterparts).

Consider a theorem such as I.5, concerning angles at the base of an isosceles triangle, the famous pons asinorum,[27] or else the favorite of Aristotle and Kant, prop. I.32 on the angle sum in a triangle (see Figure 4). Euclid first enunciates a proposition, then exemplifies it by a concrete diagram, and specifies what is intended to be proved. Now comes the proof proper: it begins with some constructions that are valid according to the postulates (or to previously proved propositions, called "problems") and then goes through a demonstrative argument that establishes the validity of what was claimed on the basis of definitions and common notions (or previously proved propositions, called "theorems"). A Euclidean proof is thus the sum of construction plus demonstration, where "demonstration" (*apodeixis*) is meant in the sense of showing that such-and-such is the case in the diagram according to explicit stipulations.

Most important is a division of labor between text and diagram to which Manders has called attention, analyzing it in terms of *exact* vs. *coexact information*. The kind of information that is inferred from the diagram is strictly limited, with some key pieces of knowledge established in the text and not the diagram. It is very natural to see this as an outcome of practical experience with arguing about diagrams: there are certain ways in which our drawings and/or our perception are not reliable and can easily go wrong. *Exact* information includes equality of lines or angles, congruence of triangles, proportionality of lines, the precise geometric nature of a line (circle, straight line, ellipse), or parallelism.

---

[27] The Bridge of Asses, its medieval name in reference to the difficulties some students experienced "because of the number of lines and angles," as Clavius explained (Heilbron 2000).

All of this is explicitly recorded in text and not taken from the diagram, which is understandable, since the diagram may easily be wrong about any of those features. *Co-exact* information is unaffected by some range of deformation of the lines in diagrams, i.e., it is equally conveyed by very good diagrams and poorly drawn ones: "paradigmatically, that one region includes another . . . or the existence of intersection points such as those required in Euclid I.1 (which is unaffected no matter how the circles are to some extent deformed)" (Manders 1995, 92).[28]

The crucial point is that all of the co-exact information, which is pictorial or diagrammatic, is not codified in axioms but taken directly from the figures. We have seen that the postulates (1, 2, and 3) constrain the construction of diagrams, but then the diagram licenses certain attributions that are imported into the text. Postulate 5 can be read as saying that under certain circumstances a diagram element of co-exact nature (intersection point) can be found, even if to make this evident from the diagram as drawn on a finite surface one needs to represent a straight line by a bent one (a procedure that is common in Euclid, and consistent with his regimentation). The proofs by *reductio*, omnipresent in the *Elements*, in fact provide a very interesting exemplification of the distinction identified by Manders. Let us see two different proofs, one direct and one indirect.

Take for instance the famous theorem I.32, concerning the sum of angles in a triangle, which is among the first depending on the parallel postulate. The construction is very simple: just prolong a straight line and draw one parallel to it. The demonstration must prove two statements, and it is done in eight steps. Most of the proof steps involve exact information, dealing with equalities of angles, and they depend on previous theorems (I.31, I.29) and common notion 2. Only two proof steps involve co-exact information, bringing in attributions read from the diagram; they are very simple: that ΑΓΕ and ΕΓΔ are parts of the exterior angle ΑΓΔ (which is therefore the sum of ΑΓΕ and ΕΓΔ), and that ΒΓΑ and ΑΓΔ are parts of a two-right angle (so that their sum is two right angles, or 180°). This kind of information—that ΑΓΔ covers or

---

[28] An alternative terminology of "diagrammatic" for co-exact, and "metric" for exact, has been suggested by Avigad *et al.* 2010.

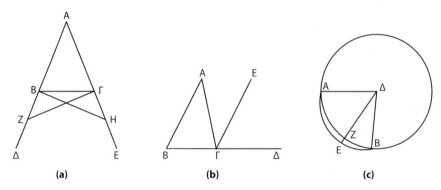

Figure 4 Diagrams corresponding to the *Elements* a) Book I. Prop. 5, b) Book I. Prop. 32, and c) Book III. Prop. 2.

comprises (i.e., is the sum of) ΑΓΕ and ΕΓΔ—is co-exact because it has to do with part-whole relations and it is not affected by deformations.

Take now theorem III.2: If two points are taken at random on the circumference of a circle, the straight line joining them will fall within the circle. The result is established by reductio ad absurdum, assuming that the segment falls outside the circle. Cases such as this one are interesting, because they exemplify *visually* some of the traits of Manders's analysis. Here, the segment joining A and B is drawn by a compass, but the text states that it is a straight line: in the diagram we find a co-exact element, merely a line joining two points; the text indicates its exact nature as being straight. The argument has nine steps, all of them exact except for one at the end, which is co-exact. It proceeds by using the pons asinorum I.5, for ΑΒΔ (going through E) by construction must be an isosceles triangle, in order to infer that angles α and β are equal. But I.16 assures us that the exterior angle ΔΕΒ is greater than the interior angle α, and is hence greater than angle β. By I.19, the greater angle is subtended by a greater side, and so ΔΒ > ΔΕ. But ΔΒ = ΔΖ. Therefore, ΔΖ is greater than ΔΕ, which means (and now comes co-exact information read from the diagram, from the inclusion of segments in line ΔΕ) that the smaller is greater than the bigger. This is impossible; hence one has to reject the assumption.[29]

---

[29] The text goes on to say that one proves in a similar manner that the segment will not lie on the circumference.

All of the above is related to the often-expressed idea that geometry is the art of rigorous conclusion established on the basis of poor diagrams. Manders (1995, 80) quotes Leibniz saying that "the circle described on a paper is not a true circle and need not be; it is enough that we take it for a circle." It is worthwhile now to pause and consider how my approach, based on joint consideration of agents and frameworks, can be applied to the case of Greek geometry. Of special interest is to consider what all this entails with regard to the role of the agents.

5.5. AGENTS, IDEALIZATION, AND ABSTRACTNESS

Let us discuss, very briefly, how to apply the scheme of Chapter 3 to geometry in the ancient form of the *Elements*, before we move on to issues related to the role of the agent in this practice.

5.5.1

In Euclidean math, one may take the language, i.e., the symbolic framework, to consist both of technical terms, such as the more than twenty-three defined notions of Book I or the eleven of Book III, and of diagrammatic characters (lines, circles), inscriptions that are ruled and regimented by the postulates. The diagram characters are juxtaposed to produce more complex geometric elements: two lines are drawn and a third element, the angle, thereby emerges; two circles or three or more lines engender another plane figure (when they are perceived or conceived as limits of the figure; see Defs. 13, 14 in Book I). A third element, crucial in practice because it establishes the links between text and diagrams, is the simple letters A, B, Γ, Δ, . . . and the conventions for their use, which are not codified explicitly but learned practically (for instance, that BΔE refers to an angle in some contexts, to a triangle in others, or that BE can refer to a segment in this situation or a square in this other, and so on).

All of this, of course, can be analyzed from the vantage point of modern mathematics; it can even be formalized.[30] But the whole point is that ancient geometric practice exploits the resources of pictorial

---

[30] See Avigad *et al.* 2010.

combination of characters in diagrams, using diagrammatic representations from which it extracts crucial information for its inferences. It does not possess or need the resources of modern symbolic renderings. Thus, the point for a practice-oriented analysis is to understand the difference between these ways of practicing, the advantages and the shortcomings of each one of them.

The technical terms defined in the *Elements* are not analyzed into primitive and derived terms—another symptom that the practice of definition in the *Elements* is totally different from the modern axiomatic practice. Some of the definitions are crucial for the proofs, e.g., when they encode exact information, as happens with Def. I.15 of a circle; it will often be inferred that certain segments are equal because they are radiuses of a circle. Other definitions are crucial, as we shall emphasize, in establishing the basic idealizations of ancient geometry. All of them are visually realized in diagrams; they correspond to simple elements or complex gestalts that can be recognized in a diagram.

To briefly continue with other components of the theoretical framework, I shall mention only statements S and questions Q. Among the statements, we have seen the author of the *Elements* distinguish carefully between *problems (q.e.f.)*, which solve certain questions about constructions, and *theorems (q.e.d.)*, which establish certain propositions.[31] Among the questions, emphases would be different depending on whether we are thinking of geometry as a well-rounded edifice presented to the student, or considering the practice as open and developing. From the latter standpoint, most important are the unsolved questions, such as the very famous problems of squaring the circle, duplicating the cube, and so on; from the former, most important are some questions that were solved by Greek geometers—such as the partial classification of incommensurable magnitudes (the legendary Book X) and the measurement of the circle or the quadrature of the parabola (Archimedes)—and other questions about volumes, about conic sections (Apollonius), and so on. However, one should not think only of highly demanding questions, because, in the practice of geometry, simpler problems can be essential and paradigmatic for the learning of the practice. Some such

---

[31] One also finds porisms and corollaries, but we need not go into details.

exemplars that are very well known include questions about figures drawn on the sides of an equilateral triangle (Gougu-Pythagoras and its extensions, I.47, VI.31), questions about the regular solids (Book XIII), and so on. The fact that a simple theorem such as I.5 has been called the pons asinorum (bridge of asses) indicates, I think, its relevance as an exemplar and its role in the learning process.

5.5.2. Objects and Agents

The diagram characters are used dually, I have proposed, both to *constitute* the objects of geometry and to *represent* them. Ancient Greek geometry is of course a science of figures, and instantiations of such figures can be found in many different situations. It affords procedures for the precise construction of certain geometrical figures from given figures and with limited (regimented) means; it establishes results concerning properties of figures, proving, e.g., that they have certain areas and volumes. It also explores the limits of what is feasible with the limited means for construction that it allows—although establishing results of non-constructability would require other theoretical means.[32]

The object of geometry can be taken to be, in Aristotelian spirit, the diagrams and other similarly shaped objects, or aspects of objects.[33] But not the diagram as it is drawn physically, i.e., as empirically given— not the token with its particularities and imperfections; the object is the diagram *as perceived and interpreted*, in light of the idealizations and exact conditions superimposed on it by the text, by Euclid's theoretical framework. Hence, the figure studied by the geometer is the ideal type, not the empirical token. The ideal type, however, does not exist outside the mind of the (reasonably good, "normal" as we have said) practitioner, and it may well not come to existence without the help of diagrams—or at least of similarly shaped figures in nature or in human works. It emerges only in the complex relational situation that involves agents and frameworks in interaction. As Aristotle said, this kind of mathematics is theoretical in that it regards its objects as "immovable

---

[32] See Hartshorne 2000 and Bos 2001 (the matter is related to Descartes and Gauss).

[33] Aspects of objects: e.g., the planar shapes that we see in a real object, say the façade of a cathedral; shapes which are drawn geometrically by the architect designing or understanding the building.

and separable," but this does not necessarily compromise an epistemological analysis of this science with the thesis that there indeed exist
"immovable and separable" entities (*Metaphysics*, Book E, 1026a 6–10).

Manipulation of the diagram is not only strictly regimented by the
postulates; it is also to be performed in light of what the conception
or interpretation of the diagram allows. Because the object is the diagram, I say that diagrams *constitute* the objects of geometry; because
only the diagram as interpreted is the object, I say that they *represent* the
objects. As instances of what the (interpreted) diagram represents, the
diagrams drawn by a geometer (or printed on a computer) may be just
as good or bad instances as any other thing: the circled shape of the
sun or the triangles or ellipses that can be recognized in the façade of
a building or in the shape of a vault.

With all this, I am underscoring the role of the agent in the kind of
knowledge obtained from the practice of classical geometry. In a sense,
what I am saying is that the old riddle about the nature and status of the
objects of geometry is indeed a riddle about human perception, conception, and knowledge.[34] Any reasonably complete theory of human
cognition will have to offer a solution to that riddle. My proposal is that
the quasi-concrete objects of geometry do not exist outside the particular concrete things (diagram, for example), and also not outside the
interaction of an agent with those particulars. And this is not merely
a physical interaction, but a semiotic one, involving idealization and
regulation. This is a matter of perception and, also, of interpretation;
of seeing the diagram but seeing it as exactly a circle or a regular solid
figure; it involves both a factual aspect and a normative one.

Although I have suggested that Euclid's geometry can be regarded
as a theoretical study of practical geometry, with it, the invention of
ideal objects (and the invention of their "theoretical contemplation")
took place. Notice that the thrust of my approach to such phenomena
is in insisting on their agent-dependency (which is not the same as their

---

[34] Macbeth (2010) has insisted that the diagrams do not instantiate (the drawn circle is
not there to be a particular example of the circle that the text talks about) but rather formulate content, so that the diagram is really an icon. I believe this oversimplifies the situation.
As I said above, the diagram is both instance and icon, because *as they are used in the practice*
(hence also in connection with the text) the diagrams both exemplify and represent, being
thus used to formulate content.

mind-dependency),[35] regardless of whether we speak of "ideal" objects or "abstract" objects. I believe the kind of approach developed here dispels some of the traditional fog surrounding famous riddles such as those about the sources of knowledge of "causally inefficient" abstract objects.

Classical geometry takes advantage of the cognitive abilities of human agents in a most noteworthy way, by relying on very robust perceptive and manipulative and conceptual abilities, which furthermore are carefully regimented so as to maximize agreement and eliminate sources of error. As a result, these abilities become the basis of a true mathematical practice that, however, is not purely linguistic or symbolic in the sense of twentieth century symbolic logic. Incidentally, one of the outcomes is the aesthetic value that characterizes this form of geometry, which adds to its knowledge value. (In this connection, let me remind you of Edna St. Vincent Millay's poem "Euclid alone has looked on Beauty bare.")

The role of the agent in geometric thinking is a vast topic that we are just beginning to explore. The question of visual thinking, its role in mathematical practice, and its possible epistemological roles in justifying knowledge have been studied in an innovative way by Marcus Giaquinto (2007). He relies on cognitive science studies of perception, and on certain philosophical views on concepts (Peacocke's), so that he can claim that perceptual recognition yields representations having a kind of conceptual content. On this basis, Giaquinto offers an analysis according to which we can deploy our perceptual abilities to obtain knowledge of geometric truths, in simple cases at least, whereas the same abilities are not reliable in the context of the calculus. Although I do not agree with some of the ideas he offers, I refer the reader to this important work.

There are many other aspects of ancient geometric practices that could be studied from a related angle awaiting detailed analysis. Consider, e.g., one very important question that we have left out here, the problem of *generality* in Euclid, that is, how a proof that ostensibly deals with a particular configuration can claim to establish a result that holds in general.[36] There are important cases (Heron and III.12) in which a

---

[35] This is not because they are "not really" mind-dependent, but because reference to the mental (like reference to abstract entities) has traditionally come with connotations that we aim to avoid; the tendency of my approach is also to question the physical/mental and objective/subjective dichotomies.

[36] See, e.g., the discussion by Manders (2008, 72–75) and other authors he refers to.

proposition is meant to apply to figures constructed in ways so different that they affect the construction and the *apodeixis*.[37] It seems highly plausible that an agent-based perspective has much to say about such cases, but the task to articulate this in detail remains for the future. Let me also mention questions about the following: the precise role of perception and other cognitive abilities in obtaining nonexact information; their role in judgments concerning possible variation of figures, which sometimes affect the proofs.

And, finally, an important issue that has been commented upon by Manders, at least occasionally, namely the role of the dialectics or dialogics of proponent and opponent in ancient Greek geometry, and how this offers the means to critically analyze a purported proof. (And its interiorization by agents; for it is quite obvious that an expert geometer—a Euclid or Clavius or Saccheri—will have interiorized it and learned to anticipate objections.) Proclus, who underscored dispute and dissent in relation to geometry, mentions some important disputes among the ancients: Apollonius on the common notions, Zeno of Sidon on prop. I.1, Proclus himself on lines enclosing a space, and also on prop. I.12, and so on.[38]

5.5.3. Idealization and Exactness

The very first page of the *Elements* confronts us with the problem of idealization. It contains the first "definitions" (*horoi*), which have long been a matter of contention. Some authors go as far as to consider them an interpolation, an element foreign to the strictly deductive structure of the work, which a later scribe added. In fact, the first few definitions are not used in the arguments and proofs; consider the well known:

1. A point is that which has no part[39]
2. A line is breadthless length
3. The ends of a line are points

---

[37] Other cases can be found: I.7 with Γ inside the triangle or to the right; juxtaposition in I.4; etc.

[38] See Proclus 1992. He also discusses "objections" in a special technical sense.

[39] According to this, as I like to tell my students jokingly, God is a point, since there is general agreement in theology that God has no parts. The number 1 would be a point, too, according to the Greek conception of numbers. The joke has a bit of interest since it is based on reading the definition as having some kind of global scope (like Frege's logic), which seems to go *against* Euclid's intentions.

What is the use of these *horoi*, literally "milestones" or "landmarks"? The third connects the concepts introduced in Defs. 1 and 2; it might thus be regarded as a claim or proposition, but in fact it is presented as a *horos*.

The *horoi* are marks that help lay out the terrain in which the geometrical game is going to be played. But what kind of a definition is offered by 1 and 2? From the standpoint of mathematics in the nineteenth or twentieth centuries, those definitions do not "define" in any serious sense of the term, but merely give an intuitive understanding of the intended meaning of some basic geometric terms. Even Aristotle understood that undefined terms are necessary before the first definitions are set up, but the geometer Euclid (regarded by some as an Aristotelian) violates this in the very first sentences of his treatise. We have a strange situation, even taking into account that most experts in Aristotle see only a tenuous connection between the *Elements* and the philosopher's views on methodology.[40] One thing at least should be clear, namely that the criteria for definitions in the practice of the *Elements* are quite different from those of modern mathematics. We find one of those hermeneutic difficulties that may contain the promise of opening new doors to a deeper understanding of the topic.

Sometimes it is said that the *horoi* express a pedagogical concern, but this seems to me an obvious understatement of their importance for Greek geometry, and of their epistemic relevance. Let us look at them from another angle, from the point of view of practice. We should not discard too quickly the possibility that the first Euclidean definitions offer something to the person who begins the study of geometry, and something peculiar. It has always struck me that the first printed edition of the *Elements*, produced in Venice, in 1482, by Erhard Ratdolt, should reproduce graphically the straight line, the point, the circle and its diameter, two adjoining right angles, etc.[41] It also presents visual

---

[40] Harari (2003) writes: "Hence, if there is a relation between Aristotle's conception of first principles and those of the mathematicians, Aristotle provides an ideal framework based on contemporary mathematical practice and which may or may not have been noticed by authors such as Euclid."

[41] This Latin version was based on a translation from the Arabic and is known by the name of Campanus of Novara, who commented it. It had many editions, some of them revised by Luca Pacioli. Similar explanatory drawings can be found in many other sixteenth-century editions, such as the famous one by F. Commandino, *Euclidis Elementorum libri XV* (Pesaro, 1572), or the first Spanish edition by R. Çamorano (Sevilla, 1576).

representations of the postulates (*petitiones*) with the parallel postulate as no. 4.[42] Clearly, back then novices needed some pictorial introduction to the subject. This is confirmed by the fact that many manuscripts also contain such diagrams.

The first definitions indeed suggest a *way of reading diagrams*, a perspective for seeing or conceiving what is implied by a diagram, and what is not. And this way of reading is not at all evident, especially if one previously knows only practical geometry. For the definitions and the reading that comes with them lead the practitioner to certain crucial idealizations. Most importantly, the definitions suggest certain forms of response (and of indifference) to some aspects of the diagram:[43] thus, the crossing of two drawn lines will be a (very small) planar region, but we are taught to disregard this and consider in the argumentation that one and only one point has thus been determined. Definitions 1–3 are part of a school of the practice of Greek geometry that teaches us to "see" merely one point as the object determined by the said intersection, a one-dimensional line as the intersection of two planes, and so on. The geometer will act (drawing) and reason (inferring) accordingly, employing the diagram not as it is empirically given, but as it is conceived. Nevertheless, the relevant informational content is not fully exhausted conceptually or theoretically; the information is still (partly at least) extracted from the visual representation.

A good example for underscoring the distance between perception and visual thinking à la Euclid is any diagram displaying a tangent, in the simplest case the tangent to a circle. By considering visually a diagram of the circle and its tangent, I am at a loss to determine whether there is one point or more than one point of intersection. Well, there is at least one point, and the simplest assumption is, then, that there is only one point. But is that the case "really"? Seventeenth-century mathematicians could "see" in such situations infinitely small segments as the intersection, and people such as Peirce could even consider this a necessary assumption regarding a true continuum. But Euclid's

---

[42] And a fifth postulate saying that two right lines cannot enclose a surface, a feature of the Theonian editions (see above), present in them as axiom 12.

[43] Indifference and responsiveness to features of representations is a topic that I have seen Manders elaborate on in an unpublished conference.

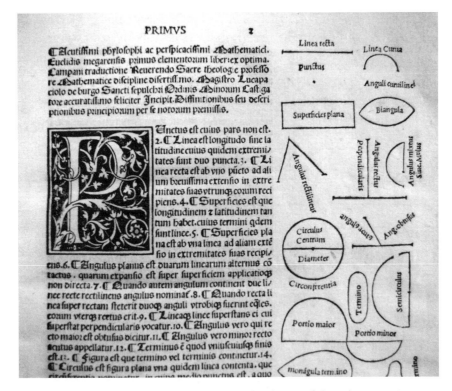

Figure 5 First page of the Paganini-Pacioli edition of the *Elements* (Venice, 1509), which features diagrams of line, point, angle, circle, etc. accompanying the Definitions in the text. It is not known whether such representations accompanied the early manuscripts at the time of Theon of Alexandria (fourth century) or earlier. (Courtesy Fondo Antiguo, Universidad de Sevilla)

idealizations, including what is said in Def. 1 and others, entail that the tangent merely "touches" the circle at a single point. The theorem that the tangent to a circle is the perpendicular to its diameter through a given point is found in Book III prop.16, its proof being based only on results from Book I, and, particularly, on the angle sums in triangles (I.17)[44] and the relations between angles and sides (I.19).

[44] I.17 is not based on the parallel axiom; it states that the sum of two angles in a triangle is less than 180°.

Consider now Def. 4, which is linked with a large body of literature on the topic of definitions of "straight," from the idea of a very tense rope to Euclid's "lying evenly" to geodesics in spaces of constant curvature.[45] It reads:

4. A straight line is a line which lies evenly with the points on itself.[46]

The problem of reaching a satisfactory conceptual characterization of the straight line has been a prominent preoccupation of authors who worried about the foundations of geometry. Those who tried to find a satisfactory solution to the problem of the parallels (the perceived need to turn postulate 5 into a theorem) often ended up realizing that it is not obvious how to characterize a straight line. It seems likely that Def. 4 can only be understood by a beginner if it comes accompanied by a diagram, or else the beginner already knows what is meant by a straight line in practical geometry. (For example, it seems natural to wonder why the perfect line of a circumference does not "lie evenly with the points on itself".)

The main point I want to emphasize is the distance that the first few sentences in Euclid's treatise creates from any practical geometry— a distance linked with the riddle of the continuum, discussed below. Abstraction and the role of these idealizing definitions can be seen from a pragmatist perspective, promoting a positive view of the role of diagrams and a nuanced vision of the links between the diagrammatic and the ideal.[47] In this pragmatist perspective, the diagram plus the definitions and postulates establish the concepts of point, line, etc., concepts that dawn upon the agent when she learns to employ the diagrams in a certain way, according to the stipulations established by

---

[45] For details, see Heath's edition (Euclid 1908), and Gray 1989, among others.

[46] The next few definitions are similar to the one already mentioned: 5. A surface is that which has length and breadth only. 6. The edges of a surface are lines. 7. A plane surface is a surface which lies evenly with the straight lines on itself.

[47] Different both from Plato's view of the drawing as a mere sensory aid to the imagination, the very first step in the scale of learning, while the "true figure" is accessible to our understanding when we rise to the contemplation of forms; and from Kant's approach, which distinguished between empirical intuition (imperfect, *a posteriori*, unable to lead to true science) and pure intuition (*a priori* source of apodictic knowledge), insisting on the view that mathematical knowledge depends only on the latter.

definitions and postulates. Notice that this kind of approach commits us to the view that geometric knowledge is conceptual, but at the same time liberates us from linguistic reductionism regarding concepts. Needless to say, much work will probably be needed to fully clarify this standpoint.

5.6. A LOOK AT THE FUTURE—OUR PAST

It would be of great interest to develop further the previous analysis by considering the rise of Cartesian geometry. With Descartes's systematic interconnection between symbolical algebra and geometry, we find a beautiful case of the introduction of new symbolic means and conceptual apparatus into an already-developed domain. According to our standpoint, it should be expected that the introduction of a new Framework shall involve the elaboration of new conceptual perspectives. Descartes was not a revolutionary concerning geometry; his approach was intended to preserve continuity with the geometry of the ancients. But he was, of course, a revolutionary vis-à-vis method, primarily concerned with the *extension* of the means for elementary constructions in geometry, with a view to augmenting the power of geometric methods. If Euclid's plane geometry was carefully restricted to circle and segment, Descartes sought a principled extension that could engender a closed and well-defined theory. His method was dual, involving analysis of a geometric situation by means of algebraic symbolism, until knowledge was obtained of the conditions for a solution (by means of algebra), and synthesis of the geometric solution by interpreting geometrically (through the available constructions) those algebraic conditions. In the process, the range of geometric configurations under study was enlarged, the field of geometry expanded, and—most important—the scene was prepared for other, more revolutionary symbolical methods: those of the infinitesimal calculus.

However, for those highly interesting developments I will refer the reader to careful work done by other writers.[48] It seems important,

---

[48] See especially Bos 2001, and also Panza 2011 and Serfati 2005. For a mathematical discussion, see Hartshorne 2000.

instead, to devote some time to a possible worry in the reader's mind. We have devoted a full chapter to old-fashioned Euclidean geometry, in the context of a book that aims to analyze mathematical knowledge. Is this not mere history? Certainly Euclid's notions and methods played an important role in the configuration of mathematical knowledge in antiquity, the Middle Ages, and the early modern period. But today's mathematics would be unrecognizable to Euler! The domination of geometry ceased in the generation of Gauss, 200 years ago,[49] and they also started a complete transformation of geometry itself. Isn't it clear that, as of today, Euclidean geometry has little relevance to mathematics? With good reason, Bourbaki-member Jean Dieudonné exclaimed "Down with Euclid!" at a conference on math education in 1959.

I must argue against such views. All of the topics discussed here have been chosen with a view to understanding the web of practices that is intertwined in the conformation of fundamental concepts and theories in mathematics today. One obvious reason is the role played by geometry in the conformation of the real number system, which plays a crucial role in the sequel; but there is more. My claim is that studying Euclidean geometry and the role of visual thinking in it is highly relevant for any attempt at understanding mathematical epistemology and cognition. Students of mathematics, more or less since 1960, are no longer exposed to Euclidean methods—yet spatial thinking of a comparable nature remains active and central to the way in which individuals gain knowledge of mathematics. Pre-university students today have a strongly algebraic-algorithmic conception of what math is, a reductionist, unbalanced view. They are not sufficiently exposed to spatial thinking, for which reason it is likely that some people who could become good mathematicians are lost, due to their lack of interest in the mechanistic application of algorithmic methods, while some people who might be good accountants, but not mathematicians, start to study university mathematics as a result of false impressions. However, it is clear that spatial thinking is nonetheless developed—in the

---

[49] Symbolically, Gauss took a saying attributed to Plato, "*ho theos geometrei*", the god geometrizes, and transformed it into "*ho theos arithmetidsei*," God does arithmetic. Geometry was no longer at the core of math; now that place was taken by arithmetic; see Ferreirós 2006.

context of fractions, of real numbers and measurement—through the heuristic use of diagrams for physics, drawing, and so on.

Indeed, there are reasons to suspect that the Bourbakian dictum "A bas Euclid" has created an educational situation in which the training of spatial thinking abilities, which are central to mathematics, is far from optimal. Let me quote the mathematician V. Arnol'd:

> Jacobi noted the most fascinating property of mathematics, that in it one and the same function controls both the presentations of an integer as a sum of four squares and the real movement of a pendulum. These discoveries of connections between heterogeneous mathematical objects can be compared with the discovery of the connection between electricity and magnetism in physics or with the discovery of the similarity in the geology of the east coast of America and the west coast of Africa.
>
> The emotional significance of such discoveries for teaching is difficult to overestimate. It is they who teach us to search and find such wonderful phenomena of harmony in the universe.
>
> The de-geometrization of mathematical education and the divorce [of mathematics] from physics sever these ties. (Arnol'd 1997, 231)

The nineteenth century was the time of arithmetization, when mathematical development came increasingly "under the sign of number" (Hilbert), and the twentieth century was the time of abstract Structuralism, of a highly abstract and formal presentation of mathematics. This promoted the "de-geometrization of mathematical education and the divorce from physics" that Arnol'd deplored.

But even in those days, against the tide in math education, geometric thinking has constantly been coming back—in math, but also in physics, biology, etcetera. Geometrization of mathematical theories and results is a common phenomenon: the theory of real functions may have become increasingly formal and "arithmetized" in the period 1820–1930, but in the study of complex functions there emerged geometric notions such as Riemann surfaces that increasingly dominated the field. With the emergence of more flexible, non-metrical geometric notions (culminating in modern topology), there arose the possibility of introducing spatial thinking in the most diverse fields of mathematics: not only in the study of functions, with function spaces of the most diverse kinds, but even in the fields of number theory and mathematical logic.[50] Banach spaces, introduced in the 1920s, are central to functional analysis, while Hilbert spaces are indispensable tools in partial differential equations, Fourier analysis, quantum mechanics, and ergodic theory. Number theory has established stronger links to geometry throughout the twentieth century: one might point to Minkowski's work around 1900, but also and especially to the line of work running through Weil to Laglands to Wiles, which has given rise to what is called arithmetic geometry.

Visual representation is clearly "part of the process of doing mathematics" (Hoffman, cited in Mancosu 2005), even if current mathematical practices emphasize "properly mathematical" justification based on the models of arithmetic and set theory, on the twentieth-century logico-linguistic understanding of proof. Jessica Carter (2010) has shown with an interesting case study the hidden role of visual and diagrammatic thinking in modern analysis: her case has to do with very modern math, a certain result on Gaussian random matrices in free probability theory, where she was able to document that a group of researchers employed diagrams to obtain certain combinatorial expressions—diagrams that were intentionally excluded from the published version, following established practice. In such cases, diagrams can offer clues to essential properties, they can be a preferred heuristic route for some lines of research, with traceable impacts on the genesis of concepts. This kind of situation is also important to take into account, insofar as it reminds us of the great difficulty we may often experience in locating and documenting relevant elements of mathematical practice, which may well be invisible in the usual written sources.

The enormous power of the introduction of geometric ideas into a given field of mathematics is a well-known phenomenon to researchers, but it remains sadly invisible to a majority of people, including

---

[50] One could give very many examples in the field of logic, starting of course with different areas of model theory, and including the whole area of topos theory, which itself arose from the idea of cross-fertilizing logic and algebraic geometry. One can find nowadays books on topology for computer scientists, which I guess defies most people's intuitive understanding about the "order of things."

many of those who have enjoyed sound basic training in science. As we have seen, in the era of Euler and Lagrange it became fashionable to avoid diagrams, and the conviction emerged that they can only serve as heuristic guides, but are fundamentally dangerous since they promote wrong "intuitive'" conclusions. Yet, in recent decades diagrams have become a subject of much interest for mathematicians, cognitive psychologists, philosophers, mathematical educationalists, and even logicians (see Mancosu 2005). Among mathematicians, partly due to the strong impact of computer simulations since 1980, there have been calls for a return to intuition and visualization, and "visual approaches" to topics such as geometry, topology, and complex analysis have been offered. Also, philosophers such as Giaquinto (2007; 2008) have analyzed basic questions about the kind of epistemic warrant that diagrammatic or visual reasoning can provide.

Looking at things from a totally different angle, consider results from cognitive science relevant to the question of spatial thinking. Classic experiments in developmental psychology show that infants eighteen to twenty-four months old employ geometric cues from the environment, while they are incapable of using nongeometric cues—an ability that develops later on the basis of experience (Hermer and Spelke 1994). Such language-independent reliance on geometric cues is shared with nonhuman apes; it includes responses to angle and length that are quite sophisticated by the age of three.[51] Work on the Amazonian tribe of the Munduruku, who do not possess elaborate geometric vocabulary, nor use maps or other similar representations, has led to striking results about seemingly universal abilities to discriminate between straight lines, angles (in particular right angles), and alignments of line and point (Dehaene et al. 2006). Notice that we are talking about metric properties that are directly relevant to the Gougu-Pythagorean result with which we started this chapter.

Researchers such as Dehaene and Spelke go as far as talking about "geometry as a universal mental construction," arguing that the basic principles of Euclidean geometry "are reflected in intuitions of space that develop progressively throughout childhood, but still appear

---

[51] See the review in De Cruz 2009, which is useful independently of the main thesis of that paper.

universal" (Izard et al. 2011, 329–330). Notice that the claim is not exactly a form of innatism, as they make room for cognitive development in infancy and childhood, obviously linked to progresses in perception and action; yet the idea that there is such a thing as a "natural geometry" whose basic metric features coincide with Euclidean geometry is no less striking. Doubtless, more interdisciplinary work on this issue is needed, relying on experimental psychology, anthropology, neuroscience, and—by no means least—the history of mathematics.[52] Yet it seems pretty clear that spatial thinking deserves to be given a special place whenever we consider the origins and the development of mathematical knowledge. And let me insist: if we continue disregarding the training of spatial thinking in the mathematics classroom, it will be at our own risk.

---

[52] Historians such as J. Høyrup, K. Chemla, H. Bos, A. Heeffer, and others have done much work on reasoning styles, roles of different types of diagram practices, algorithm practices, the interplay of such, and so on. Historical work is often irreplaceable insofar as it may prevent us from drawing unwarranted conclusions about such difficult questions as the origins and emergence of basic mathematical notions.

6

Advanced Math

The Hypothetical Conception

Around 1900, it became common to argue—at least in the context of geometry—that mathematical axioms are hypotheses. This viewpoint goes back at least to Riemann, to his famous lecture "On the hypotheses which lie at the basis of geometry," but it can be reinforced and generalized on the basis of results obtained in foundational research. This revolution was brought about, more than anything else, by the epochal shift in the conception of geometry associated with non-Euclidean geometries[1] and later with the rise of modern, abstract geometry. But once again, the point brought home by geometry was later reinforced by other developments, and should be generalized. Here I shall defend a hypothetical conception of higher mathematics.

Many philosophers of mathematics today tend to accept some form of "quasi-empiricism," and therefore the view that modern mathematics includes hypothetical elements. Such propositions or principles, according to this view, are not accepted on account of their certainty and necessity; they are not apodictic principles, contrary to Kant, but propositions adopted mostly on *extrinsic* evidence. That is to say, because they lead to certain "seemingly evident" consequences (among which the most prominent tend to be results in classical analysis, the theory of real numbers and real-valued functions), no better alternative route to these consequences is available, and no contradiction can be derived from them. Some *intrinsic* evidence is also relevant, of course, for reasons such as simplicity and the fact that the system of principles forms a

---

[1] Riemann's "hypotheses" referred indeed to the basic truths of Euclid, but more generally to the axioms he employed as a basis of differential geometry, from which he regained Euclidean and non-Euclidean geometries.

coherent whole, a cogent overall view that is arguably well understood. But it is unclear whether we have the right to speak of basic "intuitions" underlying the concepts and principles of the system, in any strong sense of the word going beyond the mere availability of a seemingly clear semantics and conceptual understanding.[2]

Nevertheless, in my view it is preferable to avoid Lakatos's term "quasi-empirical" as this adjective is obscure and even misleading. It aims to mean that the methodology of mathematics is not so different from that of the empirical sciences, but it is misleading to interpret this in the sense that mathematics is (to some unclear extent) another empirical science, or in the sense of a simplistic "reduction" of mathematical methodology to that of the natural sciences. Moreover, the idea of quasi-empiricism is often linked with the view that inductive methods are at play when the hypotheses are established, but this is a complex matter that I shall prefer to put aside. The presence of hypotheses at the very heart of mathematics establishes an important similitude with physical theory, as Hilbert himself emphasized, and undermines the simple distinction between "formal" and "empirical" sciences.[3] For all of these reasons I propose to speak simply and directly of a hypothetical conception of mathematics.

6.1. THE HYPOTHETICAL CONCEPTION: AN INTRODUCTION

The hypothetical conception of (part of) mathematics is difficult for philosophers and mathematicians alike to accept, because it runs against time-honored views of mathematics as a body of certain, evident, necessary truths. Mathematics had been regarded, for centuries and even millenia, as the paradigm of a science formed by apodictic truths; such views were hegemonic from the time of Plato and Euclid to the days of Kant and Gauss. Even after Riemann, influential authors

---

[2] This more delicate matter will be taken up in the last chapters. To clarify, I find it best to reserve the term "intuition" for strong accounts such as Kant's (space and time as *necessary* forms of sensible experience) and/or Brouwer's (time and an intuition of twoness).

[3] See Weyl 1927, Quine 1951, and Putnam 1967.

such as Kronecker and Frege tried to re-found mathematics, and resist innovations, with the aim of preserving certainty. Nonetheless, key authors in all the main tendencies of twentieth-century foundational studies came to agree on a hypothetical conception of the "classical" edifice of mathematics. Let me briefly mention four examples.

Egbertus Brouwer and other constructivists proposed abandoning "classical" mathematics because it depends on ungrounded assumptions; by refusing to accept postulates such as the Axiom of Choice and even the Axiom of Infinity, they underscored the hypothetical nature of mainstream mathematics. David Hilbert accepted this diagnosis and came to speak of "ideal elements" (Hilbert 1926) as an inescapable ingredient in mathematics; the goal of his celebrated program in proof theory was to justify adoption of such "ideal elements" by means of consistency proofs. Even a logicist like Bertrand Russell spoke of axioms accepted merely on the basis of "inductive evidence," i.e., because seemingly evident propositions can be derived from them, no good alternative route to these is in sight, and no contradiction has so far been derived.[4] To complete the picture, the radical platonist Kurt Gödel was also open about hypothetical methods being at play in mathematics:[5] e.g., in his view not all the axioms of set theory "force themselves upon us," some being upheld on the basis of extrinsic evidence of the kind adduced by Russell. Gödel acknowledged that some mathematical postulates are adopted on criteria such as their explanatory power and generality, their ability to predict results that may be "verified" independently (e.g., number-theoretical results), their simplicity, or their role in making possible fruitful extension of theories.

Two clarifications should be made immediately. The word "hypothesis" is being used here in its etymological sense, without the connotation that mathematical hypotheses may represent aspects of physical reality. A hypothesis is etymologically a supposition, "what is placed at the basis," just as the Axiom of Parallels is put at the basis of classical Greek geometry. Indeed, Aristotle used the Greek term "hypothesis" [ὑπόθεσις] for what we call the axioms of geometry; it seems that

---

[4] The unlikely place for such declarations was vol. 1 of *Principia Mathematica* (1910), in connection with one of the logical axioms (!) of Whitehead and Russell's system.

[5] See Gödel (1947), Gödel (1980), 310, 359–61, and the elaborations in Maddy 1990, 1997.

Riemann was taking it up directly from him, with the additional modern connotation that hypotheses are not certain and evident, but merely probable.[6] A possible way to emphasize the peculiar nature of mathematical hypotheses, without the above-mentioned connotation, is by saying that these hypotheses are *constitutive* and not representational, in the sense that they constitute or lay out a mathematical domain.

Second, there is no need to take this as an all-or-nothing approach: one can, and, in my view, should, accept that parts of mathematics are *not* hypothetical. Recall that Bernays and Hilbert endorsed the hypothetical conception, talking about "ideal elements," while claiming a special epistemological status for finitary mathematics. The hypothetical conception does not entail that all of mathematics is of the same character. One could certainly argue that some parts of the theoretical body (say parts of elementary geometry) enjoy a special non-hypothetical status. A rather obvious candidate for certainty is the theory of natural numbers, as embodied in the first-order Peano–Dedekind axioms, when presented in a way that avoids commitment to actual infinity; one can argue then that the Peano–Dedekind axioms are true of counting numbers (see Chapter 7). This means that one can avoid apriorisms even here, adopting an argument in line with the cognitive, historical, pragmatist approach we are advocating.

Thus, the hypothetical conception applies to advanced mathematics, not to elementary mathematics. This raises the open questions of how exactly to circumscribe the non-hypothetical domains, and how to account for their epistemological character. These difficult questions must be approached on the basis of empirical studies of human cognition, and they become particularly delicate (it seems) in the case of geometry; see Chapter 5, section 5.2 and the concluding paragraphs.

---

[6] Speaking of the geometrical hypotheses, when regarded as representing physical space, he says: "These facts, like all facts, are not necessary but have only empirical certainty, they are hypotheses; one can thus investigate their probability, which however is very high within the limits of observation." In the second part he is thinking about Euclid's parallel axiom compared with astronomical observations; in the first, he employs the word "hypothesis" in the way now usual in philosophy of science, different from our terminology above precisely because it adds the connotation that geometrical hypotheses are representing aspects of physical reality. Elsewhere, Riemann writes that, in modern times (unlike in Newton's), by a hypothesis we mean something added by our thought to the phenomena (*zu den Erscheinungen Hinzugedachte*).

To sum up, we should see mathematical propositions as parts of systems based on hypotheses, i.e., constitutive hypotheses in the sense explained above; they are true in those systems, but not true of something else.

Such hypothetical systems are not based on mere convention, but arise out of a richly embedded historical development—if you wish to avoid mentioning history, out of a richly embedded network of practices—far from arbitrariness. We talk about assumptions that are not quite arbitrary, since they arise more or less naturally in connection with previously known phenomena and results, and they come in good measure conditioned by this previous context. It is precisely at this point, by conceiving it in the context of an analysis of the interplay of practices, that the hypothetical conception of mathematics in practice we propose takes a distinctive turn. When properly situated as a perspective on the web of mathematical knowledge and practices, the hypothetical conception affords novel perspectives on issues such as the objectivity of mathematical results, discovery vs. invention, and the old realism vs. anti-realism debate.

To clarify this matter, it will be worthwhile to contrast the historically and practically situated perspective that we are proposing with the more traditional and well-known views of if-thenism and conventionalism.[7]

"If-thenism" is the doctrine that "all mathematical statements are conditional in form," a view asserted by Russell in the very first sentence of his *Principles of Mathematics*: "Pure mathematics is the class of all propositions of the form 'p implies q'" (Russell 1903, 3). The same idea was expressed in his famous sentence that, in mathematics, one does not know what she is talking about, nor whether it is true or false. Typically, this type of perspective is offered as a fully general, atemporal characterization of mathematics; Russell, in particular, aimed to emphasize the abstract and universal nature of mathematical concepts and principles. Although Russell's if-thenism was linked with his basic logicistic understanding of mathematical knowledge, others have interpreted if-thenism along the lines of a conventionalist philosophy of mathematics. Anyhow, they seem to be content with the idea that mathematicians explore the deductive consequences of (in principle) *any*

---

[7] I follow here a suggestion of Torsten Wilholt in a review of Ferreirós and Gray 2009; see *The Bulletin of Symbolic Logic* vol. 13, Sept. 2007.

consistent system of stipulations, or axioms. Were it to be true, mathematicians would be engaged in the investigation of all possible logical consequences from all possible premises or principles.

Needless to say, that is not the case. In fact mathematicians are only interested in a relatively narrow range of axiom systems, or structures. Confronted with the dizzying perspective proposed by Russell, any textbook in modern algebra should strike the reader for the tremendously limited scope of the structures under investigation. And the reasons why mathematicians have been so obsessed with a few most particular systems, such as the theory of real numbers, would be some kind of mystery: $\mathbb{R}$ with its order and topological properties is a most particular kind of field, rather irrelevant if our interest lies in fields in general. This suffices to underscore that if-thenism, although compatible with the actual development of mathematics, leaves much to be desired because it remains too general, vague, and devoid of explanatory power. It fails to offer any explanation for how mathematicians go about deciding whether a mathematical result or theorem is important or not,[8] about deciding whether a new principle should be accepted or not, and so on.

The hypothetical conception of mathematics in practice differs from if-thenism in the following way: rather than focusing on the consequentialist aspect of mathematical knowledge (the way in which mathematicians analyze the logical consequences of given principles), our approach emphasizes the issue of the provenance of the hypothetical principles. If-thenism, with its foundational or logical drive, oriented itself toward a static, monolithic perspective on mathematics. The hypothetical conception advocated here is oriented toward a dynamical analysis of the emergence of new systems. In so doing, it aims to capture a significant part of the dynamics of mathematical growth: the emergence of new practices, new concepts, new principles, and new theories, in which the interactions between mathematics and the sciences have played, and still play, a crucial role. This becomes possible, because the approach developed here is historically situated and takes centrally into account the multiplicity of theories and

[8] This, Gödel regarded as the most important question to be answered, according to Grattan-Guinness (1994), 11.

practices and their crucial interplay, which guides the constitution of meaningful concepts and restricts admissible principles. One can thus start to substantiate the claims made above about objectivity of results, show the way in which mathematical growth transcends the simplistic dichotomy of discovery vs. invention, and so on. The rest of the book is devoted to such issues.

6.2. ON CERTAINTY AND OBJECTIVITY

Although it may seem so at first sight, the above does not force us to abandon completely the ideas (and ideals) of certainty and objectivity. Let me explain why.

First, to the extent that some portions of mathematics (like the theory of natural numbers) are not hypothetical, we have certainty in our knowledge of them—whatever the epistemological reasons. This important matter is treated in Chapter 7. But second, one can achieve certainty in knowing what follows from what, that is, *conditional* certainty concerning what follows deductively from given axioms. The proofs of mathematics can be certain even if some premises (say some axioms) are hypothetical. Indeed, I would argue, in disagreement with Lakatos and many historically inclined philosophers, that the methods developed in twentieth-century mathematical logic have provided a kind of absolute rigor in that sense. Full rigor, that is, in the positive orientation, meaning that once a result has been established (e.g., that the Banach-Tarski paradox is derivable from ZFC), we can be certain that it follows from the premises in question. The negative orientation, particularly when we aim at the conclusion that a given conceptual approach cannot be implemented, is much more delicate for the simple reason that the space of conceptual possibilities may have been insufficiently determined or insufficiently explored.[9]

---

[9] Consider, for instance, the issue of analysis based on the assumption of infinitesimals. The nineteenth-century idea that the infinitesimal had been expelled from mathematics forever was insufficiently grounded: currently we have at least one solution based on classical logic (nonstandard analysis in Robinson style) and another based on intuitionistic logic (smooth infinitesimal analysis; see Bell 1998). I should add that several nineteenth-century mathematicians continued studying infinitesimals; see Ehrlich 2006.

At any rate, however, the certainties offered by modern mathematics fall short of the classical image of an edifice built on stone, founded on propositions that are true and certain. Can one make compatible this kind of hypothetical conception with the peculiar objectivity of mathematical knowledge?

The philosophically inclined mathematician and physicist Hermann Weyl (Hilbert's most impressive disciple, but also a critic of classical mathematics) wrote:

> The constructs of the mathematical mind are at the same time free and necessary. The individual mathematician feels free to define his notions and set up his axioms as he pleases. But the question is, will he get his fellow mathematicians interested in the constructs of his imagination. We cannot help the feeling that certain mathematical structures which have evolved through the combined efforts of the mathematical community bear the stamp of a necessity not affected by the accidents of their historical birth. Everybody who looks at the spectacle of modern algebra will be struck by this complementarity of freedom and necessity. (Weyl 1951, 538–39)

Such a compatibility of freedom and necessity, or, in our terms, hypothetical character and objectivity, can reasonably be attained. This is an objectivity without objects, to adopt a happy expression of Kreisel and Putnam. We are talking about a form of intersubjectivity and the associated "relative necessity," so to speak, of mathematical results. Indeed, its detailed analysis makes it possible to claim that this is a peculiarly *strong form of intersubjectivity*—very likely, the strongest there is for humans.

We shall examine in detail some examples of this, ranging from the elementary to the advanced. To put it differently, one has to accept that, if the development of mathematics seems to be endowed with necessity, to force some results upon us, on closer examination this comes down to a lack of arbitrariness—more precisely, to the existence of strong constraints on theoretical developments, in particular, on new hypotheses. Some authors, e.g., French philosopher J. Cavaillès, tried to conceive of mathematical development as a process that, in spite of being contingent and historical, is nevertheless marked by necessity. That which is obscure and somewhat paradoxical in Cavaillès, I shall try to make fully clear and precise, and also show as grounded immanently in concrete understanding of humans and their capabilities.

Adoption or rejection of a certain hypothesis is the outcome of consensus within the mathematics community. It is perfectly conceivable that the Axiom of Choice (in whatever version we formulate it, set-theoretic or category-theoretic) or even the Completeness of $\mathbb{R}$ might have been rejected by the vast majority of twentieth-century mathematicians. The present statistical distribution of "classical" vs. constructivist mathematicians could be exactly reversed from what we observe. As already remarked, that is not to say that such hypotheses are "purely conventional," that no restriction operates on the free choices that lead to general consensus. Indeed, there exist different kinds of restrictions or linkages, from those related to our constitution as human beings, to others related to the activities of counting, measuring, and constructing, to still others deriving from the role of mathematics in scientific modelling of phenomena. As a result, one must hasten to add that such conceivable scenarios (rejection of Completeness, or of Choice) are highly unlikely in view of the established practices of mathematics in the nineteenth and twentieth centuries, including the practices related to mathematical physics.

Even so, philosophically the hypothetical conception entails as an inescapable consequence the contingent nature of (parts of) mathematical knowledge. If one could define the function of mathematicians as bringing to light some preexisting structures, given in advance of the development of their activities, there would be no reason to talk about an essential historicity. Of course many philosophers have suggested that, for one reason or another, structures such as the real-number continuum are preexisting and given to us. As already mentioned, in my view such a philosophical assumption would be unwarranted, and I definitely side with Riemann and Hilbert, who denied that the structure of the continuum and/or actual infinity may be (so to speak) read off physical reality. The continuum and infinity are posited by thought, and the philosophical developments offered here are marked by the decision to put brackets around ontological assumptions such as those just mentioned. Our finding, which I hope shall be welcome as an interesting insight, is that objectivity can still be found in a form of knowledge that falls short of perfect certainty.

It is important to emphasize that the claims I am making should not be understood in terms of a simplistic dichotomic perspective. The details of my arguments make it clear that we are not dealing with dichotomies such as historicity vs. rationality, or invention vs. knowledge understood as discovery, or fictionalism vs. full-fledged realism. Such dichotomies are forced and impose a false stiffness upon our philosophical discourse; here we encounter again a classical theme of pragmatism, which has explored carefully how dichotomic thinking fails. The philosophical approach we are delineating is aimed at navigating the space between those reefs, between Scylla and Charybdis—for there is indeed such a space. One can accept the presence of conventional elements (the hypotheses we are discussing) and yet insist on the objective traits of developments based on them; assuming a strict conventional/objective dichotomy would force our philosophical views into a Procrustean bed.

But, precisely because we are not committed to the view that mathematical hypotheses represent an independent reality, they could always have been rejected by the community of mathematicians. Herein lies a strong source of historicity for mathematical knowledge, and also one (only one, but arguably the deepest) of the sources of that diversity of mathematical practices that historians and other students of mathematical practice make it their business to explore. According to the reconstruction offered here, contingency and the hypothetical are inevitably installed at the heart of modern mathematics, although the highly interconnected web of mathematical and scientific activity (past and present) acts in such a way that it becomes disguised and can even seem invisible. Needless to say, such a viewpoint, if correct, must affect also our epistemological understanding of other sciences, beginning with physics.

In fact, the introduction of all the above-mentioned axioms or principles was perfectly rational, albeit historically contingent. It was rational given the web of mathematical knowledge and practices that was in place at the time (say around 1850–70) and the problems of justification and systematization that it posed. Take the problems of mathematical analysis, and in particular that of rigorizing the theory of the real numbers: it was (and remains) rational to pursue the program based on analyzing those numbers as defined by set-theoretic means from the naturals; which led to the axioms of Infinity and Power Set.[10] In order to understand why this is so, one is forced (in my opinion) to take into account that mathematical practices do not live in a vacuum of pure thought. Contrary to what Cavaillès liked to claim (in Bourbakian spirit *avant la lettre*),[11] mathematics is not a sui generis, "perfectly autonomous" development of concepts and structures. Mathematical ideas always depend on a technical and practical background; they are constantly linked (cognitively) with non-mathematical practices. Among these I count not only technical practices such as counting and measuring, but, especially, scientific practices of modelling that are at the core of the sciences, in which mathematics plays a central constitutive role.[12] In a nutshell, a full explanation of mathematical knowledge and the web of mathematical practices makes it necessary to embed this network within the larger web of scientific practices.

6.3. ELEMENTARY VS. ADVANCED: GEOMETRY AND THE CONTINUUM

How can we locate the hypothetical principles of mathematics? This is not hard to do: by considering great debates about modern mathematics in the period 1850–1950—especially, but not only, those between constructivists and postulationists—it is easy to identify such elements in the classical edifice. Some of them come to mind very quickly, such as the Euclidean axiom of the parallels (see Gray 1989) and the Axiom of

---

[10] It was also rational to pursue the alternative program based on attempting to reconstruct analysis without such set-theoretic, infinitistic definitions. A lot of progress was made along both lines of research in the twentieth century, many interesting results were established, and the differences between both programs and their reach (so to speak) were gradually clarified. The mathematical community has opted for the "modern" infinitistic approaches because they satisfy better the constraints imposed by the web of mathematical (and scientific) knowledge and practices.

[11] After all, it is Weierstrassian spirit, reflecting the nineteenth-century ideal of pure science. See Ferreirós 1999.

[12] Let me quote Einstein here (1933, 183): "Experience remains, of course, the sole criterion of the physical utility of a mathematical construction. But the creative principle resides in mathematics." (Indeed he went as far as saying, in Pythagorean spirit: "Our experience hitherto justifies us in believing that nature is the realization of the simplest conceivable mathematical ideas.")

Choice, a quintessential example of a hypothesis (see Moore 1982). The attitudes and procedures that mathematicians have used in treating the Axiom of Choice (see section 6.5) offer much insight into the methodological strategies employed in confronting a principle that is openly and explicitly regarded as conjectural. But the above-mentioned debates offer evidence that other principles, albeit usually adopted without a shade of doubt, are also of the hypothetical kind. I am thinking especially about two of the propositions most consistently put in question by constructivist mathematicians: the Completeness Axiom for the real number system—doubted by Kronecker in 1870, rejected by Brouwer and Weyl around 1920 (see Mancosu 1998, Ferreirós 1999)—and the Axioms of Infinity (Mancosu 1998, Feferman 1998), of which even the weakest version—assuming the totality of natural numbers as a completed object—has been rejected by the intuitionists and, of course, by finitists.

I wish to emphasize that the introduction of hypothetical principles started quite early, and such principles can be found in rather elementary bodies of mathematical results. In order to see those points, and as crucial preparation for discussions in the chapters to follow, it is important to consider basic geometry.

The game of Euclidean geometry depends, according to our analysis in Chapter 4, on a constellation of ingredients that partly come out of practical geometry and partly stem from the hypotheses and idealizations built on top of it. The elements afforded by practical geometry make it handy and susceptible to manipulation—think, in particular, of the line characters (lines drawn by ruler, circumferences drawn by compass) and of the "syllables" (angles, lines with endpoints) and "words" (triangles, circles, etc.) formed with them. But the practice of Euclidean geometry requires that the mathematician handle those characters according to very precise stipulations, according to conceptual determinations introduced linguistically. This combination is ultimately the source of the careful control that makes this practice so robust and convincing.

Those stipulations are in fact hypotheses; they introduce ideal or abstract entities as transforms of the real things in the drawing. We cannot know whether geometrical points exist in the world,[13] but we

---

[13] In fact, most people today believe they don't exist, since there is nothing (physicists teach) below the limit set by Planck's constant—despite the fact that the physicists' world-pictures and worldview are replete with fields and other point set spaces.

postulate them; Euclid can prove that parallel lines exist (on the basis of "absolute geometry"; see I. 27), but he introduces a simplifying hypothesis with his postulate 5, a consequence of which is that the "parallel" be unique. We cannot know whether exact measures can be transported without alteration from region to region in space; but again—perhaps out of simplicity and sheer ignorance of the alternatives—we tend to postulate the homogeneity and isotropy of space.

Although we have not discussed it in the preceding chapter, the theory of measurement (ratios and proportions, Book V) was a central ingredient of ancient Greek geometry. Newton went so far as to say that geometry "is nothing but that part of universal [rational] mechanics which accurately proposes and demonstrates the art of measuring" (preface to Newton 1687). Not so, really: it is "the art of measuring," but it is also, before that, the art of geometrical construction in the plane and in space—construction of polygons, of conics, of the octahedron and icosahedron, and so on. Newton's characterization of geometry suggests that he was heavily biased toward seeing books V, VI, and their extensions (e.g., in the work of Archimedes) as the heart of geometry.

One crucial point concerning the relation between measuring and construction procedures deserves to be emphasized. The celebrated Eudoxan theory of proportions became necessary because the allowed construction methods, as everyone knows, engender figures that cannot be measured by number ratios (e.g., the diagonals in squares and pentagons). But notice how the idealizations also play a role here; for instance, one easily establishes that the Euclidean algorithm applied to the side and diagonal of a regular pentagon engenders a nonterminating process; but this conclusion depends on the assumption that smaller and smaller copies of a pentagon behave exactly like the initial one, that is to say, the conclusion depends on the assumptions of infinite divisibility and of points "without parts" (dimensionless) as the ultimate elements. Without these idealizations, the notion of strictly incommensurable lines would not have arisen.

Another crucial point is that existence (to use that modern term) in Greek geometry depends on the basic construction methods, not on the theory of proportions. The given segments, circles, pentagons, solids, etc. are those that can be constructed in the regimented way; there is simply no assumption of completeness (continuity of lines) beyond that. Here lies a crucial difference between the original space of Euclid and the modern "Euclidean spaces" associated with the full real number system, i.e., with analysis, a difference that, unfortunately, is too often not understood.

From that point of view, the modern problems of the continuum exist only partially in the world of ancient Greek geometry. And yet, I would like to emphasize that the riddle of the continuum was created with the idealizations introduced in the *Elements* via hypotheses. To put it plainly: with the invention of the point came the riddle of the continuum. Let me insist that the hypotheses of interest here were implemented more in the definitions of Book I and the theory of proportion, than in any of the axioms, be they postulates or common notions.

Consider again Def. 1: even if we read, restrictively and, it seems, faithfully, "A point is that [plane figure] which has no part," this is immediately prone to causing perplexities. The following question springs to mind: How many points, then, in a drawn "point"? Similarly, How many lines in a drawn "line"? One will soon come to the answer—infinitely many—from which many further difficulties emerge. The hypothetical introduction of ideal elements, such as dimensionless points and lines without breadth (purely one-dimensional), engendered the question of whether a geometric line could reduce to a multitude of points. The extant fragments of Zeno of Elea and the writings of Aristotle (especially in the *Physics* 6.9, a book devoted to the theory of the continuum) make it clear that this crucial step in idealization was made some two hundred years before the time of Archimedes and Euclid. To rebate Zeno's paradoxes, Aristotle introduced the idea—central to mathematical physics—that time intervals can be represented spatially too, and just as space is infinitely divisible, so is time—making it possible for the arrow to reach the end of the stadium in a limited time, despite Zeno's argument.

Ever since then, the antinomy of the discrete and the continuous has been a constant presence and source of trouble for mathematical thought;[14] indeed, for scientific thought in general. In a noteworthy table of antinomies, Riemann gave in the first place the contraposition

---

[14] See Bell (2010): Aristotle identified continuity and discreteness as attributes applying to the category of Quantity. As examples of continuous quantities, he offers lines, planes, solid bodies, extensions, movement, time and space; among discrete quantities he includes number and speech.

between the perspective of "finite elements of time and space" and the infinitary notion of "the continuous" [*Stetiges*], understood as a "conceptual system that lies at the boundary of what is representable" [*an der Grenze des Vorstellbaren*]. This notion of two possible, opposite conceptions of the reality underlying our notions of space, time, and the physical realm was clearly present in his pathbreaking lecture on the hypothesis at the basis of geometry (although Riemann's working option, in mathematics as in physics, was in favor of the infinitary and the continuous).

The question of the sources of infinity in mathematical knowledge, and the riddle of the continuum in particular, will occupy an important place in all that follows. In the case of ancient Greek geometry, we find a natural dividing line between the elementary and the advanced in the introduction of ideal notions such as the point and of hypotheses such as the Axiom of Parallels.

It is ironic that, when we ask ourselves how and why this all emerged in ancient times, a natural answer is the following: perhaps the reason was mere simplicity. In order to develop a theoretical analysis of geometrical constructions and measurement, the simplest assumption making possible precise study is this ideal notion of a point "which has no parts" (and correspondingly with one-dimensional lines and two-dimensional surfaces). This also seems to be a promising avenue for understanding the Parallel Axiom. The constellation of ingredients that conformed ancient Greek geometry, in particular its practical origins in techniques of geometric drawing, suggested that one can construct a line parallel to any given line through a point outside it. In fact, one can show on the basis of other assumptions—pertaining to what would be called "absolute geometry" in the early nineteenth century—that there is one parallel line (*Elements* I.31). That much being given, the simplest assumption is postulating that there is only one straight line parallel to the given line through that point.[15]

---

[15] This characteristic of simplicity, or, if you wish, degenerateness, of Euclidean geometry was to be recovered later, inside the broader theoretical contexts of Lobachevskii-Bolyai geometry and Riemannian differential geometry. Hyperbolic geometries depend on an arbitrary constant K that can take nonnegative values, and Euclidean geometry is obtained when $K = 0$ (from the standpoint of Riemannian geometry, K represents the value of the curvature tensor at each point; hyperbolic, elliptic, and Euclidean geometries are the particular cases where the curvature does not vary from point to point).

Nevertheless, the interplay between elementary geometric results and advanced developments based on hypotheses must be seen as a ground for future investigations. Our understanding of these questions is very rudimentary, as can be seen in the example of the Pythagoras-Gougu theorem. What are the pre-theoretical sources of this result? It is clear that this question must be asked, since evidence suggests that the result was found independently in quite different cultural environments. Yet, when judged as a perfectly precise proposition within "absolute geometry" (an idealized theory), it turns out that the Gougu theorem implies the Axiom of Parallels.

As the reader must know, there is a huge distance between the continuous as it entered ancient geometry and the modern problems of the continuum (particularly in analysis and set theory). This can be seen in the light of Aristotle's views on this matter, which are never plainly contradicted by the content of the *Elements*. Aristotle maintained that many quantities are continuous and infinitely divisible; he defended that physical reality is a continuous plenum,[16] but also insisted on the irreducibility of the continuum to discreteness—a continuum cannot be "composed" of indivisibles or atoms. The parts of a continuum are of the same kind, they are continuous and divisible, and it would be absurd to think of a continuum as "made up" of points. This makes it understandable how he could be a synechist and at the same time argue that only the potential infinite is admissible (*infinitum actu non datur*). The methods and results of Greek geometry made it possible to determine points on lines in many different ways, but the idea that the line might be an infinite aggregate of points never entered the game.

Despite the development of new notions of indivisibles and infinitesimals in the seventeenth and eighteenth centuries, and of limits and real numbers later, the old Aristotelian view was not forgotten even in the last two centuries. Let me give two examples. Most noteworthy is what the great analyst and set theorist Nikolai Luzin wrote in 1930:

---

[16] Continuum as having no gaps, no "solution" or interruption. Ulrich (1836) explained it thus: "Dem Raume wird Unendlichkeit, Stetigkeit (Continuität) und Gleichartigkeit beigelegt; also in Absicht auf die Möglichkeit der räumlichen Gegenstände findet keine Grenze, in Absicht auf die Folge der räumlichen Theile keine Unterbrechung, und in Absicht auf die Beschaffenheit der Theile des Raumes keine Verschiedenheit Statt."

... if we take the uniform definition of irrational numbers from the theory of Dedekind, ... we obtain the (perhaps illusory) possibility of considering the continuum as a *set* formed by rational and irrational points. This was the viewpoint on the continuum adopted *a priori* by Cantor.

The goal of the Theory of Sets is to solve a question of the greatest importance: whether one can or cannot consider linear extension in an atomistic way, as a set of points—a question which is not new and goes back to the Eleats. (Luzin 1930, 2)[17]

The set theory of Dedekind and Cantor was, in the heterodox view of Luzin, meant to answer a deep mathematico-philosophical question posed already by ancient philosophers, and redefined via the new mathematical practices of analysis and set theory: whether the continuum can or cannot be conceived in a pointillist or "atomistic" way. This is the question that Aristotle had answered negatively.

My other example comes from the University of Göttingen, from the textbook on pure mathematics by the professor of mathematics G.C.J. Ulrich (1836). In that work he explained that geometric bodies are limited by surfaces, surfaces by lines, and lines by points; but "the point has absolutely no extension, it is indivisible" while "the line, as a continuous magnitude, is infinitely divisible in its length."[18] And thus:

From the foregoing it follows that neither the line can be composed of a series of points lying next to each other, nor can a surface be so composed of lines [lying next to each other], nor a bodily space be so composed of surfaces [lying on each other],—although in a

---

[17] "[S]i nous prenons la définition uniforme de nombre irrationnel de la théorie de Dedekind, ... nous obtenons la possibilité (peut-être toute illusoire) de considerer le continu comme un ensemble formé des points rationels et irrationels. Tel était le point de vue sur le continu adopté a priori par G. Cantor. Le but de la Théorie des ensembles est de résoudre la question de la plus haute importance: si l'on peut ou non considerer l'étendue lineaire d'une manière atomistique comme un ensemble de points, question d'ailleurs peu nouvelle et remontant aux éléates."

[18] "Es giebt drei Arten räumlicher Gegenstände: Körper, Flächen und Linien. Die Körper werden von Flächen, die Flächen von Linien, die Linien von Puncten begrenzt. ... Der Punct hat gar keine Ausdehnung, er ist untheilbar. Die Linie aber, als continuirliche Grösse, ist nach ihrer Länge in's Unendliche theilbar; eben so die Fläche nach der Länge und Breite; aber die Linie, als Grenze der Fläche, ist nach der Breite untheilbar, d. h. sie hat überhaupt keine Breite." (401–402)

bodily space one can think and assume to be given infinitely many surfaces, lines, and points; in a surface infinitely many lines and points; in a line infinitely many points. (Ulrich 1836, 404)[19]

Let me mention that Ulrich, a colleague of Gauss, became a member of the Göttingen Academy on the latter's proposal.

6.4. TALKING ABOUT OBJECTS

Associated with the presence of hypotheses in mathematics, and also with other issues, is a crucial logical trait of modern (also known as "classical") mathematics: its so-called methodological platonism. Finding the right name for these notions is not simple, and probably a reference to Plato is not the most adequate move, but following Bernays (1935) it is customary to speak of "platonism."[20] This is a well-known topic that emerged out of foundational studies in the twentieth century; but it seems important to clarify our standpoint. It will suffice to summarize the following main ideas.

6.4.1. Two Kinds of Platonism

Philosophers of mathematics agree that one has to distinguish between two forms of platonism,[21] a distinction that some mathematicians find difficult, while it seems trivial to others:

1. INTERNAL, WORKING, OR METHODOLOGICAL PLATONISM. It is characteristic of modern mathematical theories that reference be made to elements that are not explicitly defined

---

[19] "Aus dem Vorstehenden ergiebt sich, dass weder eine Linie aus einer Reihefolge aneinanderliegender Puncte, noch eine Fläche aus nebeneinanderliegenden Linien, noch ein körperlicher Raum aus übereinanderliegenden Flächen zusammengesetzt sein kann, obschon in einem körperlichen Raume unendlich viel Flächen, Linien und Puncte; in einer Fläche unendlich viel Linien und Puncte; in einer Linie unendlich viel Puncte gedacht und angenommen werden können. Auch giebt es keine kleinste Linie, . . ." (404)

[20] This may diverge widely from Plato's original views. It has been claimed that French philosopher A. Lautman was the only real Platonist in twentieth-century philosophy of mathematics.

[21] Beyond the *locus classicus* Bernays 1935, references include: Jané 1993; Shapiro 1997, pp. 21–27 and 38–44; Linnebo 2011.

or definable, but whose existence is postulated; this is a methodological trait of such theories, established by logical analysis.

One of the best-known examples is the well-ordering of $\mathbb{R}$, which can be shown to exist according to ZFC, even though it has never been made explicit—and it is certainly consistent with ZFC that there is no definable well-ordering of $\mathbb{R}$ (Feferman 1965). One could thus speak, following Hilbert and Zermelo, of the *ideal existence* of mathematical objects. Different from this is

2. EXTERNAL PLATONISM, OR ONTOLOGICAL REALISM. This is the philosophical position consisting in the view that there exist independent, abstract mathematical objects, to which the theories refer. Such objects are taken to *exist* in a sense that, albeit analogous to the existence of physical objects, does not coincide completely since they are not spatio-temporal entities.

Notice that Hilbert's famous idea that logical consistency (of an axiom system) suffices for mathematical existence (of the corresponding system of objects) was, among other things, a way of underscoring the difference between mathematical and metaphysical existence. He was thus accepting platonism$_1$ but not platonism$_2$. Realism (external platonism) is one among several possible interpretations of internal, working platonism. It is, of course, the simplest such interpretation, but nothing prevents us from admitting the methodological platonism of mathematics while understanding it from a perspective that avoids the ontological commitments of realism.

The theory of natural numbers does not require strong existential assumptions, so constructivism is adequate for that part of math. The same applies to the theory of rational numbers; but the real number system cannot be defined (i.e., explicitly introduced) with the sole means of a theory of the rationals or the naturals. It is the merit of constructivists, from Kronecker to Weyl and Brouwer, to have gradually clarified this state of affairs. If one were to restrict the domain of real numbers to those that can be explicitly defined from the rationals—e.g., by concretely given Cauchy sequences—then the ensuing domain would be countable and would lack completeness. As a result, one could not prove theorems in analysis the usual way, e.g., arguing from the existence of a lowest upper bound for any bounded, increasing sequence of numbers, or relying on the existence of a limit point for any bounded infinite set of reals $S \subset \mathbb{R}$. This point, made explicit by Weyl (1918), explains why constructivistic analysis must deviate from its classical counterpart.

Platonism comes in degrees, as Bernays emphasized, so the existential assumptions of mathematics are stronger in some theories than in others. There is nothing astonishing about this situation, said Bernays (1935, 68), for it is a familiar procedure in contemporary mathematics to restrict the assumptions in each domain to those that are essential. The famous case of the Axiom of Choice is one in which the methodological platonism of modern mathematics became overtly explicit. As is well known, Zermelo (1904, 1908b) employed this axiom to prove that every set can be well ordered; in particular, the interesting case was $\mathbb{R}$. It was clear to analysts already in 1905 that no explicit way of defining a well-ordering of $\mathbb{R}$ was likely to exist, yet Zermelo had proven there are such orderings on the basis of "purely existential" postulates. (The Axiom of Choice is purely existential in the sense that it postulates the existence of certain sets without offering any means to define them, or a way to present them by means of explicit operations, or the like.)

Such distinctions were not at all clear in nineteenth-century mathematics. In fact, most mathematical contributions up to 1800 tended to be explicit, offering constructions for whatever objects had to be found. The emergence of a new mathematical style, "modern" or "abstract" mathematics, with its strong methodological innovations, subverted the traditional reliance on constructive procedures. For the traditional mathematician, those things exist that we can determine in a concrete and effective way, in a step-by-step construction. The modern mathematician (whose standpoint is called "classical," ironically) repudiates constructivist restrictions: the logical symbol $\exists x$ is employed to specify certain properties of the structure that is characterized by his or her axioms, and the only limitation to free postulation is the consistency of the axiom system. While Euclid's objects were explicitly constructed (by lines and circles) in a finite region of the plane, modern analysis postulates the existence of a limit point for *any* bounded infinite set of reals $S \subset \mathbb{R}$.

One can say, following Bernays, that modern mathematics investigates relations between elements or objects that are *assumed to be given* independently of human thought. After all, in the domain of pure thought, anything that can be thought without contradiction is admissible. Hilbert attempted to make this move fully explicit with his axiomatic approach, and to justify it objectively with consistency proofs by restricted means (which unfortunately turned out to be impossible). Nevertheless, one can in retrospect show that the notion of the real numbers as proportions between magnitudes—based as it was on the assumption that the underlying domains of magnitudes are continuous—carried the seeds of platonism (Chapters 8 and 9; see also Ferreirós 2011).

6.4.2. Mathematical Objects

The methodological platonism of modern math does not force us to adopt a form of realism. From our standpoint, the presence of a platonistic methodology is just a clear indication that mathematics (being part of human theoretical activities) is based on hypothetical assumptions. It does force us to admit that mathematical propositions should be seen as parts of systems based on hypotheses—constitutive hypotheses in the sense explained above; they are true *in* those systems, but not true of something else. And the key philosophical problem that this raises, beyond the need for its transparent critical analysis, is making clear how something hypothetical—mere postulation—can appear to be objective.

Is that a sufficient basis to justify the common talk of mathematical objects? It is, according to a good number of authors who promote what may be called deflationary platonism (see Tait 2005 and Parsons 2009). We have a tendency to objectify, just as we have a tendency to avoid considering relations that are beyond binary or ternary. After long practice with counting, we may become seriously interested in the properties of numbers, and eventually we talk about numbers as objects: there are infinitely many prime numbers! What for a long time was regarded as proportions between quantities, i.e., relations of a certain kind, eventually came to be theorized as the real numbers. After long practice with analyzing the similarities and differences between particular mathematical systems, we may become seriously interested in structure-preserving mappings (morphisms), and eventually we talk about morphisms as objects—perhaps about arrows as basic objects in a category.

The phenomenon is quite general. Whenever a theoretical framework is admissible, we talk about objects (this or that Bessel function, the photon or the Higgs boson, mRNA or the gene codifying a certain protein in the membrane of this bacteria),[22] and when the relevant theory is strongly supported by some scientific community, we see great significance in the fact that the world is well described by talking about those objects. In mathematics, there is a particularly strong tendency to *thematize* relations, correspondences, or structural elements that emerge in a certain theoretical or practical context, turning them into the objects of a new theory (see the previous examples). This is quite natural and perfectly sensible, the only problem being the philosophical naïveté that people frequently exercise by extrapolating their common-sense notions of object to the realm of advanced scientific theories.

Indeed, if you press philosophical questions about these kinds of objects, I will answer that mathematical objects such as those mentioned above are *not* like the Neanderthal found in Atapuerca, not like the mitochondria in your cells, and not even like much more elusive "entities," such as the gene FOXP2 or the space-time field of GRT. And yet we are entitled to talk about such objects, about some unnamed transcendental irrational number (say one that will not be named by any mathematician in the third millennium), about the sheaves on a certain topos, and so on.

Would you like to say that somebody's marriage is an object of the same kind as a mitochondria? No. Then, why would you want to compare the "mode of existence" of photons and prime numbers? The focus should not be on the relatively simple notion of an object, which in the end should carry no metaphysical weight—unlike subcategories of that notion, such as physical object or biological object. Whenever a theoretical framework is admissible, and even more when the framework is presently accepted, we are entitled to talk about objects. Our focus should be on the question of admissibility of theoretical frameworks.

Parsons has promoted a deflation of platonism by considering differences between the status of objects across mathematical theories (geometry in his view is special due to the role of "quasi-concrete"

---

[22] But the examples can also be more elusive for a variety of reasons: the plane of Lobachevskii-Bolyai, the ether, the Neanderthal as a species, this or that mental illusion.

objects) and, on the other hand, by insisting on the simple and clear idea that our basic framework for speaking of objects is first-order logic (FOL). This logical framework raises some demands, but still is a flexible framework and does not imply that all objects are of the same kind. Fortunately so!

"Speaking of objects just is using the linguistic devices of singular terms, predication, identity and quantification to make serious statements," writes Parsons (2009, 10). To talk about objects is to predicate on them, to indicate relations among them, to identify them: this function is Lebesgue-integrable; that number is the logarithm of this other one; $-1 = e^{i\pi}$. We do this in the logical framework of FOL, which is sufficient for the codification of mathematical results and proofs,[23] and along the way we quantify on mathematical objects. The formulation of general statements requires us to go beyond names, to introduce variables and quantify on them: e.g., not only is $49 = 7^2$ but, talking about natural numbers, $\forall x \exists y (y = x^2)$. There is nothing so mysterious in this, although it may seem so to those who are trapped in naïve ontological views of the monistic kind, say the absurd notion that all objects ought to be like ping-pong balls.

Let me put the matter in a very different way. What is predicated of numbers—which we then see as objects—in a certain FOL framework can be predicated of certain relations in a more complex theoretical framework. Gauss wrote that mathematics is "in the most general sense, the science of relations"; other authors, like Poincaré, would later agree with him. As he put it in 1831:

> The mathematician abstracts entirely from the quality of the objects and the content of their relations; he just occupies himself with counting and comparing their relations to each other. [Gauss 1831, 175–176]

Here Gauss was thinking of physical objects that have qualities, quantities, and relations to each other. The mathematician abstracts and considers basically only relations, including relations among relations. But we humans have a definite preference for object language, and so

---

[23] See, e.g., Ferreirós 2001.

the world of our knowledge and our theoretical practices is full of Bessel functions, irrational numbers, gauge fields, and so on. When the mathematician "counts and compares" those relations, she talks about natural and real numbers, about maps, about manifolds, and so on. I can share Gauss's view, adding only that our analysis of relational structures in world phenomena often embeds them into richer structures, enriched precisely by hypothetical elements added in our thought.

Let me briefly mention how this complies well, too, with views such as those of philosopher of science van Fraassen when he insists that adopting a scientific theory does not entail a serious commitment to the "real existence" of its theoretical entities—it entails serious commitment only to the empirical adequacy of the theory. There are modalities involved in the scientists' attitudes to the accepted theories, and indeed those modalities are relevant to scientific practice.[24] The same happens with mathematicians, who may well accept the theories of classical analysis but want to see them presented in more constructivist or predicativist frameworks (Bishop and Bridges 1985, Feferman 1998), or to produce a *reverse* analysis of the indispensable assumptions for certain results (Simpson 2009). Such attitudes can be found not only inside the field of foundational studies, but outside it as well.

6.5. WORKING WITH HYPOTHESES: AC AND THE RIEMANN CONJECTURE

In mathematics one finds many different kinds of hypotheses, sometimes under the guise of axioms, sometimes explicitly presented as conjectures. In fact, working with hypotheses or conjectures is an essential part of the practice of mathematics, even though it has been paid little attention in traditional philosophy of mathematics. Conjectures such as the Riemann Hypothesis about the zeros of the function $\zeta(s)$, or the Poincaré Conjecture concerning 3-manifolds (formulated in 1904, proven by Perelman in 2003), the Taniyama-Shimura conjecture (a theorem thanks to work by Wiles in 1995 and other authors in 2001), and

---

[24] An interesting historical case was that of Hertz on mechanics; see Lützen 2005.

the Langlands program—to name but a few—have played central roles in organizing research in different areas of mathematics. Such conjectures circulate from researcher to researcher, precious material that may guide their work, and they have become particularly prominent since the early twentieth century due to what appears to be a rise of the "art of conjecturing" (see Mazur 1997).

Mathematicians normally use the name "conjecture" or "hypothesis" for a statement that is regarded as provable (or refutable) on the basis of established principles of their discipline, that is, on the basis of accepted axioms. The hope is to eventually see the conjecture turn into a theorem, as happened with two of the examples given above, and as is currently expected of the most famous mathematical conjecture, the Riemann Hypothesis. Only in some exceptional cases do they use the same words for statements that are known to be independent of the accepted principles, i.e., candidate axioms (most noteworthy being perhaps Cantor's Continuum Hypothesis). There are in practice great differences in status among the various conjectures, and so there is a whole spectrum of intentional attitudes, or degrees of confidence, shown by mathematicians in working with them. Certainly, to employ a conjecture as a basis for investigating its consequences is not to make a bet on its truth; sometimes, in fact, a researcher may well be trying to establish its falsity.[25] Also, some conjectures are regarded as superficial and lacking deep implications for their field (an example may be the Goldbach Conjecture in number theory), while others are judged particularly deep and full of implications (such as Riemann's).

Notice that such language usage among mathematicians only coincides partially with ours. From our perspective, not only the conjectures indicated above, and the Continuum Hypothesis, are worthy of that name, but so are basic axioms such as the Axiom of Choice and the Axiom of Completeness. Our perspective is that these assumptions, which cannot be regarded as evident but nonetheless function as "constitutive principles" lying at the basis of analysis or set theory, can be

---

[25] This is claimed about Jack Silver's work on set theory: "Silver's original work involving large cardinals was perhaps motivated by the goal of showing the inconsistency of an uncountable measurable cardinal; instead he was led to discover indiscernibles in L assuming a measurable cardinal exists" (from the Wikipedia article on Silver, accessed in Dec. 2013)

regarded as hypotheses solidified or reified due to general acceptance. We shall continue developing this idea in the following chapters.

How do mathematicians work with hypotheses? There is abundant material on this topic, which offers ample information about the methodology employed by researchers, particularly in connection with the history of some of the hypotheses we have mentioned. Abundantly rich examples are the Completeness principle in analysis, the Choice principle in set theory, and the Riemann Hypothesis in analytic number theory; see, e.g., the well-known book by Moore (1982) in connection with the Axiom of Choice.[26]

In those cases in which the principle in question is sufficiently relevant—paradigmatically when it is a controversial axiom—one may find polemical discussion in which different practitioners express themselves for or against the hypothesis. Interestingly, Zermelo's introduction of AC in 1904 had been preceded by several instances of implicit use of the principle, but also by at least two cases in which Italian authors rejected the idea of having an "arbitrary rule" applied infinitely many times. This somehow anticipated the extraordinary debate that ensued after Zermelo's use of AC to prove the Well-ordering theorem.

More commonly, even if it is a very important conjecture such as Riemann's, one will find it presented tentatively.[27] On the way to clarifying the status of the conjecture, it is particularly important to analyze and clarify its proof-theoretic links, the theoretical web of knowledge in which it is embedded, its consequences, its equivalences, weaker propositions linked to it, and stronger propositions. In fact, the relevance of the conjecture for the mathematical community will be a function of the richness of that theoretical web.

Along the way, trying to obtain a proof of the conjecture, mathematicians may come up with important partial results. A simple example of this is provided by the history of Fermat's Last Theorem that $a^n + b^n = c^n$ has no integer solutions for $n > 2$ ($n \in \mathbb{N}$). After Fermat, and later many others, proved the conjecture for $n = 4$, efforts had to concentrate on

---

[26] I presented a summary of the main issues in Ferreirós (1999), 311–320, 333, 339, 361–367, 379–385. Zermelo's defense in (1908b) is extraordinarily interesting and witty.

[27] Few have bet on the falsity of the Riemann Conjecture, an example being J. E. Littlewood, who said: "I believe this to be false. There is no evidence whatever for it . . . [I feel] there is no imaginable *reason* why it should be true" (Littlewood 1962).

odd exponents. Subsequent developments were very gradual: Euler in 1770 essentially solved the case $n = 3$, Dirichlet (and Legendre) in 1825 the case $n = 5$, and Lamé in 1839 the case $n = 7$. Then around 1850 came an impressive step forward with Kummer's work on cyclotomic fields, which allowed him to prove Fermat's theorem for all regular prime exponents;[28] this work won him the Grand Prize of the Paris Academy of Sciences in 1857. (Note, however, that Kummer's work was much more important in its consequences for algebraic number theory than for having proved infinitely many cases of Fermat's last theorem.) The twentieth century developments would occur in the setting of algebraic geometry, again being more important as such than as steps toward establishing the conjecture—consider, e.g., the Mordell conjecture (1922) that became Falting's theorem in 1984.[29]

A very significant part of the effort devoted to a relevant conjecture is the investigation of its consequences, and in particular of its equivalences—principles provable from it that are as strong as the conjecture itself—and also the study of weaker forms of the hypothesis—consequences that are proof-theoretically weaker than the hypothesis itself. Any good book about the Axiom of Choice contains a wealth of material of this type, with equivalents of AC, such as the Well-ordering theorem, Zorn's lemma, or Tychonoff's theorem; weaker axioms such as Countable choice or Dependent choice; many weak consequences of AC in diverse areas of mathematics, e.g., the Hahn-Banach theorem in functional analysis (and also results in set theory, measure theory, algebra, topology); and even stronger variants such as Global Choice.

In the particularly polemical case of AC, it was quite important during the early stages to analyze instances of implicit use of the principle in analysis, algebra, or set theory itself (see Zermelo 1908b, Sierpinski 1916, 1918).[30] The interconnections between the axiom and

---

[28] A regular prime is a prime number $p > 2$ that does not divide the class number of the pth cyclotomic field. The nth cyclotomic field $Q(\zeta_n)$ is obtained by adjoining a primitive n^{th} root of unity ζ_n to the rational numbers ($n > 2$). The class number is the order of a finite group (known as the ideal class group) that describes the extent to which unique factorization fails in the ring of integers of an algebraic number field (in this case a cyclotomic field).

[29] On this topic, see, among others, Edwards (1977) and Corry (2010).

[30] It may be interesting to note the sections of Sierpinski (1918): I. AC in proofs of the equivalence of diverse definitions of infinite set; II. In the theory of cardinal numbers; III. In the theory of ordered sets and transfinite numbers; IV. In the theory of pointsets; V. In

important results in modern analysis, so-called classical analysis, were so rich that they came to constitute very powerful reasons for its full acceptance. It is not only that; as well known, early investigations in measure theory such as those of Borel and Lebesgue had implicitly employed the axiom, e.g., in the proof that the countable union of measurable sets is a measurable set. (This application became notorious because those same mathematicians would make objections after the AC was clearly isolated and employed by Zermelo to establish the Well-ordering theorem.) The countable form of AC was a most natural principle, employed unhesitatingly in the past, for it seemed innocuous to "extract" a sequence of points converging to a point P from a countable sequence of embedded (closed) domains converging to P. As Sierpinski established, in the absence of AC, many classical notions split up, e.g., the continuity of a function at a point in the sense of Cauchy and in the sense of Heine diverge; also, the notions of a limit-point of a sequence and an accumulation point diverge. The earliest cases of implicit but essential use of AC that were found by Moore in his careful study (1982, 14–16) are, interestingly, a result of Cantor's in analysis (on sequential continuity of functions, published in a paper of Heine) and one of Dedekind's in algebraic number theory (having to do with $\mathbb{Z}$-modules).

Evidence for a conjecture may come in several different ways. There is, of course, the partial results, if any have been established; but there is also its general coherence with known results in the field (number theory in the case of the Riemann Hypothesis, RH). If we consider in particular the RH, there is also a significant body of quasi-empirical results—strong numerical evidence obtained by the actual computation of zeros (Gourdon recently calculated the first ten trillion nontrivial zeros, in 2004). And the list would continue, including knowledge about limitants to possible results, which may even come in the form of probabilistic results (e.g., exceptions to RH must be rare if one moves away from the critical line $R(s) = \frac{1}{2}$; and it is known that more than 40 percent of nontrivial zeros of $\zeta(s)$ are simple and satisfy the Riemann Hypothesis).

---

measure theory; VI. In analysis; VII. In the construction of examples; VIII. Problems implying the existence of nonmeasurable functions; IX. The generalized AC; dependent choices.

Moreover, in cases like that of the Riemann Hypothesis, even the evidence coming from analogous conjectures in more general settings may be relevant. Thus, generalizations of the conjecture can also be very important. This is particularly the case with RH, since the Riemann zeta function is the prototype of a general class of functions, called L-functions, associated with algebraic or arithmetical objects (e.g., arithmetic varieties). These "global L-functions," as Bombieri calls them, are expected to satisfy their corresponding Riemann Hypothesis. The consequences of a Riemann hypothesis for global L-functions are important and varied: "the most important properties of the algebraic or arithmetical objects underlying an L-function can or should be described in terms of the location of its zeros and poles, and values at special points" (Bombieri 2000). The existence of this rich "superstructure," so to speak, built around a conjecture is crucial to understanding the reasons why a conjecture such as the Riemann Hypothesis is considered so important. Moreover,

> the best evidence in favor of the Riemann Hypothesis derives from the corresponding theory, which has been developed in the context of algebraic varieties over finite fields. (Bombieri 2000)[31]

Many experts, like Bombieri, are of the opinion that these results in the geometric setting cannot be ignored; the analogies are too compelling to be dismissed outright, and so they are relevant to the understanding of the classical Riemann Hypothesis.

---

[31] Hasse in the 1930s, Weil in 1940s; "Through his researches, Weil was led to the formulation of sweeping conjectures about Zeta functions of general algebraic varieties over finite fields, relating their properties to the topological structure of the underlying algebraic variety. Here the Riemann Hypothesis, in a simplified form, is the statement that the reciprocals of the zeros and poles of the Zeta function (the so-called *characteristic roots*) have absolute value $q^{d/2}$ with d a positive integer or 0, and are interpreted as eigenvalues of the Frobenius automorphism acting on the cohomology of the variety. After M. Artin, A. Grothendieck, and J.-L. Verdier developed the fundamental tool of étale cohomology, the proof of the corresponding Riemann Hypothesis for Zeta functions of arbitrary varieties over finite fields was finally obtained by Deligne. . . . Deligne's theorem surely ranks as one of the crowning achievements of twentieth century mathematics. Its numerous applications to the solution of long-standing problems in number theory, algebraic geometry, and discrete mathematics are witness to the significance of these general Riemann hypotheses."

7

Arithmetic Certainty

The claim that arithmetical knowledge enjoys certainty is obviously in need of qualification and clarification. You may know about the existence of nonstandard models of arithmetic, and also about the distinctions between intuitionistic and classical arithmetic; the computational complexity of arithmetic statements has long been a topic of discussion; also, the complexity of number theory as a discipline should be a well-known fact. How are such facts of mathematical experience to be reconciled with my claim?

7.1. BASIC ARITHMETIC

To begin, the claim must be about *basic* arithmetic or, let's say, elementary number theory—it cannot be extended to a similar claim about number theory "as a whole." Notice that number theory is a discipline, not a well-defined body of theory; it is rather more like a complex of mathematical practices and associated theoretical bodies (in the plural). The investigation of questions about the natural numbers and the patterns to be found inside their sequence has led to such diverse endeavors as algebraic number theory (where issues about the integers are studied within the much larger setting of algebraic integers), analytic number theory (where they are studied within the even larger domain of complex numbers), and arithmetic geometry or Diophantine geometry (where Wiles's work belongs).[1,2]

---

[1] See, among many others, Manders 1989, Stillwell 2002, and E. Grosholz forthcoming.

[2] Famously, in 1994 Andrew Wiles proved the modularity theorem for semistable elliptic curves, which was enough to imply Fermat's last theorem, and Christophe Breuil, Brian Conrad, Fred Diamond, and Richard Taylor extended his techniques to prove the full modularity theorem in 2001.

If the aim were to systematize "all" of number theory, one could only hope to *freeze* such developments, artificially fencing in a certain body of theory, and then look for an axiomatic analysis of it.[3] We have known for decades now that the resulting analysis of number theory will be not elementary (in the logical sense of a reduction to sentences derivable in first-order Peano Arithmetic) but higher-order. Richard Dedekind was the first to provide such an axiomatization of number theory by means of set and map theory; notice that he was aware of the feasibility of reducing methods in algebraic number theory and analytic number theory to his basic framework. (Arguably, this makes a difference between Dedekind's advanced axiomatic analysis of $\mathbb{N}$ and Peano's more elementary approach.)

One may believe that all important results having to do with patterns in the fine structure of the natural number sequence will turn out to be a consequence of the principles of basic arithmetic. The paradigmatic case of fine structure is the pattern of the erratic sequence of prime numbers, an important part of it being the law of frequency distribution of the primes described in the Prime Number Theorem. Another example is the famous Wiles-Fermat theorem, usually called the "last theorem" of Fermat. One may be convinced that this result must be provable within elementary number theory; and that conviction may be argued for by citing the fact that the Prime Number Theorem turned out to be provable in first-order Peano Arithmetic.[4] (Let me add that, perhaps typically, the elementary proof is not the simplest: arguably, the simplest known proof of the PNT is due to Newman, but it is nonelementary for its use of Cauchy's integral theorem from complex analysis.)[5] Nevertheless, such convictions are not mathematical facts; we are facing *beliefs*. They may well guide research; for us they belong in the description of certain practices, and cannot be taken at face value. Typically, while some of these beliefs may turn out right (we

---

[3] Another example: apparently no one has ever thought about axiomatizing model theory, but it could be done artificially in the above sense.

[4] See Goldfeld 2003.

[5] The elementary proof is mainly due to A. Selberg in 1948, although Erdös also contributed to the research; the proof discovered in 1980 by American mathematician Donald J. Newman is described as "direct, modern and elegant" while the "elementary" Erdös–Selberg argument is "more involved".

have mentioned the example of PNT), others will be there waiting to be confirmed or refuted (the Wiles-Fermat theorem). And we certainly know of important number-theoretic facts that are not elementary: the Harrington-Paris theorem, Goodstein's theorem, Friedman's version of the Kruskal tree theorem.[6]

So we shall need to distinguish between elementary number theory and other, more advanced, levels in the study of numbers: algebraic number theory, analytic number theory, and perhaps set-theoretic number theory (where the so-called second-order arithmetic, PA^2, is the most important level).[7] There is nothing ad hoc in this kind of distinction, which has been perceived by mathematicians throughout. As early as 1801, Gauss distinguished between elementary arithmetic (*Arithmetica elementaris*), which focuses on the basic operations and computational technique, and "higher arithmetic" (*Arithmetica sublimior*), i.e., the general study of the "proper," particular properties of the integers—this is our number theory.[8] It was Gauss himself who, thirty years later, called for an expansion of the field of "higher arithmetic" or number theory, to include the Gaussian (and other algebraic) integers. In the first half of the twentieth century, some mathematicians—most notably G. H. Hardy—believed that there exists a hierarchy of proof methods in number theory—more generally, in mathematics—depending on what sorts of numbers (integers, reals, complex) a proof requires.

7.2. COUNTING PRACTICES, AGAIN

I shall argue that the axioms of Peano Arithmetic are recognized to be true of counting numbers. Thus, we need to start by saying something about counting. According to the standpoint we sketched in Chapter 2, it is one of the basic grounding practices of mathematics (alongside measuring and drawing geometrical forms); such technical practices

---

[6] For a good popular exposition, see Stillwell (2012). See also T. Tao's lecture on primes, 'Structure and randomness in the prime numbers' (UCLA Science Faculty Research Colloquium, Jan 17 2007) in http://www.math.ucla.edu/~tao/preprints/acnt.html.

[7] See Simpson 1999. Third-order arithmetic already is sufficient to codify Descriptive Set Theory, which makes clear its heavy load of set-theoretic content.

[8] This happens in the introduction to *Disquisitiones Arithmeticae*. See Ferreirós (2006), 261.

are important root elements in the web of practices related to mathematical knowledge

This ability to count and the number concept can be developed by humans of any cultural background, apparently, but in many cases it is just latent.[9] To the best of my knowledge, it cannot be developed by our animal cousins. The number concept, in that sense, is a cultural product; it has only been developed with the help of cultural tools (oral language suffices for obtaining it); it is not innate but culture-dependent. This observation does not have any implication of relativity or a lack of clear referents for number:[10] to put it paradoxically, the number concept is a *perfectly natural* cultural product, and knowledge of number is characterized by *certainty* of a kind that has little counterpart even in other areas of math. As I wrote some time ago:

> To state that mathematics is a part of culture, and mathematical knowledge is a social product, does not imply the thesis that mathematical theories are mere cultural products, nor that the only relevant factors to explain their genesis be sociological factors. That is the distance between a soft and almost trivial version of the idea of social construction, and the strong versions defended by Bloor and others, which are based on an unproven and implausible hypothesis. I would rather suggest that the case of mathematics and its associated cultural invariants seems to show that cultures *are not* ultimate factors, self-sufficient and autonomous, without links with nature. (Ferreirós 2005, 64, Spanish in the original)

In fact, the nature/culture dichotomy is only valid on a first approximation; when talking of cognitive factors underlying mathematics,

---

[9] Namely the cases of cultures that do not have a full system of number-words: it is well known that many societies do not count precisely beyond three, four or five, and I propose to infer that such cultures have not developed or "fixed" a *concept* of number (although of course particular individuals in the society may be able to come to that point).

[10] One may want to say that numbers are abstract objects, and this can be done with the necessary provisos (Chapter 6 section 4); but the view that is more consistent with our approach is that numbers refer to *relational situations* of a perfectly objective (intersubjective) kind that involve the ingredients I) to IV) discussed in the sequel. Already Gauss had proposed that mathematics has to do with relations and relations among relations (quoted in Chapter 6, section 4.2).

cultural elements must be taken into account. Thus, my standpoint is better described as a form of pragmatism than of naturalism.[11]

Note that, since we are adopting precision and the absence of limitations of size as defining traits of the number concept, practically speaking this means that we have a simple criterion based on oral language.[12] The number-words are used for counting, and counting is a very peculiar (and highly stable) way of *doing things with words*. With counting, the imprecise innate grasp of cardinality is joined by a notion of order (ordinality), which refines the notion of cardinality into a precise concept, with the aid of specific words,[13] so that they combine to obtain the number concept.

In the process of counting, as it is done in daily experience, the following elements can be found:

I.   we deal with a certain collection of objects that we categorize "conceptually," i.e., that we are somehow able to isolate or specify;[14]

II.   we survey them in a certain (arbitrarily imposed) order, while orally producing a number sequence (via number-words or number-symbols, or even strokes like a primitive shepherd);

---

[11] If by the latter we understand—as tends to be usual—a reductionistic, scientistic approach; see Ferreirós 2010.

[12] To be supplemented, if needed, by criteria having to do with material practices such as that of cutting tallies on a stick; consider the Lebombo bone, from roughly 35,000 years ago. In principle, nothing would prevent members of a human group from developing a number concept without corresponding words, e.g., by using tallies. Suffice it to say that this abstract possibility seems not to occur in practice.

[13] We might say that they are "technical words" but now in a broad sense. "Eleven" or "twelve" are highly technical expressions, due to their role in the technique of counting, which explains why they resist change so much more than other words of English (notice the contrast with "thirteen" and "fourteen").

[14] I follow Frege (1884) in associating the class or collection with a concept, but I put this word in brackets because there is a long cognitive story to be told, underlying our use of words as "concepts" (with significantly different stories for concepts of different kinds—such as dog, chair, prime number). This, however, is not the place to enter into that topic. Also, I use the word *collection* to avoid "set" and its connotations; by a collection I mean those we engage with in everyday life: a collection is *not* a single object, and the operation "collect" is not iterated (a set is a single object, different from its elements and from the collection of the elements; the operation "set of" is iterated, even transfinitely). Cf. Chapter 9.

III.   by so doing we establish a concrete one-to-one correspondence (an effective pairing up) between the number
sequence and the counted objects; and

IV.   the number that we finally reach in the counting process
is employed to determine the cardinality of the collection
(invariant for alternative ways of counting, i.e., for different
orderings of the objects).

I will be pleased if all readers consider this analysis a triviality, because
my analysis aims to be naïve.

Counting so considered is nothing but a technical practice of man-
kind, one of the oldest, and one of the earliest we consistently teach
our children, generation after generation. Due to the restrictions of
constructivity and effectiveness that I have indicated, it is obvious that
such counting is only possible for finite collections of objects. Counting
typically requires visual perception, language-related abilities of cat-
egorization, memory of the spoken numerals in their standard order,
practical knowledge of their (recursive) formation rules, the ability to
order a collection according to certain rules (no duplications, no missing object), and command of the practice of coordinating body movements—normally pointing a finger toward one object of the category
at a time—with the production of numerals. Say, for example, that a
child is counting toys or beans: beyond having sufficient control of the
practice of producing numerals according to a combination of memory
and grammatical rules (". . . ten, eleven, twelve, thirteen, . . . , twenty,
twenty-one, . . . , thirty") and the practice of establishing concrete one-
to-one correspondences,[15] he or she must be able to correctly perceive
and categorize the objects.

Similar remarks apply to the case when a native of Papua New Guinea
is using body parts to count. Even if no oral language were being used,
perception, deeds, the conceptual skills that are typically mastered only
with language, and the semiotic element of representation (with body
parts standing for ordinals and cardinals) are nonetheless present.

---

[15] In the typical case of counting a collection of visualizable things, the concrete correspondence is actually between a gesture simultaneous with an utterance on the one hand,
and a thing or event on the other; that is to say, between "finger+word" and "one."

The cognitive complexity of counting is certainly very high, incomparably higher than that of subitizing, but the most noteworthy aspects of this practice are its stability, reliability, learnability, and intersubjectivity. I dare to suggest the idea that mathematics, in its most elementary strata, may be the best expression of that which we humans have in common, merely in virtue of being human.

Mathematical practices, properly speaking, like reckoning arithmetic and other more sophisticated forms of arithmetic, are of course linked with the concrete practice of counting. But they go far beyond it. It is possible to reinterpret counting practices from an advanced mathematical standpoint, but this can only be done from the vantage point of modern math with its characteristic assumptions. The advantages are bought at the price of assuming the hypotheses on which advanced mathematics is founded (see Chapter 6), the conceptual clarity and precision of the analysis is paid the price of foundational difficulties. A particularly simple case is offered by the analysis given above of counting naively conceived and Dedekind's structural approach to the natural numbers.[16]

Adopting all the artillery of the theory of sets and mappings, and looking at counting from above, so to speak, Dedekind offered the following reinterpretation:

Practical Counting	Structural Analysis
—collection of objects,	—a set of entities
—recursively generated number sequence	—the structure $<N, 1, \sigma>$, with σ successor function
—effectively given pairing up	—a one-to-one mapping.

The intuitive appeal of Dedekind's theory comes from the way it links with this common everyday life experience of counting. Interestingly, Dedekind was so steeped in the modern set-theoretic way of thinking that he did not realize the distance between the two; he went as far as calling his analysis "naïve."[17] What I did above was to adapt freely his

[16] Dedekind (1888), especially § 71.
[17] See the 1872/78 draft in Dugac (1976), appendix; see also the first preface.

approach, making it concrete and constructive, i.e., kroneckerizing it and eliminating its abstract framework.

7.3. THE CERTAINTY OF BASIC ARITHMETIC

What is basic arithmetic?[18] In agreement with Gauss and Peano, we shall take it to be the study of the most general and basic properties of the natural numbers, coincident with the analysis and study of the basic operations and order relation among the naturals. It is the study of the natural number sequence $\mathbb{N}$, in their order of succession, where $n' = \Sigma(n)$ is the successor of number n, and of the ordering relation $<$ and operations $+$, $\cdot$, and $\uparrow$ (exponentiation). Expressed in formal language, this is just the first-order theory of Peano Arithmetic, PA, with $\uparrow$. There is no principled reason why the Prime Number Theorem ought to be provable in this first-order theory, although of course that result is a highly relevant fact about PA as a formal system.

The axioms of Peano Arithmetic can be asserted to be true of counting numbers. Thus they are axioms in the old traditional sense—true assertions known with certainty—not merely in the sense of "defining conditions" (characteristic of axioms for algebraic structures). Counting, as we said, consists in an agent recursively producing the sequence of natural numbers while he or she surveys a group of objects or phenomena (of any kind) and establishes a correspondence between successive numbers and the different objects or phenomena. Think of yourself counting the lines on this page so far. The correspondences we are talking about here are not abstract mappings, but concrete, effectively established relations; most often, establishing the correspondence is aided by bodily movements: pointing fingers, producing eye movements, and so on. Now, the counting numbers are simply the numerals employed for counting, or any symbols employed to represent them (i.e., the Latin symbols or the decimal positional symbols), as determined by the rules for their generation. This is a potentially infinite

---

[18] Notice that proof theorists nowadays tend to use "elementary arithmetic" for the system EA or EA_0 that results from the system PA$\uparrow$ by restricting induction to Δ_0 formulas (having only bounded quantifiers). See Avigad (2004).

sequence of words or symbols, beginning with "one," where there are unequivocal rules that fix the successor of a given counting number.

Consider the Peano axioms as he formulated them:

1. 1 is a natural number.
2. For every natural number n, $n' = \sigma(n)$ is a natural number.
3. For every natural number n, not $n' = 1$ (no natural number has 1 as its successor).[19]
4. For all natural numbers m and n, if $m' = n'$, then $m = n$.
5. If $\varphi(x)$ is any property such that i) $\varphi(1)$ and ii) for every natural number n, if $\varphi(n)$, then $\varphi(n')$, then $\varphi(x)$ for all x. (If the base case i) and the induction step ii) are valid, then the predicate or concept φ applies to every natural number.)

These axioms can be formulated in FOL with identity, i.e., with a signature including the constant symbol 1 and unary function symbol σ. To these, one should add six other axioms, the recursive clauses that define addition, multiplication, and exponentiation. But for our present purposes we can skip these technical details. We can also skip the important technical point that, within a *first-order* presentation, the idea of "property" in Axiom 5 must be replaced by the precise, recursively defined concept of *open sentence* in the language of PA.

I have said that the axioms are true of the counting numbers, that is, by reflection on the practice of counting we can realize that the axioms are correct, true statements. It will be clear that the crucial difficulty here is with Axiom 5 of Mathematical Induction (MPI), which in fact is an axiom scheme, and hence represents schematically an infinity of different principles, one for each concrete open sentence. The associated informal reasoning would in essence be the same if, instead of a schematic first-order formulation, we employed a more general formulation: given any property of numbers that we may eventually come to acknowledge (and this surpasses the open sentences $\varphi(x)$ in a given formalism, to include sentences expressed in enriched systems), if that

---

[19] Of course, one might just as well have defined 0 to be the first natural number, in which case $\sigma(0) = 1$. But nothing essential depends on this convention (axiom 3 is then: For every natural number n, not $\sigma(n) = 0$).

property applies to 1 and is also hereditary (step ii) in axiom 5), then all natural numbers share the property.[20] The difficulties this reflection raises are also entangled with the interpretation of the quantifiers, and we shall have to say something about this.

The first three Peano axioms are rather trivial; they merely capture basic features of the process of recursive generation of number-words; as a matter of fact, a precise statement of the generative rules we employ for numbers in any vernacular language is much more complicated than the above set of axioms. The fourth axiom just says that whenever two numbers are different, $m \neq n$, their successors are also different— and this is also a simple fact based on the generative rules.

The Axiom of Induction may be used to establish facts about the basic operations of addition and multiplication, e.g., to prove that $n + m = m + n$ (for any n, m) and that $m \cdot (n + p) = m \cdot n + m \cdot p$ (for any m, n, p). It may be used to prove that the sum of the first one hundred numbers is $101 \cdot 50$, or, more generally, that the sum of the first n numbers is $\frac{1}{2}(n + 1) \cdot n$; this statement involves a quantifier. And it may also be used to prove the Fundamental Theorem of Arithmetic, namely that each natural number admits one and (except for the order of factors) only one decomposition into a product of prime numbers. In this case, the proof that for each number there is one decomposition would use induction—actually an improved form of it called *strong induction*: we prove that n has a decomposition on the assumption that each number smaller than n has one. As one can see, the complexity of the open sentence referred to above by $\varphi(x)$ can be as great as you wish, and in particular the sentence may contain any (finite) quantity of quantifiers.[21]

Since the domain of quantification here is infinite as a consequence of Axioms 1–4, the situation may seem challenging. (Think of the

---

[20] Notice that the set-theoretic, "full" semantic interpretation of SOL is *not* to be regarded as *standard*. The logic of SOL can be captured by semantics ranging from the Henkin or "general" semantics to proof-theoretic versions.

[21] Perhaps it is relevant here to add that the notion of finiteness need not be mathematically or logically defined: it suffices to say that, in Mathematical Induction, $\varphi(x)$ represents a *given* sentence—therefore it will have finitely many symbols, in particular, finitely many quantifiers. The qualification "finite" (in brackets in the main text) would be dispensable if logicians had not become used to consider not-purely-formal systems such as infinitary logics—sentences in infinitary languages are abstract mathematical objects, not concrete elements of mathematical practice!

domain Δ as a set: it is nonempty by Axiom 1, closed under σ by Axiom 2, which is an injective mapping by Axiom 4, and 1 is not in the image $\sigma(\Delta)$ of the domain by Axiom 3—hence Δ is infinite.) The interpretation of quantifiers must be such that it does not overstep the limits of basic arithmetic; as we are considering the truth of the Peano Axioms w.r.t. the sequence of counting numbers, we must respect whatever limits may be implied in the latter. In particular, it should be clear that actual infinity must not be presupposed: we may speak of the domain Δ as comprising all possible numbers, we may regard it as potentially infinite, but we should not employ at any point the assumption that the domain is a set, an actually infinite object.

This is done most naturally (it seems to me) by interpreting the quantifiers dialogically. The MPI is employed to establish generalizations, universal sentences of the form $\forall x\, \varphi(x)$, where the variables x, y—be they explicit or implicit in the practice—range over natural numbers. We consider a game or situation in which player A proposes one such generalization, while player B will play the role of the opponent. Consider, then, any well-defined property of integers $P(x)$ that complies with the two assumptions of the induction argument. Opponent B may question the validity of the universal sentence $\forall x\, P(x)$, in which case she may request from A a proof of the correctness of $P(x)$ for any given number. If so, B must indicate a number n, a concrete counting number; this is the important point in the dialogical interpretation I propose: B offers A a particular number, and asks A to validate his or her claim— this is the meaning of a universal statement of the form $\forall x\, \varphi(x)$.

Let's say, without loss of generality, that B picks $n = 1001$. Now A must prove $P(1001)$ correct, which she might do explicitly—absurdly and tediously—by applying assumption (i) once, and assumption (ii) one thousand times. Instead of this, of course, A will point to a few applications of the clauses, say up to establishing $P(10)$, and then will ask B to agree that indeed $P(1000)$. "You see?" Any normal agent B who understands this situation and these practices will agree. As a matter of fact, our cognitive functioning leads us to generalize and "see" that the argumentative pattern applies equally well to 1001 and to any other number. Already around the year 1000, when the first proofs by Mathematical Induction were given, the certainty and generality of the method were clearly, albeit implicitly, reflected in the practice of the mathematicians involved. The earliest proofs proceeded by example and insofar cannot be regarded as fully explicit: they provided proofs for the first few numbers, leaving it to the student to see the perfect analogy, and thus to recognize the generality involved.[22,23]

Knowing, as you do, a lot of mathematics and sophisticated modern logic, you might think that the above argument is deceiving. You may sense that it proves nothing, for an implicit application of mathematical induction must be lurking in the background. After all, there are infinitely many natural numbers, so $\forall x\, P(x)$ is an assertion about infinitely many cases. It is like an infinite conjunction; hence only an application of induction can establish it *in general*. I have cleverly deceived the reader by inviting B to pick a concrete number, while the argument was meant to be general. Not so! It is crucial to understand that, at the level of elementary arithmetic linked with counting numbers, quantifier $\forall x$ is not interpreted functionally to mean "for each and every one of the (infinitely many) numbers"—it means "*for any given* number."[24]

---

[22] The earliest implicit proof by mathematical induction was given around 1000 in a work by the Persian mathematician al-Karaji (*Al-Fakhri fi'l-jabr wa'l-muqabala*, which means *Glorious* [treatise] *on algebra*), where he proved a result on the sums of integral cubes already known to Aryabhata. He gave his proof for $n = 10$, going down to the base case $n = 1$, and he clearly implied that other cases can be treated the same way. Thus it was left to the student to see that the property in question can be proven for any number provided that it has been established for its predecessor ($\varphi(n')$ if $\varphi(n)$), and to infer from this the general nature of the procedure and the result. (The work of al-Karaji is known thanks to quotations given by al-Samaw'al al-Maghribi in *al-Bahir fi'l-jabr* (*The brilliant in algebra*). See Rashed 1972 and 1994 and Katz 1998, 255.) The modern rigorous and systematic treatment of the principle came only in the nineteenth century, with Hermann Grassmann, Giuseppe Peano, and, above all, Richard Dedekind; but even Grassmann did not provide a general formulation of the principle as such.

[23] In the Maghreb of the twelfth century, al-Samaw'al came closest to a modern proof by mathematical induction in premodern times, which he used to extend the proof of the binomial theorem and the "Pascal" triangle previously given by al-Karaji. Katz remarks that "Al-Samaw'al's inductive argument was only a short step from the full inductive proof of the general binomial theorem." Two centuries later, the Jewish scholar Gersonides wrote *Maaseh Hoshev* (*Work of Calculation*, 1321), where he used mathematical induction to prove certain simple combinatorial identities.

(Compare all of this with the state of historical knowledge a century ago, when Cajori (1918, p. 197) wrote: "The process of reasoning called 'Mathematical Induction' has had several independent origins. It has been traced back to the Swiss Jakob Bernoulli, the Frenchman [sic.] B. Pascal and P. Fermat, and the Italian F. Maurolycus.")

[24] The situation is very significantly unlike that of infinite games, which are strong mathematical idealizations—the jump from potential to actual infinity is always the crucial difference.

But wait. The specification of counting numbers by saying, first, that *Num*(1) and, second, that if *Num*(x) then *Num*(σ(x)) is surely incomplete. We should add an extremal clause to the effect that only things such as 1, 2, 3, . . . effectively generated by the first and second rules will be held to be numbers. Yet this clause by itself does not do any theoretical work of restricting the number domain![25] Of course, one option would be to take the MPI as an explicit statement that encapsulates what the extremal clause aims to say. But then we are employing mathematical induction to ground the MPI!

What if we lived in a world created by Descartes's evil genius, where there *are* nonstandard natural numbers, and B might actually choose a nonstandard α? To wit, consider α such that $1 < \alpha$, $2 < \alpha$, $3 < \alpha$, and, in general (for all your counting numbers n), $n < \alpha$. If opponent B questions the validity of the universal sentence $\forall x\, P(x)$, and requests from A proof of the correctness of $P(n)$, and then *picks* $n = \alpha$, then surely proponent A is doomed. She will fail to establish the correctness of $P(n)$ in this case!

The objection is clever, but wrong. In fact, it is either incoherent or self-refuting. Let me explain why. When we established that, in general, for any counting number n, α is such that $n < \alpha$, we already conceded to A what she needs. We admitted to having antecedent knowledge of what a counting number is, and not just knowledge that 1 and 2 and 3 are counting numbers, but knowledge of what a counting number in general is—any possible number n. So the objection is self-refuting; or, otherwise, it is incoherent; for, assuming what I just claimed is wrong in the second sentence of the previous paragraph, nothing prevents me from choosing to instantiate "$n < \alpha$" with $\alpha < \alpha$!

The point should be obvious. The practice of counting is a concrete human practice: it does not live in an abstract world of mathematical entities. The counting numbers are generated by agents according to explicit rules. Once these rules are given, and agents employ them properly, we *do not need* to stipulate that 1 is not the successor of any given number (to be precise, to postulate $(\forall x \neg (1 = \sigma(x))$ as an axiom). That the base number 1 is not the successor of any counting number

---

[25] See Parsons (1992).

is a fact about the practice of generating counting numbers.[26] We then idealize and consider any possible outcome of a counting process, since we are interested in finding universal statements that hold for all numbers. But a number α such as above is not a possible outcome of a counting process.

7.4. FURTHER CLARIFICATIONS

We have not discussed how exactly to circumscribe the properties to which our arguments can be applied. We have talked, rather ambiguously, about well-defined properties. Being prime is a very precise and easy-to-define property, being a "curved number" or a "colored number" is not; more interestingly, "n is a feasibly computable number" is not well defined. The property of being a perfect number happens to be well defined, because by tradition (handed down from the ancient Greeks) we define "perfect numbers" in terms of sums of their proper divisors. Perhaps it is worthwhile saying that this ancient Greek "perfection" may be different from the Platonic (real) perfection of numbers living in the autonomous realm of mathematical entities. With this little joke I merely mean to say that, were it not for tradition, "perfect number" would be as ill defined as "colourless number."

Does the above account of basic arithmetic and its certainty commit us, perhaps implicitly, to some restrictions on the admissible properties? For instance, do we have to restrict the properties that instantiate $\varphi(x)$ in applications of the MPI, on pain of losing the certainty that we have been reclaiming? One might imagine that this is so, e.g., that only predicatively acceptable properties can be considered.[27] Such an option would mean that full, impredicative Peano Arithmetic is no longer elementary, but advanced (in our sense of the term, section 6.3). Notice that the realm of well-formed formulas in the language of PA is large, and allows for extraordinarily complex properties; thus the arithmetic

---

[26] As already stated, we leave 0 out (we may think of it as an integer), though of course nothing would prevent us from having, as usual, a formal system PA with 0 as the base number.

[27] See Feferman and Hellman 1995, who give concrete arguments for the view that Peano Arithmetic can be established on strictly predicative foundations.

hierarchy has long been a topic of study, with origins in the work of the Hilbert proof-theoretic school during the 1920s. It will be worthwhile devoting some space to this topic.

Consider a few interesting number-theoretic results. Lagrange proved that every positive integer is the sum of four squares. This is easy to formalize:

$$\forall x\, \exists y\, \exists y'\, \exists y''\, \exists y'''\, (\, x = y \uparrow 2 + y' \uparrow 2 + y'' \uparrow 2 + y''' \uparrow 2 \,).$$

A related but much more advanced result is the Hilbert-Waring theorem, proposed in 1770 by Edward Waring, that, given any natural number k, there exists a(n associated) positive integer s such that every natural number is the sum of at most s kth powers of natural numbers (for example, every number is the sum of at most nine cubes, of nineteen fourth powers, and so on). The situation we find here is typical in number theory: numbers are the objects of study, but they are also employed to state the relevant property, i.e., numbers are being enumerated. This is the source of great formal complexity.[28] At the same time, however, in many cases that arise naturally, the relevant properties and sentences can be formulated with only bounded quantifiers. The Lagrange result can be stated as:

$$\forall x\, \exists y < x\, \exists y' < x\, \exists y'' < x\, \exists y''' < x\, (x = y \uparrow 2 + y' \uparrow 2 + y'' \uparrow 2 + y''' \uparrow 2 \,).$$

Expressing the Fundamental Theorem of Arithmetic purely formally is not very simple, and here too the number of primes involved depends on the target number; but it is perfectly doable, and it requires only bounded quantification.[29]

A nice feature of relying on Δ_0 sentences (where quantifiers are all bound) is that the finiteness of the relevant decompositions is thereby asserted. Δ_0 sentences are at the root of the arithmetic hierarchy, and the Peano axiom system with induction restricted to Δ_0 sentences is

---

[28] The Hilbert-Waring theorem is a rather complicated sentence, but it can be formulated as a Π_1 assertion thanks to an explicit bound on s obtained by Hardy and Littlewood in the 1920s. See Avigad 2004 for further details and references.

[29] Consider this formalization, which of course is not optimal, where $P(x)$ represents the open sentence stating that x is a prime number:

$$\forall x\, \exists y' < x\, \exists y'' < x \ldots \exists y^{(n)} < x\, \exists z' < x\, \exists z'' < x \ldots \exists z^{(n)} < x\, (P(y') \wedge P(y'') \wedge \ldots \wedge P(y^{(n)})$$
$$\wedge\, x = y' \uparrow z' \cdot y'' \uparrow z'' \cdot y''' \uparrow z''' \cdot \ldots \cdot y^{(n)} \uparrow z^{(n)}.$$

a proof-theoretically weak, but still surprisingly powerful, system. (We may call it EA_0; for more on this topic, see Avigad 2004.) Weyl (1949, 51, 56) once made a contrast between two definitions of "even number," one by simple recursion,[30] the other through a general proposition: n is even iff there is a number x such that $n = 2x$. He commented that the second definition "appeals to the infinite generality of numbers" and presupposes the number domain as "a domain of absolute existence, which is 'not of this world'. . . ." This strikes me as rather wrong. After all, the second definition can be refined into a sentence with bounded quantifiers: n is even iff $\exists x < n$ such that $n = 2x$. There is nothing infinitarian here, nothing "not of this world."

I would actually claim that the general notion of a counting number involves the idea of an *initial segment* of the number sequence: a counting number is always conceived as obtained through a process of recursive generation of numerals; "thirteen" comes after "twelve" and before "fourteen," just as 13 comes after 12 and before 14. (Let me add that, at the level of counting practices, it does not make much sense to prioritize either the cardinal or the ordinal aspect of numbers, since both are essentially involved in this kind of practice.) Notice also that the conception of an initial segment of the number sequence is much clearer than quantified sentences of some complexity. For these reasons it is quite satisfactory to rely on bounded quantification while studying number theory; in essence, our previous renderings of results relied on the initial segment $Zn = <0, 1, 2, 3, \ldots n>$ for the study of the properties of n.

In the above sense, I claim that the Dedekind-Peano axioms are true of the counting numbers. By reflection on the practices of counting and of expressing arithmetic properties, we realize that the axioms are correct, true statements. (If you regard it as necessary, you may qualify my statement by restricting the admitted properties to some level of the arithmetic hierarchy—say, Δ_0 sentences, or Δ_1 or Δ_2; albeit fixing some particular level strikes one as somewhat artificial and conventional, except for level zero.) For this reason I find it natural to maintain, as many proof theorists do, that the general scheme of induction,

---

[30] 1 is uneven; n' is even or uneven, depending on whether n is, respectively, uneven or even.

$$P(0) \wedge \forall n \, [P(n) \to P(\sigma(n))] \to \forall n \, P(n),^{31}$$

is most naturally taken to be open-ended "in the sense that it is accepted for any definite property P of natural numbers that one meets in the process of doing mathematics, no matter what the subject matter and what the notions used in the formulation of P" (Feferman 2009). Feferman goes on to add: "The question—What is a definite property?—requires in each instance the mathematician's judgment."

This provides us with *certain* knowledge of the PA axioms and all the theorems deducible therefrom. The claim applies to basic arithmetic. How far up in number theory this certainty extends, is a different problem.

7.5. MODEL THEORY OF ARITHMETIC

Concerning the issue of nonstandard models of arithmetic, we have already advanced one important idea. One must distinguish between strictly formal systems—which are meant to stand alone, to employ the usual metaphor—and the symbolic frameworks of mathematical practice—which are made to stand on the shoulders of agents, to employ a new metaphor. Formal systems were invented by Frege around 1880; they were employed in the modern way by Hilbert and his followers around 1925. Thus, we should expect symbolical arithmetic to have different properties when it is regarded as part of mathematical practice from those it has when transformed into a strictly formal system.

Nonstandard models of arithmetic are relatively old, as their origins lie around 1930 in the work of Skolem (1934); in fact, the earliest extant consideration of sets of numbers with nonstandard elements is in the work of Dedekind (see the famous letter to Keferstein in van Heijenoort 1967). Anyhow, the important point is that the issue of nonstandard models belongs in model theory, the study of relations

---

[31] My comings and goings with the base number of arithmetic (first I took it to be 1, now 0) should cause the reader no special trouble. With counting numbers, the natural starting point is 1; when you are doing "scientific" arithmetic or number theory, considerations about simplifying and generalizing the laws recommend starting with 0.

between set-theoretic structures and formal axiom systems. The viewpoint can be presented informally the following way: a particular axiom system is regarded as a "selector" of structures, from among the immense variety of set-theoretic structures that we regard as given (within the framework of set theory); a certain structure Σ is picked out when there exists an interpretation of the axiom system into Σ that validates the formal axioms, while structure Σ' is left out when there is no such interpretation.

The reader is likely to reflect that the very notion of nonstandard number is quite alien to the common practice of arithmetic, not to mention how absolutely foreign it is to the practice of generating numbers for counting. Counting practices and the common reckoning that is associated with them are both *constructive*; because the sequence of numbers is just a potential infinity, and cannot be completed, it is obvious that all counting numbers are "standard" (strange as expressing it this way may seem). The same applies to basic arithmetic, the way we have presented it in the previous subsection. In other words, the problem of nonstandard elements emerges only within the top-down perspective of set theory, and will never arise from the bottom-up perspectives of counting or even basic arithmetic.[32]

My thesis is thus the following: nonstandard models arise *only* as a result of the complex of hypothetical assumptions that underlie set-theoretic structural mathematics; it is only when we submerge a formal system (inspired by the counting numbers, reckoning operations, and mathematical induction) within the seas of set-theoretic structures that this whole problem arises. Prima facie, this has nothing to do with our previous considerations that showed the Peano axioms, adequately interpreted, to be true of basic arithmetic practices. Nonstandard models belong in what might be called *logical* number theory (in parallel to algebraic number theory and analytic number theory).

Here, the word "logic" refers to mathematical logic, and in particular to the disciplines of model theory and set theory—the theory ZFC

---

[32] In this connection, the reader may find instructive the paper by George and Velleman (1998): they compare two approaches that they call the *build up* definition and the *pare down* definition of the naturals (BU in the style of Poincaré-Lorenzen and PD in that of Dedekind-Frege); we might also say *generation* of numbers and *specification* of structure $\mathbb{N}$.

being a necessary basis for model theory, the main structural frame-work from which the latter draws resources.[33] Just as algebraic number theory or analytic number theory is not directly relevant to our analysis of basic arithmetical knowledge, neither is logical number theory.

Another related aspect of this transition to advanced mathematics is that the PA axioms cease to be regarded as true statements, and become axioms in the algebraic sense. It seems to me that much of the discussion on arithmetic during the last century has been complicated by the fact that advanced logical results (belonging to logical number theory) have been taken to possess great significance for the basic understanding of arithmetic. But, as we have seen, it ain't necessarily so.

7.6. LOGICAL ISSUES: CLASSICAL OR INTUITIONISTIC MATH?

In relation to arithmetic, and to pure mathematics in general, it is natural to pose questions concerning the choice of basic logic. The last century of development of logical theory has raised deep and difficult questions concerning the justification of classical logic—the logic of Boole, Frege, Whitehead-Russell, Hilbert, and . . . well, of mainstream modern mathematics. In classical logic, a statement or proposition can be either true or false, *tertium non datur* (no third option exists, the Law of Excluded Middle). Whenever a proof by reductio ad absurdum is established, we rely on classical logic: we assume a statement as premise, derive a contradiction, and, from this, conclude the falsity of the initial statement; the overall argument relies on the schema

$$(p \rightarrow q), \neg q \Rightarrow \neg p,$$

---

[33] Surprisingly, this is not emphasized too often, and there is even a tradition among philosophical logicians to obscure it. (A referee remarks that my previous statement may be too strong, that the centrality of set theory is a case of "common knowledge" that "goes without saying", wrongly insofar as this misleads students and nonexperts. I still have the impression that, in the mid-twentieth century at least, philosophers interested in logical semantics intentionally downplayed the role of set theory, as if it were negligible. The matter probably deserves careful scrutiny.) For related considerations about the historical importance of set theory as a source of structures and models, see Ferreirós 1999, 119–23.

which depends on *tertium non datur*. Many logical systems have been devised that behave differently from this, but they have not affected mainstream mathematics (except for the community of people working on topos theory). Not even the community of set theorists has been affected, despite the fact that a good number of philosophers believe such alternative systems to be the answer to the well-known paradoxes.

Should elementary arithmetic, built on top of the practice of counting, be classical arithmetic or intuitionistic? And, what about the theory of real numbers (Chapter 8)? These problems are relevant to the philosophy of logic and the foundations of math. In this respect, I offer a line of argument justifying the logic of modern "postulational" mathematics on the basis of the goals and values linked with the mathematical practices it embodies—in particular, those practices that go *beyond* the area of foundations. Let us begin with arithmetic, and then move to the general case.

7.6.1

Is the principle of excluded middle valid in general, in particular with quantified sentences? Should we accept or reject that $\forall x\ \varphi(x) \vee \exists x\ \neg\varphi(x)$? Let's assume a predicate $\varphi(x)$ such that we are not able to prove $\forall x\, \varphi(x)$, and also unable to indicate a concrete n for which $\neg\varphi(n)$, i.e., a witness of the truth of $\exists x\ \neg\varphi(x)$. It is very well known that there are such predicates (and we establish this fact constructively, effectively, not by reductio): one such is the Goldbach conjecture that any even number $n > 2$ is the sum of two prime numbers; call the corresponding predicate $S2p(x)$. Should we accept or reject $\forall x\, S2p(x) \vee \exists x\ \neg S2p(x)$ as a logically valid sentence?[34]

An intuitionist only regards a statement as true when it has been conclusively established, either by proof or by counterexample (with explicit witnesses). To establish a disjunction conclusively, you must have established one of the disjuncts; hence you need either a proof of $\forall x\, S2p(x)$ or else a counterexample such as $\neg S2p(n)$ for explicitly given n. There is no guarantee that this can be done for each predicate. From

---

[34] There is empirical evidence that the Goldbach conjecture is true, as it has been computationally confirmed for large numbers (empirically) up through 4×10^{18} (see http://sweet.ua.pt/tos/goldbach.html). But this is irrelevant to our question.

the intuitionist point of view, then, one must reject the principle of excluded middle: $\forall x\, \varphi(x) \lor \exists x\, \neg\varphi(x)$ is not valid in general.

A classical mathematician has a less stringent—and less computational—notion of truth. The property $S2p(x)$ is well defined and perfectly clear: it just says that x is an even number, $x > 2$, and that $x = p' + p''$ where p', p'' are prime numbers.[35] If you ask me whether $S2p(1000)$, I happen to know that the answer is yes, even though I don't know which concrete prime numbers witness the fact that $1000 = p' + p''$ (it would be easy to check; nor do I know, right now, in how many different ways 1000 can be written as a sum of two primes). Assume we have checked that $S2p(x)$ is true of the first one thousand integers. I claim that either this goes on forever, applying to all possible counting numbers, or there are numbers n such that $\neg S2p(n)$. Furthermore, there will then be a *first* number n such that $\neg S2p(n)$, i.e., such that $S2p(m)$ for any $m < n$. (The well-ordering principle for natural numbers is a truth about the counting numbers.) The reasoning is based on mere conceptual possibilities, and does not depend on our actual ability to compute.

It is for such reasons that in the foregoing sections we have accepted the validity of classical logic for elementary arithmetic. Even a constructivist like Weyl (1918, 9) admitted the laws of classical logic—in particular, the quantificational form of *tertium non datur*, $\forall x\, \varphi(x) \lor \exists x\, \neg\varphi(x)$—for the domain of natural numbers. This, he argued, is a consequence of the fact that the naturals form a "closed" domain, or "category of individuals," in the sense that each one of them as an individual is characterized by specific, distinctive properties, and there is a generative procedure that successively generates the individuals in this domain, presenting each one of them in fully specific form.

It is not at all evident that the same can be claimed about the real numbers, and certainly there exists no generative procedure of that kind even for the reals in a unit interval (of the form $0,abc\ldots$ with a, b, c digits). In connection with this, some authors have suggested that the $\mathbb{R}$ structure calls for a weaker logic, say intuitionistic first-order logic. The same would apply a fortiori to a domain such as $\wp(\mathbb{N})$, and therefore to the treatment of set theory. This is the standpoint from which work on constructive set theory has been developed (see, e.g., Crosilla 2009).

---

[35] If you wish to be more explicit, write this last property in first-order logic, with $\exists x\, \exists y$ and using the property *Prime* (x).

Before we continue discussing the logic of mathematics in general, let me remind the reader of something else: even if you take the other horn and go intuitionistic, it does not matter much at the level of arithmetic (at least from the standpoint of formal consistency, i.e., of Hilbert's metamathematics). It has been known since 1933, thanks to Gödel and Gentzen, that the formal system PA is consistent relative to intuitionistic arithmetic, HA (initials for Heyting arithmetic). The proof is simple, employing a translation of statements of PA into statements of HA, such that a statement φ is derivable in PA iff the corresponding φ' is derivable in HA (see Gödel 1986, vol. 1, 282ff). This has the effect that any contradiction in PA would translate into a contradiction in HA, so that, assuming HA is consistent, so must be PA; it turns out that "intuitionistic arithmetic and number theory is only apparently narrower than the classical ones, and in truth contains them" (Gödel, op. cit., 295). As Bernays and Troelstra remark, Gödel's 1933 paper taught the members of Hilbert's school that intuitionistic principles went beyond finitism, and that there were alternatives to finitary reasoning as a basis for metamathematics (op. cit. 284).

Those results reinforce our previous conclusions about the certainty of arithmetic. Beyond that, one can say that intuitionism introduces genuine restrictions at the level of the real numbers, analysis, and set theory. Needless to say, no such result of relative consistency has been obtained, reducing the consistency of classical analysis to that of constructive analysis (Bishop and Bridges 1985; Brouwer's intuitionistic analysis is controversial, as it conflicts with *tertium non datur*).[36]

7.6.2

According to Weyl, the classical reals do not form a "closed" domain; they are not well individuated, and therefore are not a "category of individuals." One can see, e.g., that there is no generative procedure corresponding to them, nothing remotely similar to the generation of natural numbers. Moreover, a typical real number, in the sense of one taken at random, will be a transcendental real that has never been studied

---

[36] Let me add that analogous relative consistency results have been proved, using "formally intuitionistic" systems of second-order arithmetic, and intuitionistic set theory IZF—which are based on intuitionistic logic but do not respect other principles of intuitionism (see Troelstra's remarks in *op. cit.* 285, the references therein, and Crosilla 2009).

and probably will never be "named." Hence Weyl suggested that classical logic would be inapplicable in this case. This suggestion, which I find quite interesting, amounts to the view that the domain of real numbers is not "closed" but rather an "open horizon," an incompletely determined totality. In such cases, intuitionistic logic would be applicable.

As one can see, from a purely mathematical point of view there are reasonable arguments to the effect that classical logic should not be taken for granted in all cases. How, then, shall we explain that the vast majority of mathematicians do not even care to understand properly the subtleties and complexities involved? It is my conviction that this cannot be explained purely or a priori, ahistorically, merely at the level of foundations. However, let me argue very briefly that the logic of modern "postulational" mathematics is natural relative to the goals and values associated with the web of mathematical (and scientific) practices in which it is immersed.

The crucial idea is simple. Classical logic, the logic of *tertium non datur*, is the natural one when human agents are using language or models to try to represent physical phenomena. That is, when we use declarative sentences (such as "It is raining" or "A positron has crossed the cloud chamber") or physico-mathematical models such as Kepler's for the solar system (or Bohr's for the hydrogen atom, Dirac's of the spinning electron, and so on). In such cases, either the sentence or model depicts the physical situation correctly (within certain margins of error) or it does not; one only needs to consider these two alternatives. Mathematicians have simply inherited this way of thinking, this practice of arguing and inferring, which has been theirs throughout the very long historical period (starting before antiquity and up at least to 1850; and, for many subcommunities, up until today) when mathematics and natural science have gone hand in hand and can hardly be differentiated.

In a detailed historical narrative of this process, the role of astronomy and astronomical models (which have been a paradigm for all of physics) would have to be underscored. As the reader probably knows, astronomy was regarded as a branch of mathematics from antiquity to the early nineteenth century. Think of the great Gauss, undoubtedly one of the finest mathematicians of all times; from a social-professional point of view, he was an astronomer, director of the Astronomical Observatory in Göttingen.

It is interesting how the case of Brouwer fits this picture and clarifies it (see also section 8.7). He developed an alternative logic in which reductio proofs are not valid, claiming that this step was a necessity when we moved from finite domains to (potentially) infinite domains, i.e., to the continuum and analysis. In fact, Brouwer's conception was marked by conscious rejection of the strong analogy that Cantor and Dedekind had established between finite and infinite. By contrast, Brouwer would agree with Weyl in the view that "inexhaustibility" is inherent in the very concept of infinity; hence no such analogy can be set at the foundations of mathematics. What marks an infinite set—something (from the constructivist standpoint) defined and determined by mathematical thought—is that we cannot survey all its elements *unless* there is a concrete, well-established defining property.[37] Thus, he argued for the necessity of abandoning the law of excluded middle if we are going to establish analysis as a purely mathematical theory, wholly autonomous and independent of natural science.

There is some consensus today that logical systems—or, if you wish, logical practices—face much more flexibility and freedom of choice than mathematical systems, although the reasons remain somewhat obscure. Historically, this is quite a noteworthy phenomenon: in their time Frege and Russell would have insisted that logic is much more solid, stable, and unchangeable than any mathematical theory; but their invention of formal systems was ultimately linked to a process that ruined that conviction! My view that explaining the allegiance of mathematicians to classical logic must invoke the impact of scientific practices upon the historical and the present configuration of mathematical knowledge suggests also an explanation for the looseness of current conceptions of logic. That came about as logicians became more and more disconnected from natural science and mathematics, establishing stronger links with linguistics, philosophy, and computer science.[38]

---

[37] That leads to rejecting as incoherent the idea of an arbitrary infinite set.

[38] This also explains why logicians who are mainly mathematicians (such as experts in set theory) cling to classical logic, while this is not the case for so-called philosophical logicians.

8

Mathematics Developed

The Case of the Reals

The interplay of previous practices and new ones, of earlier knowledge and new frameworks with their corresponding hypotheses, will be a driving force in the development of mathematics. It is also the driving force in the learning of mathematics: with all due modifications, aimed at rationalizing and simplifying the material, the process of learning must present again some of the crucial ingredients that determined the growth and the current configuration of the mathematical edifice. Surprisingly, this has not been sufficiently taken into account in recent years (under modernizing pressures, sometimes vanguardist ones, to radically change the curriculum).

This interplay we are postulating can be seen at work in all the crucial shifts of mathematical knowledge, even when we observe the emergence of structures that may appear to be radically new—and also in times when mathematicians thought that they were inventing nothing, but merely understanding more properly and completely the traditional material. In this chapter we shall touch briefly, without any intention of providing a complete history or philosophy, on two of those crucial shifts.

What are the most crucial structures in the world of mathematics? If one has to be extremely selective, to the point of oversimplifying, two systems stand out as core structures of mathematics: the natural numbers $\mathbb{N}$ and the real number system $\mathbb{R}$. As Feferman has written, "by a kind of miracle of synergy, $\mathbb{R}$ has proved to serve together with the natural numbers $\mathbb{N}$ as one of the two core structures of mathematics; together they are the *sine qua non* of our subject, both pure and applied" (Feferman 2009, 188). Though this statement may be slightly biased by a foundational orientation, I believe from an epistemological point of view it is essentially right.[1] In any event, nobody can claim to have a basic grasp of mathematics without mastery of the central elements in the theory of both number systems.

Yet from our point of view there is an ocean of difference in their status. The natural numbers, we have claimed, enjoy a peculiar status since our knowledge of them is so basic and so strongly grounded in core elements of human cognition that one can even speak of certainty. On the other hand, it is our contention that the real number system is a paradigm of advanced mathematics, and hence of the presence of hypotheses in mathematical knowledge.

All the classical conceptions of the real numbers, from Stevin and Descartes to Dedekind and Cantor, bear the mark of previous geometrical conceptions and inherit or, one should say, exacerbate their hypothetical nature.[2] Even if we are willing to simplify, we should at least distinguish the seventeenth and eighteenth century conception as articulated along the lines of a theory of magnitudes, from the modern (nineteenth and twentieth century) view of $\mathbb{R}$ linked to arithmetic and set-theoretic notions. These related theories and conceptions will be at the center of this chapter.

8.1. INVENTING THE REALS

The concept of real number is a true crossroads of mathematics, where all kinds of stem-ideas meet and combine to give a fruitful outcome. One might compare it with the central square of a large city, or with a knot of highways where all the main routes of a country meet. It is a perfect "*mixte*" or *mikton* ($\mu\iota\kappa\tau o\nu$) in the sense of French philosopher

---

[1] Some theoretical physicists (Wigner, Penrose) might answer $\mathbb{C}$, yet although this structure is a perfect "mixte" in Lautman's sense, it hardly poses epistemological difficulties that may not have emerged already for $\mathbb{R}$. The Bourbaki (1950) might answer algebraic structures (such as fields and vector spaces), topological structures (e.g., Hausdorff space), or order structures; yet, precisely a combination these three types of structures is present in the paradigmatic case of $\mathbb{R}$.

[2] There exist other, more recent and advanced, notions that set free of geometric preconceptions, for instance, the ideas of Cantor space and Baire space, which (conceived as Polish spaces) play a central role in descriptive set theory. See Moschovakis 1980 and 1994; but we cannot enter into that matter here.

Lautman, an idea that brings to combination and perfection several previous concepts—which, regarded from this vantage point, may even appear to be imperfect partial glimpses of a more perfect idea that was waiting to be discovered.

This is not the place to attempt a history of the notion of real number, which after all would have to cover much of the history of mathematics between (say) the years 900 and 1900. However, it seems necessary to say something about the great variety of ideas and practices that were linked with the invention of the real numbers.[3] The roots of the invention included: measuring practices, fractions, and ratios; numerical and algebraic symbolism as a driving force; conceptual difficulties surrounding the very idea of the number concept; and analysis as a source of tensions. If the ancient Greeks regarded arithmetic and geometry as the two—separate—main springs of mathematics, with the invention of the reals they came together; this expanded number system absorbed previous ideas about measuring, subsuming concepts of fraction and of proportion; in so doing, it connected with systems of calculation and algebra, opening up new possibilities of development that led to analytical geometry and to the calculus; and in this way it became the central element for the richest host of applications of mathematical knowledge to the modelling and understanding of real phenomena.

How difficult it is for a modern mathematician to make sense of mathematics in the past without anachronism can be shown with the example of Morris Kline and the Greeks. In a discussion of the Greeks' "limitations," Kline comments on their "inability to grasp the concept of irrational number" (!!), which "obscured the intimate correspondence between arithmetic and geometric concepts and operations" (Kline 1972 vol. 1, Chapter 8, 173). This can only be said from a perspective that takes for granted and a priori the modern conception of arithmetic and the idea that there are numbers measuring the incommensurable magnitudes of geometry. One could analogously talk about Newton's inability to grasp the concept of a field or the notion of a Riemannian manifold, but in the case of physics, people seem to be more

---

[3] I shall take the liberty to use modern symbolism and to "reconstruct rationally," since my aim is to underscore key conceptual issues, not (here and now) to write history.

aware of the fact that such ideas are not present around the historical actors, waiting for them to open their eyes and see.

Such naïve comments underscore how central the real number system and its presuppositions (or hypotheses) are to us! The concept of irrational number was not out there, waiting to be grasped, but has to be seen rather as a bold step away from the very sound and rigorous Greek conception of arithmetic.[4] The Greeks did have a perfectly clear and well-defined concept of number, unlike we do today,[5] according to which arithmetic and geometric concepts and operations do not correspond at all. (We may judge that the modern reconception of those issues was a wonderful step forward, and think that mathematicians should be totally absorbed by it. But by presenting it dogmatically as an article of faith one does not really help anyone understand the fabric of mathematical knowledge.) Kline also lamented the "failure" of the Greeks "to construct a deductive algebra"—as a consequence of which "rigor" was identified with geometry up to 1800—and even their "failure to comprehend the infinite" (!!). Such comments are abusively anachronistic; they come as a surprise from a man as learned as Kline was. But, again, they serve the purpose of emphasizing how central for us today is the symbolic, equational, algebraic conception of mathematics (that was so foreign to our Greek ancestors).

To underscore the wrong nature of Kline's standpoint and the fact that the concept of irrational number was not out there, waiting to be grasped, I shall speak of the "invention" of the reals. A good example of a deep conceptual shift, this invention cannot be dated—it occurred somewhere between 1000 and 1700. A traditional idea, rooted in Renaissance times, is that the new European mathematics was basically determined by the Greek legacy. However, the two key ingredients that we have underscored—(1) the notion of real number and (2) the dominance of symbolic method, i.e., the great expansion of the number

---

[4] A step, by the way, that was started by mathematicians in Asia and the Middle East in the centuries from 1000 to 1500.

[5] Readers who may feel insulted are challenged to offer an analytical definition of number, under which all the basic number systems that we use today (including the complex and the transfinite) may fall. Due to the bold step away from the "Greek limitations," as Kline called them, the number concept became a perfect example of meaning guided merely by family resemblance, in the sense of Wittgenstein.

concept and the preeminence of algebra—were both foreign to ancient Greece. Their roots can be seen more in the Indian and Arabic civilizations, and so one could say that the river of the new European mathematical tradition (seventeenth and eighteenth centuries) stems from two springs or affluents: the Greek and the Indo-Arabic.

With Brahmagupta in the sixth century, Mahavira in the ninth, and Bhāskara in the eleventh, Indians contributed the positional system of decimal numeration with zero, clear rules for calculation, the negative numbers (in analogy with "debit and credit"), and even the decimals, which "are calculated like the integers."[6] They initiated a thousand years of development of polynomial equations, seeing, e.g., that quadratic equations have two roots if one admits negative and irrational solutions (which they tended to treat uniformly). "People do not approve of negative solutions", said Bhāskara, only to practice the opposite; and the irrationals were not a "normal" thing, but these pioneers kept moving forward.[7] All of this promoted the trend toward considering both fractions and incommensurable quantities as *numbers*, in an extreme deviation from Greek tradition.

Acquainted with both the Greek and the Indian traditions, Arabic mathematicians would criticize the latter for its lack of rigor; but the fact is that their reception and development of the Greek heritage transformed the hierarchies. Arithmetic began to be treated on equal footing with geometry, on the basis of the new tendency to establish a perfect correspondence between them (using rational and irrational numbers; see Katz 1998, 272–273), and the preference for algebraic methods began its slow rise. The ensuing loss of rigor was compensated by a strong creative drive, and the era of exploration of the new number domains and their possibilities of symbolic exploitation (by trigonometry,[8] algebra, and other new methods) began.

---

[6] Notice that one is dealing here with truncated decimals, and thus with decimals as tools to approximate quantities, not with decimals as infinitary objects (see below).

[7] After all, we can write $\sqrt{2} \cdot \sqrt{3} = \sqrt{2 \cdot 3} = \sqrt{6}$, calculating with the symbol $\sqrt{}$ (introduced in 1525) as if it worked as parentheses or as a + sign; and the outcome is quite satisfactory. But Dedekind would famously claim (1872) that the truth of $\sqrt{2} \cdot \sqrt{3} = \sqrt{6}$ had never been scientifically proven before him.

[8] Developed by Aryabhata using the semichord associated with an arch (primitive notion of *sine*), later by Abu-l-Wafa in the tenth century. Cosine and other notions were also employed, and tables of sines were calculated using trigonometric identities (in a way similar

Since the acceptance of irrational numbers was strongly promoted by the availability of methods of calculation and approximation, especially by means of decimal fractions that "are calculated like" whole numbers, it is instructive to consider decimal fractions more closely. Historically, they started to be used as early as the tenth century by al-Uqlidisi (*Kitab al-fusul fi al-hisab al-Hindi*, part IV), though merely in the context of simple operations like division. They kept reappearing in the coming centuries, e.g., with al-Samawa'l al-Maghribi in 1172 there is full understanding of decimals in the context of approximation of rationals or quadratic irrationals.[9] The Persian mathematician Jamshid al-Kashi, around 1400, who believed to be the discoverer of the idea, employed them systematically, e.g., in his famous determination of π to sixteen decimal places.[10] He had a good notation of the kind 13|163. Rashed puts al-Kashi's important contribution into perspective by emphasizing[11] (1) the analogy between the systems of sexagesimal fractions employed by astronomers, and the system of decimal fractions, and (2) the usage of decimal fractions not only for approaching irrationals of the type of the *Elements*, Book X, but also for π.

It is well known that Simon Stevin employed them in *Disme, the Art of Tenths* (1585), a treatise that enjoyed wide circulation, so that their use—and the concomitant idea of an intimate correspondence between arithmetic and geometric concepts and operations—became widespread in Europe during the seventeenth century. The "system of tenths" had occasionally appeared in Europe before Stevin, for instance, in a Byzantine text brought to Venice in 1562. Stevin linked magnitude and number inextricably, saying: "Number is that, by which the quantity of any thing is explained."[12] He attributed numerical properties to continuous magnitudes and continuity to numbers—with measuring

---

to Ptolemy's, but by exploiting them *in algebraic form*). On the need to resist temptations of retrospectively seeing algebra in the ancient Greeks, see Unguru (1975) and the volume *The History of Early Mathematics—Ways of Re-Writing* edited by Netz (2003).

[9] Book 3 of his treatise *al-Bahir fi'l-jabr* studies essentially Euclid's Book X, going no further.

[10] See Katz (1998), Chapter 7, especially 242–243. For further details, see the books by Saidan (1978) on the arithmetic of al-Uqlidisi, and Luckey (1951), (1953) on the work of al-Kāšī or al-Kashi.

[11] Rashed (1994).

[12] "Nombre est cela, par lequel s'explique la quantité de chacune chose." (Stevin 1585b, 495).

processes being always crucial, and in the *De thiende*, a treatise with a very clear practical bent.[13] Decimals were a key symbolic tool for this, but the practical orientation of Stevin imposed some limits on his number concept.

Up until 1800, the usual conception of mathematics defined it as "nothing more that the *science of magnitudes*" (Euler 1771, 15), and this immediately points to the real world. Mathematics develops concepts for analyzing physical magnitudes and investigating their relations, e.g., to analyze functions, but magnitudes themselves are not "purely" mathematical. They are given to us in reality as lengths or volumes or times; therefore the "foundations" of mathematics was not itself a topic for mathematics, but rather for philosophy. Hence the habit of talking about the "metaphysics of the calculus" and so on; even Gauss used to speak of the "metaphysics of number" or the "metaphysics of space." The continuum of the real numbers is a necessary ingredient of math because the physical magnitudes are (assumed to be) continuous—beginning with the continua of space and time.[14]

Stevin compared radical numbers to fractions, arguing that both are numbers, equally clearly defined and understandable, and therefore fully legitimate: he claimed that "any radical is a number" and "there are no numbers which are absurd, irrational, irregular, inexplicable, or surd."[15] In his view, all numbers are on a par with each other and they form a continuous domain, because what they measure varies continuously: "a continuous number corresponds to a continuous magnitude." He presented this notion on the basis of metaphorical description rather than by mathematical or philosophical argumentation (Malet 2007, 501). Thus, he argued that number cannot be separated from magnitude, as wetness or humidity cannot be separated from water. He employed a distinction between arithmetic and geometric numbers,

---

[13] Moreno-Armella and Waldegg 2000, 186–187. Cf. Goldstein, Les fractions décimales: un art d'ingénieur? (2010); see http://webusers.imj-prg.fr/~catherine.goldstein/publis.html

[14] Considering things this way, we find the funny situation that traditional mathematics contrasts with modern math in a way reminiscent of the contrast between Empiricism and Idealism (although such empiricism was based on an ontological assumption that can be doubted). This would be emphasized by Kronecker and by his French followers or semi-followers (Baire, Lebesgue).

[15] *Thèses mathèmatiques*, in *L'Arithmétique* (Stevin 1585b).

"arithmetic" being whole numbers and their fractions, "geometric" numbers those that "are expressed" by powers and roots of any order. Here we can see that Stevin's number domain is not explicitly required to be complete in anything close to the modern sense: further technical and conceptual developments would be needed to reach that point. Equivalently, he lacked the means to make clear what it meant to say that the magnitudes are continuous.

But we must move on to ask a different question—namely, when can we assume mathematicians to have conceived of decimal fractions as infinitary objects?—because there is enormous difference between decimal expansions understood as tools to approximate positions given on the line (constructed geometrically) and decimal expansions understood as defining numbers or the continuum.[16] The second view involves considering *all possible* (finite or infinite) decimal fractions as "given" and somehow determined, defining all the possible elements in an arithmetic continuum, which corresponds to the (complete) geometric line of abscissas. It seems safe to assume that only the first viewpoint of approximation was attained in the Middle Ages and the Renaissance, while the second definitional viewpoint was adumbrated late in the seventeenth century.

Notice that the first approach can be regarded as a relatively minor modification of the Greek viewpoint: given is what emerges in geometric constructions, and then some conceptual devices—fractions, ratios, irrational numbers—are introduced to analyze and determine it (perhaps approximatively, as happens with decimals with al-Kashi or Stevin). The second approach, which to us may seem a simple development of the old idea, is in fact a wholly different conception: given are all possible decimal expansions, and corresponding to them we have the points on a line or any other continuum of magnitudes (areas, volumes). In this connection, it is interesting to see what Stifel, an important algebraist, had to say in 1544:

---

[16] Malet (2007) argues that, while Stevin extended the realm of arithmetic, since all geometrical magnitudes are assumed to be expressed by numbers, in another sense arithmetic was now tacitly grounded on geometry. Geometrical magnitudes allow Stevin to construct a new conceptualization of number, but they are defined by the properties and results the *Elements* endow them with.

Just like an infinite number is no number, so an irrational number is not a true number, because it is so to speak concealed under a fog of infinity [*lateat sub quadam infinitatis nebula*]. (*Arithmetica integra*, 1544; quoted in Ebbinghaus *et al.*, 1991, 33)

In the practice of a particular mathematician, however, this difference may eventually seem small. After a long period, perhaps centuries, during which several mathematicians have established the use of decimal expansions to approximate any given geometrical quantity, a novice may be trained in this practice and soon start to think that, after all, the continuum of quantities is there abstractly, defined as an image of the infinitely many possible decimal expansions.

That second point of view was adopted by a number of mathematicians after the mid-seventeenth century. A clear example is Wallis, who regarded the Greek ratios as merely "a peculiar kind of quantity," believing that there is a natural correspondence between any possible relation of proportionality and a given number. He thus seems to have assumed the existence—the givenness—of an arithmetic continuum, even though the underpinnings of such an assumption would not be fully articulated until much later. (Ratios can be added, subtracted, multiplied, and divided, despite the fact that they are merely *relations* between magnitudes, and this provided the basis for a clear analogy between (real) numbers and (continuous) magnitudes. On this basis, which underscores the dominance of symbolic forms, mathematicians let these new kinds of numbers overflow and erase the borderlines of the "pure" number concept—the natural numbers of the Greek.) People like Wallis, "moderns," were opposed by traditionalists who kept insisting that "ratios are not quantities,"[17] and no numbers are waiting out there to represent all possible relations of proportionality.

We may tend to sympathize with Wallis, but this is for the simple reason that his views were close to present beliefs. Wallis was bridging the way toward the modern arithmetic continuum (without fully understanding the conceptual complexities of this new ground), while Barrow

---

[17] This was the emphatic position of Barrow, for instance in his *Lectiones*; see Mahoney (1990) and Jesseph (2000), 88–89. If we choose to adopt the naïve viewpoint of Kline, discussed above, it will become impossible to see the "modernist", somewhat visionary, *unstable* nature of the bet of Wallis.

was remaining close to the classical understanding of arithmetic and geometry—which conceptually and foundationally is very sound. Reading their debate from a twentieth century foundational point of view, one may say that those on Barrow's side remained in the camp of constructivism and avoided the assumption of *arbitrary infinities*, which is the mark (and Achilles's heel) of Wallis's camp. Dedekind and Hilbert would be heirs of Wallis, but even they were not fully aware of the conceptual and foundational complexities of their preferred standpoint.

8.2. "TENTHS" TO THE INFINITE: LAMBERT AND NEWTON

The notion of a real number is completely foreign to the well-delimited Greek conception of arithmetic, but we can find its first germs around the year 900 CE. From the combination of the Greek and Indo-Arabic heritage there arose a tension between conceptual purity and the interest in calculation and algebra. It would take almost a thousand years to systematize completely the real number concept: in 1900, Hilbert published his well-known axiom system, anticipated to a large extent by Dedekind in 1872. Newton's sophisticated definition of number in *Arithmetica Universalis* was clearly designed to include irrationals, but would not cover negative or complex numbers:

> *By* Number *we understand . . . the abstracted Ratio of any Quantity, to another Quantity of the same Kind, which we take for Unity.* And this *is threefold, integer, fracted,* and *surd: An* Integer *is what is measured by Unity, a* Fraction, *that which a submultiple Part of Unity measures, and a* Surd, *to which Unity is incommensurable.* (Newton 1720, 2)

This approach is advanced and proceeds top-down, as it introduces a most general concept of (real) number and distinguishes within it three special types: the integers, the fractions, and the irrationals or "surds."[18] The process by which that most general number concept was fully secured was tortuous and full of difficulties.

---

[18] It does not advance step by step from the basic notion of "natural" number to the introduction of wider number domains, as would be done in the nineteenth century by Ohm, Weierstrass, and Dedekind (see Ferreirós 1999). On the shortcomings of the Newtonian

For a preliminary periodization of this millenary history, I would suggest that:

1. the period 900–1600 constitutes a kind of prehistory of the real number concept, when there was a growing tendency to identify magnitude and number, and new symbolic tools were developed, in particular, decimal expansions, which came to facilitate adoption of the new kind of number;

2. the next period might be called the *protohistory*, 1600–1800 roughly, as one can find the reals treated as quasi-objects (even though the main objects of mathematics are still magnitudes) and the parallel between infinite decimals and infinite series becomes a powerful driving force in their promotion;

3. and the *history* properly, say 1750–1950, when they finally emerged as arithmetic objects, indeed for many authors the central objects of pure mathematics, the number system.

Most important for us is the shift in the status of the real numbers, from tools to handle magnitudes (as in proportion theory) to independent mathematical objects; this shift, linked to the question of givenness underscored above, is to some extent related to the crucial innovations in the seventeenth century. Yet there are reasons to think that, if we are to consider centrally the idea of the real numbers as given independently and in their full system, this can only be found in the nineteenth century. The nineteenth century—the time of the emergence of modern math—saw a shift from mathematics based on the concept of variable magnitude to, in the second half of the century, the static perspective on number systems characteristic of "arithmetized" mathematics.

Malet (2007) has spoken of a Renaissance "arithmetization" of geometric magnitudes in connection with the first period; but notice that around that time it was basically only quadratic irrationals that were considered (besides π). Meanwhile, the protohistory will advance substantially toward full acceptance of infinitary objects as numbers (e.g., π will be conceived as an infinite decimal), new transcendental numbers

---

definition, it is highly instructive to read the comments made by W. R. Hamilton in his (1837), p. 4.

(such as *e*) will be introduced, and the idea of distinguishing algebraic and transcendental irrationals will emerge. As late as 1637, Descartes still perceived the realm of geometry to be linked with rigor (and synthesis), and algebra as linked with heuristic methods of problem solving (and analysis). But the success of symbolic methods, with the "analytic" geometry of Descartes and the calculus of Leibniz and Newton, eventually led to a new inversion: by 1788 Lagrange was equating rigor with symbolic, algebraic treatment!

The real numbers form a complete or continuous system of numbers, and this of course is an essential trait that distinguishes them from other number systems. It is inherited from the classical idea of a geometric line (Chapter 6, section 3) and is the source of great foundational difficulties (we shall come back to that). Another, more basic but related trait that distinguishes the real numbers is the role of infinity. While natural and rational numbers are conceivable as finitary objects, the real numbers pose the additional difficulty of being infinitary. Considering this essential role of infinity, and thinking back on the history of decimal fractions, it is natural to ask when the notion of a real number as given in its infinite decimal expansion emerged. For until we find clear signs of this type of notion, we may doubt whether decimal expansions are being employed merely as a tool.

To make this fully explicit, decimal expansions employed as tools pertain to *approximation* mathematics; they are in effect truncated decimals, and they are in essence just a handy form of fraction. It is hardly ever mentioned that the whole point of Stevin's famous *Disme* (1585) was to work with *finite* decimal expansions—which is perfectly consistent with the utilitarian orientation of this work. *Infinite* decimals can also be taken as the basis for introducing the real number system, but this involves a shift in perspective to a different kind of mathematical practice.

The question is elusive. If we request fully explicit formulations, apparently that is only to be found in the nineteenth century. Weierstrass's definition of the real numbers, first presented in 1864, is in fact a rather simple development of the idea of a decimal fraction, exploited, under an infinitary formulation, in the service of pure mathematics.[19]

---

[19] See Ferreirós (1999), Chapter 4. By that time, we find an even clearer and slightly earlier reliance on the infinite in Dedekind's work.

Nevertheless, the turning point had been crossed before this time, even if the formulation was less than totally explicit, and the really interesting issue is to try to throw some light on that earlier period.

It is natural to conjecture that it was only simultaneously with the acceptance of *infinitary processes*, in the context of the infinitesimal calculus, that infinitary decimal expansions in the strict sense were accepted as objects or quasi-objects of mathematics. If so, we should come to the conclusion that only around 1700 did the characteristic notion emerge of a real number as an object associated with an infinite decimal expansion.

Proofs of irrationality are a clear sign of a mature conception of the real numbers, especially proofs of the irrationality of a transcendental number. The first such were given by Euler in 1737 (for the number *e*, incomplete) and by his colleague at the Berlin Academy J. H. Lambert, in 1766, employing new methods of representation by continued fractions.[20] To avoid the muddy waters of explaining what Euler did, let us focus on Lambert's rigorous proof that π is irrational. The first surprising element is that it took so long to first come to this result, but perhaps by now this is not so astonishing for the reader: we are realizing that this was related to a revolutionary shift in the conception and the handling of irrational numbers.

Lambert presented his proof in a Mémoire in French, in 1766, and in a chapter of his book of "contributions to the use of mathematics and its application" (*Beyträge zum Gebrauche der Mathematik und deren Anwendung*, vol. II, 1770).[21] Let me consider the latter, a semipopular and witty exposition. He began making fun at the lack of knowledge of "circle-squarers," who abounded at the time (including Leistner and a Merkel, whom he mentions by name, 1770, p. 137), and proceeded to

---

[20] Although the concept of continued fraction can be linked with the Euclidean algorithm, it is essentially foreign to classical Greek mathematics, and experienced great development in the eighteenth century (especially with Euler, in a paper of 1737 giving the mentioned proof that *e* is irrational, and especially in Chapter 18 of his *Introductio in analysin infinitorum*, 1748). As in the case of infinite series, we have to do here with infinitary processes linked with the real numbers and real-valued functions.

[21] The ironic title is 'Preparatory Knowledge for Those who search the Quadrature and Rectification of the Circle' [*Vorläufige Kentnisse für die, so die Quadratur und Rektifikation des Cirkuls suchen*]; redacted in Oct. 1766 according to M. Bullynck (see www.kuttaka.org/~JHL /Main.html).

show how to obtain bigger and bigger fractional approximations to π from representations of this constant by continued fractions.

Then he moved on to present a continued fraction expansion for the function $tan(x)$,

$$tan(x) = \cfrac{x}{1 - \cfrac{x^2}{3 - \cfrac{x^2}{5 - \cfrac{x^2}{7 - \ldots}}}}$$

from which, taking x = $1/n$, one gets (Lambert 1770, 150)

$$tan\left(\frac{1}{n}\right) = \cfrac{1}{n - \cfrac{1}{3n - \cfrac{1}{5n - \cfrac{1}{7n - \ldots}}}}$$

This result allowed him to deduce that the function $tan(x)$ only takes rational values for irrational x, and, vice versa, the tangent is irrational for every rational value of x. This was the crucial result from which the irrationality of values such as $\pi/4$, or π itself, was established (consider that $tan(\pi) = 0$).

Lambert did not emphasize the infiniteness of the decimal expansion of π, yet his presentation was clear enough on this point. He emphasized that the continued fractions established above "continue forever" [*dieser Bruch immer fortgeht*], which in fact is crucial for establishing the irrationality of the value of $tan(1/n)$. He also discussed several calculations of the decimal digits of π, in particular what he called the "*Ludolphische Zahlen*," which he wrote as

3,141592,65. . . up to 35 ciphers,

and the "*Lagnysche Zahlen*" with 127 ciphers.[22] Notice that he writes "Ludolphian numbers" in the plural, so he seems to be referring to the list of 35 ciphers, not to π itself. But more important than this is that he was emphasizing the link between finite decimal expansions and fractional representations, and the fact that the latter can always be taken

---

[22] As we know, the last fifteen are incorrect!; see Lambert 1770, 139, 146. Referenced are Ludolph van Ceulen, 1610, and Thomas de Lagny, 1720.

further (with larger and larger numerator and denominator). This implies a clear awareness of the fact that the *precise* decimal representation of π would be infinite. Thus it was clear, for those who read and understood, that a precise decimal representation for a constant such as π could only be infinite. (Lambert took care to record his doubts that he would be read, or understood, by the "circle-squarers" on p. 135.)

As we see, the matter emerged this way in connection with relatively sophisticated results about continued fractions, and in any event with *explicit infinitary processes* in the context of analysis (another example would be infinite series). As a matter of fact, the case of Isaac Newton would seem to confirm this reconstruction. In his *Universal Arithmetick*, after defining "Number" as "the abstracted Ratio of any Quantity, to another Quantity of the same Kind," he goes on immediately to introduce what "are called *Decimal Fractions*" (Newton 1720, 3) and to exploit them in his presentation of the "threefold" kinds of numbers: "*integer, fracted,* and *surd.*" However, this work was relatively elementary and most of the "surds" (irrationals) he presents are of the kind $\sqrt{a}$. Thus if we want to find out what his deeper views on the matter were, we must consider works of a more advanced character dealing with the calculus of fluxions.

In *De Quadratura Curvarum*, published in 1704 as an appendix to the first edition of the *Opticks* (but redacted in 1691), Newton starts by saying:

> I consider mathematical Quantities in this Place not as consisting of very small Parts; but as describ'd by a continued Motion. Lines are describ'd, and thereby generated not by the apposition of parts, but by the continued Motion of Points; . . . [same for surfaces, solids, angles]; Portions of Time by a continual Flux: and so in other Quantities. These Geneses really take Place in the Nature of Things, and are daily seen in the Motion of Bodies. (Newton 1691, p. 122 in Whiteside edn, vol. VIII; p. 165 in the edn of *Opticks*)

Once again, the emphasis is on a general notion of number, immediately applicable both to rational and irrational quantities. As in *Universal Arithmetick* (1720), he is not interested in the idea of "natural number" as a collection of units, nor more generally in numbers that "consist of very small parts," but tries to reach directly the *continuous* system of all real numbers. This is "mathematical Quantities . . . as describ'd by a continued Motion," and Newton insists on the view that this is no fiction or idea in the mind: "these Geneses really take Place in the Nature of Things," i.e., in the real world. The objects to which mathematics ultimately refers are real things in the real world, and *continuous motion* is a quintessential element of this reality—hence the continuity of time and space, and of many other domains of quantities.

What has this to do with decimal expansions, and with infinity? Considering numbers on the one hand as expressed by decimal fractions, and on the other hand as given continuously in the manner just described, leads to the conception of the continuous totality of real "quantities" or numbers as corresponding to all possible series of decimals. Think of the quantities that correspond to the unit interval [0, 1], and conceive them as "generated" by "a continued motion." Analyzing what the effect of this continuous process is on the decimal fractions that would represent the positions on the unit interval, we are led to the idea that all the possible decimal expansions $0, a_1a_2a_3 \ldots$, with a_i any cipher from 0 to 9, will be obtained. Obviously, one will pass through the sequence $0.141592654\ldots$, which corresponds to $\pi-3$, and $0.4142135623\ldots$, which corresponds to $\sqrt{2}-1$, and in between through 0.25 and infinitely many others. All the infinite sequences have to be considered, with all possible combinations of ciphers 0–9 at the nth decimal place, for all n.

During the eighteenth century, there was abundant discussion of the "law of continuity," and selected passages in the relevant literature may remind the modern reader of issues such as the justification of the intermediate value theorem. But this is anachronistic reading: those discussions in the hands of Bernoulli, Euler, or Boscovich were mainly on a problem of metaphysics—whether and why to assume that *natura non facit saltus*, that all action in Nature is free of jumps. And this was assumed to have the net effect that all real functions (translated into our language) are continuous, and, more, smooth—in direct contrast with so-called "classical" analysis.

8.3. THE NUMBER CONTINUUM

The real number system is essentially characterized by some principle of continuity or completeness. Yet, although this has been acknowledged from a very early stage in the development of the concept of real number, detailed analyses of the notion of continuity or completeness itself came only in the nineteenth century. Even as late as 1800 or 1850 most mathematicians were unable to offer a precise analytic definition of continuity, and among those who had something to offer, it was common to mistake denseness for completeness. But already al-Kashi or Stevin had the idea of establishing a fundamental parallelism between the realm of continuous magnitudes (paradigmatically, geometric segments) and the number domain. Yet the careful formulation of what this continuity consisted in could only come much later. It seems likely that the reason was twofold. On the technical side, a deeper knowledge of irrational numbers and their variety was needed: new procedures leading to them, such as logarithms, infinite series, etc.; distinctions like algebraic vs. transcendental, and more examples of transcendental numbers; and new methods for studying them, such as continuous fractions. Above all, what was needed was the decision to stop introducing irrationalities one by one, sporadically and as the need arose, to go instead for a characterization of the *system* of reals as a whole (see Dedekind 1872). On the conceptual side, mathematics had to advance from being mainly concerned with processes of calculation or approximation to the study and clarification of the basic properties of mathematical domains, and the crucial shift had to be made through which the question of the continuum stopped being regarded as "metaphysical" to be considered properly "mathematical" (see Cantor 1883).

As the example of Newton's *De Quadratura Curvarum* suggests, the context of infinitesimal analysis was crucial for those moves. We have seen that the traditional approach was to consider as given certain quantities obtained by explicit constructions, but in the context of analysis it was no longer possible to restrict givenness of quantities to particular types of constructions (say geometric constructions or logarithms of previously given numbers, or the like). The transition from the algebraic methods of analytical geometry to the new calculus of infinitesimals (or fluxions) did not take a long time, but it represents a veritable inversion that created the drive for a complete number system. In the limiting procedures that already Newton (like D'Alembert or Cauchy later) saw as basic to the new calculus, one presupposes a generality of validity that can only succeed when a continuous or complete system of numbers lies behind it. What exactly this means, and to what extent it holds, were not at all clearly understood until the second half of the nineteenth century.

In analysis, one reasons in a general or abstract way every time that, instead of considering a particular series or a concrete sequence (for which the relevant limit could be explicitly determined), we consider properties such as the following:

- any Cauchy-convergent series has a certain real number as sum,
- every increasing bounded sequence has a real number as limit,
- or any function f which is continuous in $[a, b]$, such that $f(a)$ is negative and $f(b)$ is positive, must have a zero therein.

If one were to consider only particular series or sequences, carefully controlled in their buildup, the full completeness of the real number system would not need to be presupposed. The convenience of a more general way of speaking is bought at the high price of postulating completeness—which, however, for reasons discussed below, did seem to be a natural assumption in the seventeenth, eighteenth, and nineteenth centuries.

Yet many conceptions of the continuum are possible, among them the nonstandard, the intuitionistic, and the smooth infinitesimal continuum.[23] The "classical" conception of Dedekind, Cantor and Hilbert finds a "common measure" between the discrete and the continuous on the basis of what may be called a pointillist conception of the continuum (see Chapter 6 section 3): the line as an infinitary object made of points is a nuclear assumption for Dedekind, Cantor, Hilbert, and mainstream twentieth-century mathematics. The problem of bridging the gap between discrete and continuum has been one of the nuclear themes in the whole development of mathematics; we have mentioned Riemann's antinomy in Chapter 6 section 3.

---

[23] See Bell (2010), Brouwer (1927), Bishop and Bridges (1985), Bell (1998), and Feferman (2009). Leaving apart atomized conceptions (of discontinuous time, space, and movement), one has *synechist* views such as those of Aristotle, Brouwer, and Thom; the *pointillist* views of Cavalieri, Dedekind, and Cantor; and what might be called "*overabundant*" views in Peirce informally, technically associated with Robinson's hyperreals or Conway's surreals; see Salanskis and Sinaceur (1992) and Ehrlich (1994).

It can be said that the conception of continua as sets of points belongs to the core of the set-theoretic program, being a nuclear assumption for Dedekind, Cantor, Zermelo, and all their successors. (Read again Luzin's explanation of the purpose of set theory in section 6.3, p. 169.) As I shall argue, central axioms of set theory—the axiom of infinity, the axiom of choice, the axiom of powersets in its standard, quasi-combinatorial interpretation—have their source precisely there. But the pointillist conception of continua was not the main approach in the classical period of the seventeenth and eighteenth centuries: it had some representatives—of which perhaps the Italian Cavalieri is the foremost representative—but it was a minority approach, heavily criticized for inconsistencies, and never transformed into a full systematic proposal.

Most important is the contrast between the *static*, "purely existential" approach characteristic of the modern Dedekind-Cantor viewpoint and the *dynamic* conception of continua, coupled with an approximative view of the number system that was typical of the previous period. The static ingredient of the modern approach and the peculiar meaning of its existence claims—understood as statements about given elements, which do not in the least depend on explicit definition or the constructive resources available to mathematicians—was discussed quite often in the era of the "foundational crisis," and most explicitly in connection with the Axiom of Choice and the meaning of proofs that depended on it.[24] By contrast, during the classic period the characteristics of the real number system were assumed to supervene on physically existing magnitudes that flow continuously. If you allow me to use, metaphorically, some philosophical language for describing the shift, I could say that the static "being" of real numbers (or points), given beforehand or platonistically, stands in sharp contrast to the "becoming" of variable quantities that were the focus in the age of Newton and Euler (think of Newton's "fluents").

The real number (be it understood as an otherwise undefined element, or as an actually infinite set, or set of sets) became the central

---

[24] An outstanding example is Zermelo's Well-ordering theorem; see, e.g., his second paper on the topic (1908), which contains a lengthy and very interesting methodological discussion.

object in the modern period, but—I surmise—it had a completely different status in the classical period, the seventeenth and eighteenth centuries. In relation to the dynamic understanding of continua that was then typical, real numbers (like all numbers) were basically understood as *tools* for the determination, or even the approximation, of relations in continuous domains. So the contrast between tool and object, which has been used to discuss important features of category-theoretic math (Krömer 2007), is also of great importance here. Probably important, too, would be the idea that the number concept was open in the classical era, while the careful determination and bounding of various number systems is characteristic of the transition to the modern era.

It has been said sometimes that modern mathematics operated a "discretization" of the continuum. This is a rather imprecise, or even incorrect, way of reading what the "arithmetization" of mathematics was about; to be precise, what happened was a reduction to numbers *and sets*, or even to sets alone, and set theory is by no means "discrete." (In the set-theoretic paradise *V*, continua live happily, and are as much first-class citizens as discrete structures are.) However, one could still say that the continuum was typically the paradigm in the classical age, while the discrete (numbers) became the paradigm for modern math. This has to be understood in the sense that the concept of the continuum was formerly taken to be an irreducible starting point, a primitive notion in and of itself. By contrast, the moderns immediately started searching for definitions of continua: Dedekind's in 1872 (denseness + cut property), Hilbert's in 1900 (Archimedean + maximally complete), Cantor's metric-topological one in 1883 (connectedness in a metric sense + perfect set property), and so on, to the usual topological definition that emerged around 1950.[25]

Another important difference between the classical magnitude-based continuum and its modern set-theoretic counterpart is the following. With the reals based on magnitudes, as the number concept needed to deal with a continuously varying domain of magnitudes,

---

[25] This is compactness + connectedness. A topological space E is *compact* if every open cover of E has a finite subcover. (In other words, if E is the union of a family of open sets, there is a finite subfamily whose union is E.) A space E is *connected* if any two points in E can be linked by a curve lying wholly within E.

some conceptual underpinnings differ from what emerges from the abstract notion of $\mathbb{R}$ proposed by Cantor and Dedekind. Notice that the abstract Dedekind-Cantor set $\mathbb{R}$ is a unique structure that is (statically) given to us in a well-defined way, arising from the set of natural numbers that is assumed to be uniquely given too. In particular, "1" is just the natural number one or the real number +1 that is identified with 1; it is conceived to be a unique object. This is not the case with the magnitude-based real numbers, for the value of "1" is arbitrarily chosen in each particular application of the concept.[26] This might seem to be a minor issue, but there is an interesting conceptual ramification.

As a result of the conventionality of the unit of magnitude, the difference between rational and irrational magnitudes appears to be artificial (and the same happens with algebraic and transcendental magnitudes). This is what I mean: you know how to construct an irrational magnitude once a given unit is chosen; pick the unit $u = 1$, and then construct a square or a circle, or a pentagon, to determine magnitudes whose irrational value is $\sqrt{2}$ or π, or $\frac{1}{2}(1 + \sqrt{5})$, respectively. This, however, does not imply any deep conceptual or ontological difference between our magnitudes u and π. To see this, just remember that the choice of unit is conventional. Change the basic unit, choose the particular magnitude whose value was π as the new unit $m = 1$; now it turns out that u is an irrational magnitude (measuring $\frac{1}{2\pi}$).

This, I believe, was the precise idea behind Stevin's thesis that "there are no numbers which are absurd, irrational, irregular, inexplicable, or surd."[27] What appears "absurd" or "irrational" on one choice of units is perfectly rational on another. Given this conventionality, the reality to which numbers refer presents no such distinctions.

Such things are not at all the case in the abstract Cantor-Dedekind continuum, where the difference between rational and irrational (or algebraic and transcendental) numbers is invariant. The set of algebraic numbers is what it is—1 and $\sqrt{2}$ belong to it—while the set of transcendental numbers is what it is—π and e belong to it: no choice,

---

[26] That is, one chooses a length of one meter as unit (after 1800), or lengths of one yard (in the modern British system or the old one); or considering old measuring systems, one Irish mile, one Spanish legua, and so on and so forth.

[27] *Theses mathèmatiques*, in *L'Arithmétique* (Stevin 1585b).

no convention involved. Perhaps that means that we should actually distinguish between the traditional magnitude-based continuum and the modern Cantor-Dedekind continuum as two different ideas.

8.4. THE REINVENTION OF THE REALS

The idea that pure mathematics has strictly arithmetical roots is quite recent, emerging only in the nineteenth century, and it is usually identified as the ideal of arithmetization. (By contrast, the idea that pure mathematics is fully autonomous from the physical world is ancient, typical of the Neoplatonists. German mathematicians of the nineteenth century combined this kind of rationalist-Neoplatonist orientation with the shift from geometry to arithmetic.[28]) From around 1820, Gauss came to insist on the empirical character of Euclidean geometry, separating it from arithmetic or pure mathematics; this for him included all numbers up to the system of complex numbers, studied not only algebraically and analytically, but also number-theoretically, or topologically. At about the same time, Martin Ohm proposed the idea that pure mathematics is a purely intellectual concern that emerges from the concept of natural number by a step-by-step expansion; Dedekind and Weierstrass later shared this vision, which Cantor learned from his teacher Weierstrass. This led to Cantor and Dedekind publishing their definitions of $\mathbb{R}$ in 1872, which conceptually were very new, but technically not so much. Just as these mathematicians arithmetized analysis, Dedekind and others worked hard to arithmetize the new conceptions in algebra, and Hilbert in 1899 proposed an arithmetization of geometry.

The configuration of $\mathbb{R}$ that is better known for mathematicians today is the product of this radical transition that occurred approximately from 1850 to 1870. The result was an interpretation of the real numbers as elements of a purely numerical system, fully autonomous from the physical world, independent of the notion of magnitude. Not all mathematicians

---

[28] See the first prologue of Proclus's *Commentary*. The philosophy of Proclus was crucial to Kepler, and the ideas presented by the young Gauss in 1808 are clearly reminiscent (see Ferreirós 2006). On arithmetization, see Petri and Schappacher (2006), pp. 343–374.

of the period shared this interpretation, of course, but some like Dedekind and Cantor were very clear exponents. This new vision of the real numbers can be presented (genetically) as a set-theoretic construction based on the set of natural numbers $\mathbb{N}$, or else independently as an axiomatic definition of $\mathbb{R}$ in Hilbert's style. Yet both of these approaches have the same origin, and epistemologically there is little difference between them. Even though the story of the reinvention of $\mathbb{R}$ is well known, let me say a few words about it, especially since today it is uncommon for students of math to come across an exposition—which I regard as an unfortunate shortcoming of current vogues in higher education.

Weierstrass had a complicated way of presenting the number system, developed in his lectures since the mid-1860s: he regarded "numbers" (*Zahlgrössen*) as infinite aggregates of units and aliquot parts of units (which could be real or complex), aggregates that are compared according to the sums of finite series of elements. This in effect amounts to a definition of the reals in terms of infinite series.[29] Cantor found two ways of simplifying the definition: first, he restricted it to the real numbers (leaving aside the complex field) and decided to start from the system $\mathbb{Q}$ of rational numbers, taken as previously given;[30] second, he chose to consider infinite sequences instead of infinite series—obviously to each Weierstrassian aggregate conceived as a series one can associate the sequence of partial sums of the series. Thus emerged his definition of the real numbers as "fundamental sequences" of rational numbers,[31] which he presented in lectures of 1870 and in a paper of 1872. Once the abstract system of all fundamental sequences in $\mathbb{Q}$ has been introduced, one can prove that it is a *complete* system with the usual order and algebraic properties.[32]

[29] See Ferreirós (1999), 124*ff*. That chapter reviews the history in more detail. A very clear systematic exposition can be found in Chapter 2 of Ebbinghaus *et al.*, 1991.

[30] How to derive $\mathbb{Q}$ from $\mathbb{N}$ was clear both to Dedekind (who did it in the usual way, introducing integers as pairs of naturals, and rationals as pairs of integers) and to Cantor, who would later (1883) refer to the work of Grassmann on this topic, *Lehrbuch der Arithmetik* (1861).

[31] Cantor (1872). A *fundamental* or Cauchy sequence is a sequence $a_1, a_2, a_3, \ldots$ of rational numbers, such that the difference $|a_{n+m} - a_n|$ becomes as small as one wishes for sufficiently great values of n. This can be defined rigorously in the typical Weierstrassian ε-δ style, as Cantor did.

[32] Cantor (1883), section 9. A neat presentation would proceed to define equivalence classes of fundamental sequences, since many of them coincide (defining the same number), but Cantor never took this step.

Dedekind also gave a set-theoretic introduction of the reals from the field of rational numbers; he assumed $\mathbb{Q}$ as given, carefully explaining that it is a number field endowed with a total order, and is dense. Also, he was more emphatic than Cantor about the supersession of magnitudes and the purity and autonomy of pure mathematics. His definition had been found in 1858 and exposed in some lectures.[33] Rather than relying on well-established notions of analysis, properly reinterpreted, he preferred to find a simpler starting point in the idea of a Dedekind cut.[34] Some cuts on the rational numbers $\mathbb{Q}$ correspond to a rational number (e.g., they define rational reals) but some do not correspond to any rational, that is to say, they define irrational numbers. He then proved that, by adding to $\mathbb{Q}$ a new set of irrationals corresponding to those cuts, the ensuing number system is a field, totally ordered, dense, and complete—in the sense that a cut in this new system always corresponds to an already given element of the system.

Perhaps it will be a good idea to review more explicitly the argument in Dedekind's paper—whose reading anyhow I recommend. First, he emphasizes that $\mathbb{Q}$ and $\mathbb{R}$ (like the complex numbers $\mathbb{C}$) are instances of number fields. But second, $\mathbb{Q}$ is a linearly and *densely* ordered field, homomorphic to the geometric line. Third, there are points on the line that do not correspond to numbers in $\mathbb{Q}$; i.e., the phenomenon of incommensurability shows that denseness is not enough for continuity (a point that had escaped many previous mathematicians, Bolzano included). Fourth, the cut property is enough to secure continuity when added to denseness.[35] Fifth, Dedekind employs cuts on the rational numbers to fully define the set of real numbers, its order, and its operations; and he shows that $\mathbb{R}$ so defined has the cut property (or, what amounts to the same thing, the abstract system of all cuts on $\mathbb{Q}$ is

---

[33] See Dedekind (1872) and Dugac (1976), 203, which includes the first draft of the paper, giving further details. Denseness is the property that, whenever s < t, there is an element r such that s < r < t.

[34] A set of rational numbers is a *cut* iff whenever r ∈ C then every s ≤ r belongs to C, and C is bounded above. (More precisely, for Dedekind a cut in $\mathbb{Q}$ was a pair of sets, C and its complement C'.)

[35] A linearly ordered set S has the *cut property* if, given any partition of S into lower and upper parts A_1 and A_2 (nonempty), such that always $a_1 < a_2$, there is $p \in S$ such that p is either the lowest element in A_2 or the uppermost in A_1. That is, any cut on S corresponds to (is engendered by) an element of S.

"a continuous manifold"[36] and can be given the structure of a totally ordered, dense, complete field). Finally, Dedekind proves that Cauchy sequences of real numbers do have a limit, and that monotonically increasing (or decreasing) bounded sequences have limits too.

The radical novelty of Cantor and Dedekind was that, instead of beginning with the assumption that a continuous system of magnitudes is given (metaphysically or physically), they started from the rational numbers and some infinitary "operations" on the rationals (sequences viz. cuts). Initially, the idea of real numbers and operations on them lacks any meaning, and it is only reductively, through the system of all fundamental sequences (viz. the system of all Dedekind cuts), that the reals are defined and the relations and operations on them are established. For mathematicians today this is a very natural way to proceed, but at that time it was novel and many had difficulties understanding it. Cantor and Dedekind went on to insist that it is only after the "purely arithmetical" introduction of the reals that notions such as the continuity of the geometrical line, or the continuity of time and space, can be satisfactorily explained.

The definitions of Dedekind and Cantor are different, of course, as can be seen from the fact that they generalize differently. Cantor's approach through fundamental sequences presupposes a metric structure (though it can be generalized to topological spaces); they are ω-ordered countable sets, sequences, and insofar traditional objects of analysis. Dedekind's tool, the cut, presupposes a totally ordered structure; it was a novel but very simple object, lacking inner structure, and a pure set (although defined by an order condition). Cantor's tools for defining the reals can be linked with the later ideas of Cantor space (the set of all sequences of 0 and 1) and Baire space (the set of all sequences of natural numbers), while Dedekind's tools are more directly linked to the general idea of powersets—the set of all subsets of a given set.

---

[36] See the letter to Lipschitz, June 1876, quoted in Ferreirós (1999), 134. Dedekind spoke insistently of the "creation" of irrational numbers, and much ink has been spilled discussing this; notice, however, that this "creation" is strictly regulated by set-theoretic operations.

8.5. SIMPLE INFINITY AND ARBITRARY INFINITY

Feferman (2009) has emphasized that the notion of a sequence is cognitively more basic and primitive than the notion of a set. This is likely to be true; indeed, the most basic idea of a sequence is a relatively simple overdevelopment of the series of natural numbers. Consider the sequences of odd numbers, of Fibonacci numbers, or of primes: one has a rule, in each case, by which a first and a second and a third member, and so on, can be determined. That is, one has a precisely defined rule that, applied to a counting number, produces a member of the sequence—a process which, of course, may involve previously produced members. This might seem to speak in favor of Cauchy sequences as a basis for defining real numbers; yet, as we are now going to see, the necessary move to arbitrary sequences—not given by a rule—erases the distinction with arbitrary sets (or even makes sets simpler).

Set theory was characterized by the inversion of regarding general sets as more primitive and fundamental than sequences. This inversion of perspectives had significant foundational and philosophical consequences. Here, I want to emphasize one, associated with the introduction of the real numbers by Cantor and Dedekind. They relied heavily on the assumption of actually infinite sets, the first mathematical contribution that rested on such assumptions explicitly.[37] But it is important to distinguish two facets of the infinitary assumptions involved.

There are levels in the introduction of infinitary assumptions, of which (it seems to me) three are particularly noteworthy at least on a first approximation. I shall term them "simple" infinity, "arbitrary" infinity, and "large" infinities (the last referring to large cardinals, a topic of great importance in set theory).[38] The first two levels were

---

[37] The same happens with Weierstrass's numbers, presented in lectures from 1863 onward, with the difference that Weierstrass always took care to insist that his aggregates are only compared finitarily, through finite parts of them.

[38] Large cardinal hypotheses (inaccessible cardinals, measurable cardinals, and so on) are independent from the axiom system ZFC. Some of them can be justified on the basis of reflection principles, some are stronger. They have played a central role in set-theoretic research since 1950, but we shall not enter into the topic here. See Drake (1974), expositions in many good textbooks (e.g., Jech's), the very detailed treatment in Kanamori (1994), or the now-standard reference M. Foreman and A. Kanamori (eds.), *Handbook of set theory* (Springer, Dordrecht, 2010).

involved, simultaneously and unclearly, in the reinvention of the real numbers due to Cantor and Dedekind. Two kinds of infinities were stipulated at the same time, albeit confusedly, apparently without recognizing that there is great difference between them from a foundational viewpoint.

Simple infinity is the actual infinite of the natural numbers, the assumption that $\mathbb{N}$ is given as a completed totality. In the context of ZFC, every infinite set has a countable subset, a subset that is equivalent to $\mathbb{N}$. Once you take the step of assuming simple infinity, reasons of consistency force you to admit other similar totalities: examples are the set of all odd (or even) numbers, the set of multiples of any given number (say 17), and the set of all primes (or its complement, the set of all composite numbers). After that, some surprising properties appear, especially the fact that one can pair up a set and its proper part—e.g., $\mathbb{N}$ and each one of the sets mentioned before (such as the set of even numbers or the set of primes). In the previous setting of a mathematics of magnitudes, this seemed unacceptable: one of the key principles of the theory of magnitudes was that the whole is greater than its parts. Thus, famously, Galileo turned that funny property into an argument against the ability of our minds to grasp infinite magnitudes. In the new 1870s setting of a supersession of magnitudes, Galileo's argument no longer worked: a set is not a magnitude! Radically, Dedekind turned that "funny property" into the very definition of infinite sets.[39]

Why do I say that consistency forces us to admit other sets once $\mathbb{N}$ has been admitted? Because the previously available theory of the natural numbers has the means to define certain subsets of $\mathbb{N}$, and if $\mathbb{N}$ itself is admitted as a set, it would hardly be consistent to deny sethood of such *definable* subsets. (The crucial notion, then, is definability—and definability depends on a given language and theoretical means, which must be fully specified in advance. This is related to the famous semantic paradoxes, especially paradoxes of definability, such as Richard's; a change in language may entail significant differences as to the corresponding

---

[39] Dedekind (1888): a set S is (Dedekind-)infinite *iff* there is a one-to-one correspondence between S and a proper part $T \subset S$; if there exists no such correspondence, S is finite. In ZFC, one can prove that this notion agrees with the traditional notion of infinite set, defined in terms of "exhausting" all natural numbers.

definable sets.) In Dedekind-Peano arithmetic, even and odd numbers, multiples of 2 or 17 or any given number, prime numbers, squares, and a long etcetera, are definable properties that determine subsets of $\mathbb{N}$. Notice, however, that many other subsets of $\mathbb{N}$, acknowledged in set theory, do not correspond to properties definable in the usual theories of natural numbers. Thus, one distinguishes definable subsets from arbitrary subsets, which may not be definable.

Arbitrary infinity is, thus, conceptually a further step into the Cantorian paradise: one takes as given all possible subsets of $\mathbb{N}$, in the sense of arbitrary subsets (i.e., disregarding any requirement of definability), and, furthermore, one takes as given the actual infinity of subsets of $\mathbb{N}$ as a new totality or set. More generally, for any given set S, one assumes to be given the actually infinite totality of all its subsets, $\wp(S)$. This is the Powerset Axiom of ZFC, understood in the sense of quasi-combinatorialism. Quasi-combinatorialism, or simply combinatorialism, is another name given to the rejection of definability requirements in set theory (the name comes from viewing infinite sets as the outcomes of combinatorial processes, where infinitary processes would be exactly analogous to finitary ones).[40] The Axiom of Choice is a natural expression of the quasi-combinatorial standpoint; from this point of view, explicit definitions of specific functions, sequences, and sets are merely ways of picking out (or naming, specifying) an object that exists independently of, and prior to, the construction.[41]

The difference between $\mathbb{R}$ and the simple sets (similar to $\mathbb{N}$) discussed above is that the latter are definable sets, while the introduction of $\mathbb{R}$ requires us to abandon the requirement of definability. It is when one abandons the requirement of definability that we speak of arbitrary infinities. And this was the characteristic standpoint adopted by Cantor and Dedekind from 1872 onward. It characterizes the main tradition of set theory, continued by Hilbert and Zermelo, to this day.

---

[40] Bernays (1935, 260): "one views a set of integers as the result of infinitely many independent acts deciding for each number whether it should be included or excluded. . . . Sequences of real numbers and sets of real numbers are envisaged in an analogous manner. . . . The axiom of choice is an immediate application of the quasi-combinatorial concepts in question."

[41] This standpoint was emphasized already by Zermelo (1908a). See also Wang (1983) and Maddy (1988; 1997).

Notice, however, that it is perfectly coherent to accept simple infinity while rejecting arbitrary infinity. Many foundational systems, beginning with predicativism, and also constructive set theories, take this route.[42] Such alternatives are probably the best way to articulate doubts about the Axiom of Choice, which are often expressed by mathematicians. French mathematician R. Baire was perhaps the first to clearly see the possibility of these kinds of alternatives. Thus, it is perfectly possible to argue for $\mathbb{N}$ and against $\mathbb{R}$ (classically conceived) as a set. The ultimate reason for this is the difference between simple infinity and arbitrary infinity. Arbitrary infinity is involved in the postulation of $\mathbb{R}$, but also in the postulation of powersets $\wp(S)$, or what is called Baire space in descriptive set theory.[43]

Another observation that is crucial to an analysis of what Cantor and Dedekind did: the assumption that $\mathbb{Q}$ (the set of rationals) is given as a completed totality does not go beyond the assumption of $\mathbb{N}$; that is to say, it is just another instance of the postulation of actual infinity in the *simple* form that we have been discussing. To put it differently: it would be incoherent to give arguments for the assumption of the set $\mathbb{N}$ while at the same time arguing for denial of sethood to $\mathbb{Q}$.[44] Where does the difference between $\mathbb{Q}$ and $\mathbb{R}$ lie? Most simply, in the fact that each element of $\mathbb{Q}$ can be explicitly defined from the natural number arithmetic, while this is not the case for each element of $\mathbb{R}$. It is very simple to program a computer to produce an exhaustive list of rational numbers, e.g., using Cantor's technique to enumerate the rationals; each member of $\mathbb{Q}$ is then introduced by explicit definition. (The same happens for *many* real numbers, such as $\sqrt{2}$ and $\sqrt{3}$, and, in general, all algebraic numbers, and even transcendentals such as π and e. But *most* real numbers are not like this—a very important insight contributed by critics of set theory such as Kronecker. As one can easily prove from the standpoint of set theory, only countably

[42] For predicativism see Feferman's entry in Shapiro (2005); for constructive set theory see Crosilla (2009).

[43] In the sense of descriptive set theory, Baire space is the set of all sequences of natural numbers, ω^ω that can be given a topology (the product topology) as a perfect Polish space of dimension zero.

[44] A different matter is whether the rationals are going to be called numbers, as one can consistently argue for restricting the notion of number to the naturals.

many reals are definable from the theory of the naturals or the theory of the rationals.[45])

Consider again the fundamental sequences of Cantor. One could make his theory very specific and concrete by introducing a definability restriction. Since we start from the set of rational numbers and their theory, let us consider only fundamental sequences that are definable in this theory. This, however, will have the effect that only *some* real numbers will be defined: $\sqrt{2}$, $\sqrt{5}$, and, in general, algebraic numbers, and transcendental numbers such as π and e—but by no means all real numbers. If you consider Dedekind cuts, this is also exactly the case. Making his theory very specific, let us consider only cuts definable from the theory of rational numbers—or, equivalently, from the theory of naturals. (It is particularly simple to specify the cuts corresponding to quadratic irrationals such as $\sqrt{2}$ or, in general, $\sqrt{N}$ for a non-square natural number N.[46]) This will have the effect that *most* real numbers cannot be introduced, leaving a highly discontinuous set of reals, an impoverished "purely arithmetical" substitute of the set $\mathbb{R}$.[47] In such a structure, basic theorems of analysis cannot be obtained: one does not have a limit for every monotonically increasing (bounded) sequence of real numbers, nor is the Bolzano-Weierstrass theorem valid (that every set of infinitely many reals within an interval has an accumulation point).

Thus, in both cases, the crucial topological property of *completeness* for the set of reals can only be attained if we abandon the requirement of definability. It is necessary to allow for arbitrary fundamental sequences in the case of Cantor, or arbitrary cuts in that of Dedekind. This is actually what both mathematicians were presupposing in their writings, but apparently they did not realize how strong the foundational

---

[45] For those who may be wondering: calling $\mathbb{A}$ the set of all real algebraic numbers, which is a field, $\mathbb{A}$ is merely countable. And also expansions like the field $\mathbb{A}[\pi,e]$ are countable.

[46] For readers who are not mathematicians: C would then be the set of all rational numbers whose square is less than 2 ($x^2 < 2$), and C' the set of all rationals whose square is greater than 2 ($x^2 > 2$). It is not difficult to show that this cut does not correspond to any rational number; proofs were already known to the Pythagoreans. Dedekind (1872) gives a general proof for $\sqrt{N}$ as above.

[47] This observation was already made by Weyl in his book *Das Kontinuum* (1918), and it seems to correspond to Kronecker's views in the 1880s. Brouwer developed an intuitionistic version of Dedekind cuts in papers published in the late 1920s; see Troelstra and van Dalen (1988).

requirement was. Aware that they were adopting an infinitistic stand-point, they did not yet understand the difference between the two kinds of infinities they were stipulating, simple infinity and arbitrary infinity.

Modern mathematics incorporates a double rupture with Greek orthodoxy: the acceptance not just of actual infinities, but also of arbitrary infinities, as if they were actually given. Retrospectively, the roots of this second assumption can be found in traditional views about real numbers, e.g., as given by decimal expansions (Ferreirós 2011). Yet, for this to happen, the change of perspective associated with what I have called the reinvention of the reals was necessary.

8.6. DEVELOPING MATHEMATICS

Most of the time the development of mathematics has nothing to do with the introduction of new object-fields like the real numbers, or n-dimensional spaces, or the universe of sets.[48] There are many other issues at play that keep most mathematicians busy most of the time. Here are some of the main categories of issues, exemplified in cases that have been historically significant for the development of ideas about the real numbers.

PROBLEMS: Mathematicians often propose problems, and this is a very important aspect of their activity. "In mathematics, the art of posing questions is more consequential than that of solving them"; such was the extreme view defended by Cantor on the occasion of his doctorate. Some thirty years later, Hilbert agreed that problems are the lifeblood of mathematics, the crucial impulse that keeps the discipline moving.[49] Most mathematicians are fascinated by problems and by the challenge of finding a solution, which for them has the same fatal attraction that a high summit has for the alpinist. Any philosophy of mathematics that forgets this aspect of the practice will be seriously deficient.

Notice that a problem takes shape only within the context of some (more or less developed) conceptual or theoretical views and/or

---

[48] The word "field' is used in this section merely with its common-sense meaning, not with the algebraic one.

[49] Cantor's *Abhandlungen*, 30. Hilbert (1900), introduction.

methods of study. Let me mention, very briefly, some important problems linked in one way or another with the real numbers. The oldest, and an extremely famous one, is—how to square a circle? We do not know who originated the problem, but it is very natural in the context of ancient Greek geometry, and it grew into a great challenge as it became clear how difficult it was.[50] Another famous problem—that of duplicating the cube—seems to have been related to issues of religious ritual (specifically, the construction of an altar). In modern times, the problem of squaring the circle was transformed into the question, what kind of number is π? This was solved in 1882 with the proof that π is a transcendental number, which established also the impossibility of squaring a circle by means of ruler and compass (i.e., by straight lines and circles).

Another important question is, how do we express the value of π formally? Leibniz offered a beautiful series for this purpose; of course, one has corresponding problems for any other given real number (such as e, the basis of natural logarithms). Also: How do we classify the different kinds of irrational magnitudes? This is the topic of the complicated Book X of the *Elements*, which Stevin (1585b) famously called "the crux of mathematicians." In modern times, with a much more subtle picture of irrational numbers in place, there emerged a sophisticated understanding of them, more adequate than the one available to ancient Greek authors. The main classification is into *algebraic* numbers (those that are solutions to polynomials with rational coefficients) and *transcendental* numbers (which are not roots of such polynomials). The existence of transcendental reals was first established by Liouville in 1844, showing how to define or specify particular transcendental numbers. Then, Cantor, in 1874, proved their existence in abstracto, by a method that showed that every interval in $\mathbb{R}$ contains transcendental numbers.[51]

One also has questions such as, how to show that a given number is transcendental? Hermite started modern methods for transcendence

---

[50] It is linked with the seemingly straightforward problem of how to measure the length of a circumference, to which already the Bible offers a (deficient) solution (I Kings 7).

[51] Cantor's methods would establish that almost all irrational numbers are transcendental. In fact, Dedekind showed that the set of algebraic numbers is countable, while Cantor later proved that the set of irrational numbers is bijectable with $\mathbb{R}$, and thus uncountable. The set of transcendental numbers must also be uncountable and have the cardinality of the continuum.

proofs when he proved in 1873 that e is transcendental; and, with similar methods, Lindemann established the result for π in 1882. There are many open questions of this kind; e.g., it is not even known whether $\pi^{\sqrt{2}}$ is irrational, let alone transcendental; it is not known whether $\pi + e$ or $ln\ \pi$ are irrational or not; and the same applies to Catalan's constant, a number that is important in combinatorics, to the Euler-Mascheroni constant, and so on. One final example of a related question—jumping ahead to a topic that we shall consider in another chapter—is, how do we determine the difference between $\mathbb{N}$ and $\mathbb{R}$ as infinities?

METHODS: In order to solve problems, mathematicians employ tools which we normally call methods, either established methods or new methods. Archimedes employed established methods to show that the circle can be triangulated, so that its area is equal to the area of a rectangular triangle whose sides are the radius and the circumference of the circle. Thus the problem of squaring the circle was transformed into the problem of constructing a length of measure π (if this was solvable, the next step presented no difficulty).

A very important moment in the history of the reals was the introduction of a new method, the use of infinite series for dealing with real numbers and functions. This of course is an integral part of the history of the infinitesimal calculus, and we find good examples of the use of series in work by Newton and Leibniz. Their study was well established by the time of Euler, and was the main reason for speaking of a "calculus of the infinites" at that time. It became necessary to find methods for the summation of series, which enabled mathematicians to greatly expand their ability to solve traditional problems. The formal theory of series was intensively developed in the eighteenth century by the Bernoullis, Taylor, Euler, d'Alembert, Lagrange, and others. One should not forget that many important contributions had to do with methods of *approximating* the value of a series.

Naturally, new methods lead to new problems, and also to the introduction of new concepts. Eventually, the twin notions of convergence and divergence were introduced, and the problem was posed of finding methods for determining the convergence of a given series. (In the eighteenth century, divergent series were used, though it was not completely clear whether the operations carried out on them were legitimate.) Series representations were also expanded when they started to employ not only power series but also trigonometric series (this was due to Fourier especially). Riemann speaks of another veritable "turning point" [*Wendepunkt*] in the conception of infinity when Dirichlet discovered that changing the order of summands in a series may affect the value of the sum, or even its existence:

> The discovery of this fact, that infinite series and multiple integrals [split into] two classes [depending on whether or not the limit is independent of the order] constitutes a turning point in the conception of the infinite in mathematics.[52]

The exact theory of series was created in the nineteenth century on the basis of the notion of a limit by Gauss, Bolzano, Cauchy, Dirichlet, Abel, Weierstrass, Riemann, and others.

SOLUTIONS: When one speaks about solutions in mathematical practice, it is not merely a matter of finding solutions, but much more: criticizing solutions, offering counterexamples, setting bounds on solutions, rigorously establishing solutions. Work on these issues may involve the emergence of new concepts or new tools. In connection with all this, it is natural to refer to the classic work of Lakatos, *Proofs and refutations* (1976).

An obvious example, related to our discussion above, is the solutions obtained by Euler for operating with infinite series; these were criticized by Abel and others from the standpoint of a new notion of convergence of the series, and the rejection of divergent series. Abel wrote:

> In general, the theory of infinite series is at present very ill founded. One applies to infinite series all the operations as if they were finite; but is that permitted? I believe not. Where is it demonstrated that one obtains the differential of an infinite series by taking the differential of each term?[53]

One further interesting example would be provided by issues related to uniformly convergent series, a concept that was only introduced in the 1840s by Weierstrass and Seidel.[54]

---

[52] See Laugwitz (1999), 293*ff*. The quote comes from notes for a lecture, and is taken from p. 297–298.

[53] See, e.g., Jahnke (2003). Later, in the twentieth century, divergent series would be reintroduced in the world of analysis by Hardy, Rey Pastor, and others.

[54] See Lakatos 1976, appendix, but compare with later historical work.

Another good example is the tragicomedy of trigonometric representations, a theatrical play featuring Bernoulli and Fourier and Dirichlet, among others. In the eighteenth century, Bernoulli stated that any (piecewise) continuous function is representable by a series of sines; Euler and d'Alembert criticised this idea, in the context of the long and famous discussion of vibrating strings. Later, in the early nineteenth century, Fourier refined the idea to employ sines and cosines, which amended Bernoulli's solution, and he added the most noteworthy improvement of an explicit determination of the coefficients; this started an important chapter in the history of modern analysis. Next, Dirichlet added very significant rigor by carefully proving that result, adding precise stipulation of intervals for the representation, dealing rigorously with double integrals, and so on.

THEORIES: Both methods and solutions eventually lead to proofs, at least in the context of Western mathematics. Methods and proofs, and even solutions, involve key concepts and (so to speak) key points of passage en route to the goal. Different proofs of the same result may involve different points of passage, and even quite different presuppositions—for these reasons, they are of great interest to the research mathematician. Eventually, all of these things require systematization, both for pedagogical purposes and in the interest of reflection on what has been achieved and what could be achieved thereafter.[55] Thus, the problems, concepts, methods, solutions, and proofs are arranged in the form of a systematic presentation, usually guided by ideals of simplicity and economy of means. The result is a mathematical theory.

The work of mathematicians is not only, as the naïve tend to think, to establish a result once and for all through an irreproachable proof. Numerous proofs are found and compared, for reasons that have already been suggested; theorems and results are judged for their importance, and for the worthiness of their re-exposition. And theories are constantly getting reshaped, rearranged, and presented again, often on the basis of new principles or new concepts. In many cases this has led to axiomatizing the theories; in many others, to extending theories into new domains of application. All of this would deserve much more careful and detailed study, and I hope other authors will add to it.

---

[55] One finds both elementary expositions, say the *Elements'* early books, and advanced expositions, say Hartshorne's *Algebraic Geometry*.

META-THEORIES: It is important to add, however briefly, that in recent times the activity of mathematicians has risen to another level. Since the introduction of the new axiomatic techniques late in the nineteenth century, given the explicit emphasis on structures and analysis of their interplay, mathematicians ceased to work merely inside theories to start treating theories as objects of study. New object-fields emerged, in which theories or structures became particular objects, and it became possible to study different forms of relations among theories, progressions of theories, transfer methods that work among theories, and so on.

This has happened not only in areas of mathematics where that move is obviously explicit, by which I mean mathematical logic with its metamathematical methods, model theory, and proof theory. Another important example of such new methods and tools is category theory and its developments, which offer something like an eagle's view of the structures and theories established in mathematics up to the twentieth century. The road to category theory started with the comparison of structures and the study of questions in which the interconnection between different kinds of structures is featured prominently. Emphasis on mappings and morphisms ensued, and eventually new reflections on such interconnections at a higher level led to the introduction of notions of categories and functors.[56]

These contemporary developments, however, cannot be the focus of the present work, if only because the earlier configurations of mathematical knowledge (say up to 1950) already offer us abundant food for epistemological thought.

8.7. MATHEMATICAL HYPOTHESES AND SCIENTIFIC PRACTICES

Mathematicians' images of their own discipline have, of course, changed over time. In particular, the perception of mathematics in its connection with the natural sciences has seen rather strong changes. By 1800 it was generally assumed that mathematics was part and parcel of natural science, so much so that mechanics and astronomy were still

---

[56] For the history, see Corry 1996, Krömer 2007, and Marquis 2009.

regarded as parts of mathematics. During the nineteenth century, a new "puristic" perception of mathematics triumphed, promoted above all by German mathematicians and resisted initially by French and British authors. In this connection, a famous incident is the polemics of Jacobi against Fourier: in his important *Théorie analytique de la chaleur* (1822), Fourier wrote that "profound study of nature is the most fertile source of mathematical discoveries," insisting on the idea that the main aim of mathematics is to advance in the explanation of natural phenomena. The young German mathematician Jacobi, in a letter to Legendre, protested: "a philosopher like him should have known that the sole aim of science is the honour of the human mind [*l'honneur de l'esprit humain*], and that under this title a question about numbers is worth as much as a question about the system of the world."[57] Of course, echoes of similar debates can be found at all times, among the ancients in the polemics of Platonists versus Aristotelians (and even more so the followers of Democritus), in the Renaissance, and so on.

The vision of pure mathematics as totally independent and autonomous from science became dominant in the early twentieth century, embraced, e.g., by the Bourbaki tribe among many others. Yet such visions are constantly changing.[58] Late in the century the balance shifted again, and a "return to Fourier" has been saluted by experts like Kahane. An interesting polemical position was marked by Vladimir Arnol'd (1998, 229) when he insisted that "Mathematics is a part of physics . . . a part of natural science." Arnol'd regrets that in the mid-twentieth century "it was attempted to divide physics and mathematics," with "catastrophic" consequences as whole generations of mathematicians grew up "without knowing half of their science." Today there is a clear tendency to prioritize applications, though this is not always (yet) reflected in the university curriculum.

One reason for reminding readers of all that is that philosophers of mathematics are inevitably affected by the images projected by

---

[57] Reproduced in Jacobi's *Werke*, vol. 1 (1881), 454–455. Quoted, e.g., by Dieudonné, and also by Ferreirós (2007, 224), where I discuss the context for this view. Jacobi to Legendre, July 2,1830: "mais un philosophe comme lui aurait dû savoir que le but unique de la science, c'est l'honneur de l'esprit humain, et que sous ce titre, une question de nombres vaut autant qu'une question du système du monde."

[58] See Bottazzini and Dahan-Dalmedico 2001.

mathematicians. As a result, during the twentieth century there have flourished philosophical visions of mathematics that emphasized its absolute autonomy as an intellectual enterprise. But, for the indicated reasons, this may be a partial view promoted by the purism of the early twentieth century mainstream. This issue must sooner or later be confronted, explicitly or implicitly, by philosophers of mathematics. My own perspective is that one cannot deny the impact of scientific practices—very especially of mathematical physics—upon the historical and the present configuration of mathematical knowledge.[59] This is unlikely to come as a surprise to attentive readers, since it is in accord with the pragmatist and historical nature of the ideas I have been proposing throughout this book.

Notice that, for all its radicalness, twentieth-century purism was not very successful in promoting deviant conceptions of mathematics. A large part of mathematics, indeed some of its key core elements, has been inherited from pre-1900 math—which was marked by astronomy, mechanics, and mathematical physics in general. In this section I would like to explain how, in my view, this is related to the classical continuum, as a result of the importance of continuum-based representations of the physical world. In a previous chapter (Chapter 7) we have discussed how the same factors have influenced the underlying role of classical logic.

Some of the stronger hypotheses that characterize today's mathematical knowledge (to be considered in more detail in Chapter 9) can be justified on the basis of the classical continuum, but hardly without it. By this I mean the Cantor-Dedekind conception of the real numbers explained above, the continuum understood as a set of points. By this is not meant just the so-called "genetic" or set-theoretic introduction of $\mathbb{R}$ on the basis of $\mathbb{Q}$: the very same conception is taken up in Hilbert-style axiomatic system for the reals, that is, the conception investigated by analysis; and the same is true of descriptive set theory. We have seen that this understanding of the continuum is not really "classical," historically speaking, as it was rejected by Aristotle and by many mathematicians

---

[59] An important forerunner of a similar viewpoint is Riemann. See my (2006). Needless to say, one could also mention other relevant mathematician-philosophers who shared the core of Riemann's views: Poincaré and even Hilbert.

in antiquity, in the Renaissance and in the early modern period. It was especially with the development of infinitesimal concepts associated with geometry (as is the case with Cavalieri and even his master Galileo, and later with Newton and Leibniz) that the pointillistic continuum became deeply entrenched in the scientific worldview. It underlies physical models from the free fall of Galileo to Einstein's space-time, and even the spaces of states employed in quantum theory.[60]

But, of course, nothing ties a fully independent mathematics to this particular kind of structure. Some deep mathematical thinkers have dared to present alternative views of the continuum, most importantly, perhaps, Brouwer's idea, but such deviant concepts have not been taken up at all by the mainstream.

Let me say a few words about Brouwer's continuum. In a sense, it can be regarded as a modern and precise development of Aristotle's position: the idea that the continuum is an actually infinite aggregate of points is rejected; the continuum appears instead as a primary medium in which points can be determined, but which is not a sum of points. Early on, in the 1900s, Brouwer regarded mathematics as founded on a pure intuition of time, basically in accord with Kant; later, starting in 1918, he developed technical concepts that allowed him to provide a careful analysis of the continuum. These are mainly the concepts of *choice sequence*—a potentially infinite sequence of numbers (or other mathematical objects) chosen one after the other by the mathematician—and *spread*, which can be thought as a collection of choice sequences and thus as defining intervals. The crucial difference between intuitionistic and "modern" mathematics is that the latter regards any sequence as given and fully determined (atemporally, so to speak), while intuitionism insists on the idea that such things are determined by the mathematician (and not given in themselves), and underscores the temporality of human action and choices.[61]

---

[60] The Schrödinger equation depicts a continuous process that many people tend to interpret as representing what happens inside atoms; although this is not the only possible interpretation, the point is that even the physics of quantum discontinuities employs continuum-based representations.

[61] Notice that a "choice function" in the sense of set theory's Axiom of Choice is likewise an atemporally given object, a thing in itself, conceived as if it exists independently of the mathematician. Thus to speak of "choice" was not really a good terminological choice; the usual connotations of the word are better preserved in Brouwer's "choice sequences."

A choice sequence may be partly determined by a general law and partly by free choice, and it is only fully given up to the point that it has been precisely given. If the infinite sequence is fully determined by a law, then it has been given in principle; at the other end of the spectrum, we can think of arbitrary, free-choice sequences, and here temporality applies strongly. Real numbers (points on the continuum) are identified by Brouwer with choice sequences satisfying certain conditions, and one cannot assume that every real number has an infinite decimal expansion. Other differences with the classical Cantor-Dedekind continuum that Brouwer emphasized are the following: the intuitionistic continuum is not ordered by the usual less-than relation, and his real numbers are not separable.[62] An interesting edifice of intuitionistic analysis can be erected on this basis, as Brouwer explained in detail, but it deviates greatly from classical analysis.[63]

As I said, nothing ties a fully autonomous mathematics to the classical Cantor-Dedekind-Hilbert structure of $\mathbb{R}$. Brouwer's continuum must strike us as a very natural, and deep, way of pursuing the notion of a fully independent development of pure mathematics. But then, why has it not been taken up by the community of "pure" mathematicians?

I believe the explanation for this noteworthy phenomenon cannot be established purely and a priori (as the puristic image of a fully autonomous math would require), but only upon consideration of the historical configuration of mathematics around 1900. By that time, analysis formed the main bulk of mathematics, and mathematical analysis—whether we pay attention to how it was employed by physicists, or to its modern developments around 1900—was (and remains) based on the pointillistic continuum. In fact, the high time in the development of analysis and its core applications to physics, roughly 1660 to 1860, was a period when the image of mathematics was colored by the applied: Fourier represents the usual views then, Jacobi does not; math was regarded as *a science* (not merely a language) at the core of natural science. One would be hard pressed to classify the work of any of the key

---

[62] The classical set $\mathbb{R}$ is separable in the sense that it has a countable subset $\mathbb{Q}$ such that between any two real numbers there is a member of $\mathbb{Q}$; this does not apply to the intuitionistic system $\mathfrak{R}$. On all of these matters, see, e.g., the classic Heyting 1956, or Troelstra and van Dalen 1988. See also van Atten 2007.

[63] Most famous is the continuity theorem (Brouwer 1927), which states that every total function $f\colon [0,1] \to \mathfrak{R}$ is continuous and even uniformly continuous.

figures, from Newton to Laplace, purely under the label of mathematics or of physics, if we understand these words in their post-1850s sense.

Interestingly, Brouwer's attitudes to science and technology seem to offer a confirmation of my thesis concerning the link between modern math and natural science. In his work he always underscored the aim to reconceive mathematics as a pure product of the mind, fully autonomous and independent; "intuitionistic mathematics is inner architecture," as he beautifully said in (1948, 96). Unlike most mathematicians, he was also explicit—particularly in the 1900s—in *rejecting* modern technology and the associated forms of natural science, deploring the influence they have exerted upon mathematics. This was because, in his view, the aim of technology is to dominate nature, thus following intrinsically negative values; he also regarded natural science as deeply contaminated by such values since its beginnings in the Renaissance and in the early modern period. As a result, Brouwer could only consider mathematics as an activity worthy of a healthy human being when it was freed of such negative influences, developed fully autonomously and independently from the so-called applications.[64]

---

[64] See Van Dalen's biography, and very especially Brouwer's book *Life, Art and Mysticism* (1905).

9

Objectivity in Mathematical Knowledge

The celebrated objectivity of mathematical results led many authors to believe that the theorems of mathematics are apodictic, necessary, a priori. From the beginning, we have remained uncommitted to this aprioristic view of mathematics, moreover we have defended the position that advanced mathematics is marked by the presence of hypotheses at its roots. How can the idea of objectivity be rescued in this setting? Precisely by considering the interplay of knowledge and practices that takes place in mathematics. Let us see how to make this work.

I have argued that a mathematician (either a beginning student or a master) possesses working knowledge of several different practices and of their systematic interconnections; this involves interplay of different strata of mathematical knowledge. Our understanding of sophisticated abstract geometries makes contact with the more intuitive three-dimensional geometry of Euclid, and this in turn with practical geometry on the plane. Advanced mathematical knowledge is placed in a complex network of intertwined knowledge and practices; it is linked systematically with the elementary, and thus with its strong cognitive roots. *This establishes links that restrict the admissible,* guiding the formation of new concepts and the adoption of new principles, leading to objective results.

Although I have not emphasized this point in the previous chapter, we have already seen an example: the completeness of the real number system is a highly subtle and sophisticated property, a hypothetical principle for which we have no guarantee arising from our knowledge of the world. This principle can be seen to have roots in the idealizations of ancient Greek geometry—its depthless planes which have "length and breadth only," its lines "without breadth" and points "without parts"—but it is also true that it never acquired prominence before the advent of the calculus. As long as geometrical reasoning remained constructive, in the sense of ancient geometry, there was no need to demand completeness; when such reasoning became involved with infinite processes (indivisibles or infinitesimals, infinite series, continued fractions, limits) in the new calculus or analysis, ensuring completeness became a pressing issue.

The simultaneous presence of hypotheses and constraints in mathematics has consequences for some traditional riddles. It has often been asked whether mathematical concepts are invented or discovered; while much of the number system appears intuitively as a matter of discovery, it is hard to admit that the same is true for axiomatic set theory. As we shall see, the processes that drive objectivity in mathematics are such that one should rather speak of invention *cum* discovery: it is the case that mathematicians freely adopt hypotheses and insofar invent collectively, but at the same time they are forced to admit certain features of such propositions, or certain consequences or related results, due to the interplay between knowledge and practices. That is, due to the systematic interconnections with previous layers of mathematics and the constraints they impose. Stated differently, the invention/discovery duality is too simplistic for adequately analyzing the processes of knowledge development that we find at work in mathematics. This is an observation that accords very well with the tradition of pragmatism.

Similar remarks apply to the question whether mathematical principles and results are contingent or necessary. While the natural numbers seem to have been a matter of necessity (and in my opinion the same applies to parts of elementary geometry), one can hardly accept that the same may be true for sophisticated structures such as Polish spaces, or even, as we have seen, complete Archimedian ordered fields ($\mathbb{R}$). In fact, the processes that we shall now review warrant speech of contingency cum objectivity: confronted with certain hypotheses, the mathematician is by no means forced to admit them (e.g., constructivists and predicativists do not admit $\mathbb{R}$ in its classical interpretation as part of pure mathematics); but once he or she admits, certain features and results are enforced. To see how this comes about, we shall devote this chapter to a case study exploring some of the basic axioms of set theory.

How can there be intersubjective communication regarding the infinite, the arbitrarily large? This is one of the main questions we shall answer in the early sections of this chapter. The last two sections will explore more delicate ramifications of the main issue in advanced contexts: Cantor's ordinal numbers and the Continuum Hypothesis.

9.1. OBJECTIVITY AND MATHEMATICAL HYPOTHESES: A SIMPLE CASE

Although there are many other topics of interest in mathematical practice, our current question leads us to becoming particularly interested in how entirely new levels of knowledge are developed in the practice of advanced mathematics. Situations in which the mathematician detects new axioms, and in which they become generally accepted, are by no means run-of-the-mill; but they are the cases that pose a most difficult problem for an approach to the objectivity of mathematics like the one I am proposing. More commonly the mathematician will find a solution to a problem, perhaps by establishing a new method (think of Fourier and his new way of handling trigonometric series), and then a critical task will emerge of subjecting the solution to criticism, perhaps to finding counterexamples, checking the new method and its range of validity, establishing rigorously the validity of the method for certain types of cases, and so on.[1] In such situations, the underlying principles will be relatively stable and shared, even if they may be calling for analysis or refinement (the study of Fourier series was perceived as related to some deep conceptual difficulties in the principles of the calculus, e.g., in the treatment of integrals).

How objective results are attained on the basis of shared and stable principles is a matter that has been satisfactorily studied by logicians. One very important form of objectivity we find in math is of the form "this follows from that" ($\Sigma \vdash p$, $\Sigma \Rightarrow p$), and this is something we understand very well thanks to the impressive development of modern logic. Although some philosophers interested in the history of mathematics have claimed that there are no ultimate standards of rigor and there

---

[1] This is just an example: mathematicians may also detect new relevant structures and devote themselves to their analysis, perhaps by adjusting to them methods inspired by previous fields. An inventory of the full range of different situations should be expected from the current trend of work in the analysis of mathematical practices.

never will be (Lakatos was a prominent example), I dare say that we *do* possess ultimate standards for all of those cases when proofs in a theory are satisfactorily modelled by formal logic. This happens with first-order systems for the theories of the natural numbers, elementary geometry, the real numbers, and set theory—hence derivatively also for group theory, Galois theory, topological spaces, and function spaces— essentially all the mathematics normally studied at university level. This important topic is rather well known and I have nothing essentially new to add to it,[2] which is why I avoid going into more details. But there is more to mathematical objectivity than merely $\Sigma \vdash p$, "this follows from that." (As I have already explained, the approach we are exploring goes far beyond if-thenism; see Chapter 6 section 1.)

Beyond objectivity based on shared principles, there are subtler forms of objectivity—unexplored until today, as far as I know—and the most prominent example I can give is objective features in the introduction of basic mathematical hypotheses. Consider the infinitarian hypotheses that lie at the basis of set theory. As we have seen (Chapter 8 section 5), there is at least one important distinction to make here, the contrast between simple infinity and arbitrary infinity. To the extent that this distinction is reflected in the formal system of the Zermelo-Fraenkel axiomatic set theory (see Ferreirós 2011), it can be found in the axioms of Infinity and Choice, and it is linked with the interpretation of the axiom of Powerset. In more colorful terms, $\mathbb{N}$ represents the *tame* infinite, while $\wp(\mathbb{N})$ in its standard interpretation represents the *wild* infinite; in fact, the issue of Powersets is linked with the most difficult unsolved questions in set theory, beginning with the Continuum Hypothesis.

In what follows, I shall take for granted that sets are defined (analytically) by two key properties: *objecthood* and *extensionality*. When we say that T is a set, we mean that T as a whole is as much a mathematical object as its elements (for instance, $\mathbb{N}$ is no less a mathematical entity than 3), and as a result we imply that T can be an element of

---

[2] I provide clarifications about how to understand it properly, touching on the philosophy of logic and the proper way to understand the link between formal logical systems and informal mathematics, in Ferreirós (2001), (2011), and (forthcoming).

a set.[3] This notion of objecthood was underscored by Dedekind, and accepted also by Cantor; in this connection Russell spoke of being a class-as-one. Interestingly, that key idea is not frequently emphasized by mathematicians when explaining set theory, simply because there is no "axiom of objecthood," but the idea is central nonetheless; as we have seen (Chapter 6 section 4.2), we set it in place by employing first-order logic as the setting for our theory of sets. (Let me add that it would be a good practice, in informal presentations of set theory, to actually add an "axiom of objecthood" to the other ZFC axioms, because otherwise a key point is lost due simply to most students' lack of understanding of logical subtleties.) Extensionality is very well known, and it provides identity conditions for sets—T and S can be different only if they do not have the same elements.

Considering the matter abstractly, accepting the postulate that all natural numbers form an actually infinite collection, the set $\mathbb{N}$, is totally independent from accepting that all arbitrary subsets of $\mathbb{N}$, definable or not, are given—let alone whether or not they form a set $\wp(\mathbb{N})$. Admitting a "simply infinite" set $\mathbb{N}$ does have implications, as I have underscored (Chapter 8 section 5), due to the systematic interconnections between the new postulate and a large previous body of theory regarding the natural numbers. This body of theory has concepts for odd and even numbers, prime numbers, etc., since it allows us to explicitly define properties such as *being prime*. As a result, on pain of incoherence, admitting that $\mathbb{N}$ is a set one should admit that the primes, the odd numbers, and other *definable* kinds of numbers form sets. (Since all of these are definable sets, and, moreover, countable, I think it is entirely reasonable to keep talking of simple infinity; in fact, the same applies to $\mathbb{Q}$.) The incoherence I am talking about is not logical inconsistency, but more of a conceptual kind: to sustain adoption of $\mathbb{N}$ and rejection of the set of prime numbers, one would need to develop some theory about differences between both kinds of number, or between one way of "aggregating" into a set and another. Here, I'll just leave it at

---

[3] Readers probably know that this is essential for the emergence of the paradoxes, and that one can avoid the paradoxes by declaring certain problematic collections to be *classes* in the sense of Bernays (which cannot be elements of sets) instead of sets. Let me add that I use the term "objecthood" without any special ontological connotation, to mean simply that sets are treated as mathematical objects.

that, challenging readers to try to offer theoretical justification for such distinctions if they think they are reasonable or even plausible.

Granted that point, I can give a very simple example of the way in which new postulates, coupled with previous knowledge, lead to objective results. Once we accept $\mathbb{N}$ and the sets E of even numbers, S of squares, or P of primes as new mathematical objects, the established theory of the natural numbers also supplies *correspondences* between them. The arithmetical operation of doubling provides a function, $x \mapsto 2x$, that establishes a one-to-one correspondence between $\mathbb{N}$ and the subset E, and the operation of squaring provides the function $x \mapsto x^2$ that does the same with $\mathbb{N}$ and S. Also, the famous theorem on the infinity of primes can easily be turned into a method for enumerating successive primes; hence a one-to-one correspondence between $\mathbb{N}$ and subset P. Infinite sets can be put in one-to-one correspondence with their subsets; we obtain the "Galileo paradox," a result accepted by Cantor and turned by Dedekind into the very definition of infinite set.[4] Invention or discovery? Neither, strictly speaking, but invention *cum* discovery.

Let me put it this way: it is a discovery about the implications of inventing $\mathbb{N}$, in the context of previously accepted mathematics.

Do all definable subsets of $\mathbb{N}$, taken collectively as a class, form a set in their turn? This is already an independent question from that of the existence of $\mathbb{N}$ conceptually; of course, it is also formally independent from the Axiom of Infinity. And even more so is the question of whether all subsets of $\mathbb{N}$, *definable or not*, form a set $\wp(\mathbb{N})$. To begin with, we should regard as open the question whether any nondefinable subset of $\mathbb{N}$ exists at all! But if things are this way when one considers the matter abstractly, in complete independence from any other consideration of mathematical (or physical) theories and results, then how could Dedekind, Cantor, Zermelo, and many others agree in admitting both $\mathbb{N}$ and $\wp(\mathbb{N})$?

The answer is simple: because they were *not* approaching the matter abstractly, and consideration of other mathematical (and physical)[5]

---

[4] Dedekind's idea of turning that feature into the axiomatic definition of infinite sets materialized in 1872 at the latest, independently of Cantor's work. See Ferreirós 1999, 108–109.

[5] In the case of Cantor, physical and biological considerations were prominent; see Ferreirós (2004).

theories and results played an essential role in their work. They developed the new theory on the basis of *systematic interconnections* with previous knowledge and practices, so the interplay that I am claiming was clearly at work. As a matter of fact, simple infinity and arbitrary infinity entered the history of mathematics together and at the same time, in the year 1872 (Chapter 8 section 5), the reason being that they appeared in the context of a new analysis of real numbers—and in the wider context of further considerations where sets of real or complex numbers were prominent.[6]

9.2. CANTOR'S "PURELY ARITHMETICAL" PROOFS

One clear indication of the interplay at work was Cantor's concern, during the early years of his exploration of actual infinity, to provide what he called "purely arithmetical" proofs of his results. This is a constant worry in his correspondence with Dedekind, whom he relies on as the sparring partner who can detect weaknesses in the proofs. The whole period from 1874 to 1884 was characterized by that concern, which has precise and concrete methodological implications. Cantor develops new set-theoretic results, but the methods employed in his proofs belong to the more traditional style of mathematics developed in analysis and algebra by Gauss, Weierstrass, and others. Much later (especially in the *Beiträge* of 1895 and 1897) one will find characteristic new set-theoretic methods being employed, but this is not the case in the first decade of Cantor's pathbreaking work.

Let us see a couple of examples. Consider the proofs in the epoch-making paper "On a property of the collection of all real algebraic numbers," a joint work by Cantor and Dedekind (Cantor 1874; see Ferreirós 1999, 176–186). This contained two theorems, a proof that the set of all (real) algebraic numbers is countable and a proof that the set of all real numbers is uncountable. Both results were proved

---

[6] Point sets and their topology in Cantor's work on the uniqueness of representation by trigonometric series; number sets in Dedekind's algebraic and number-theoretic work, plus considerations of topology that he developed in connection with work of Riemann and Dirichlet.

by traditional methods, which Cantor regarded as "purely arithmeti-cal": the former result depended on consideration of the polynomials associated with algebraic numbers and the finite number of their solu-tions (strictly speaking, a weaker proposition than the Fundamental Theorem of Algebra). Let me add that this result may seem trivial to you today once you have taken a course in set theory; but typically such results are highly nontrivial when a wholly new way of practicing math-ematics is being developed, as was the case in 1874.

The other theorem, the first nondenumerability result, was proved by exploiting the Bolzano-Weierstrass principle of completeness, which establishes the existence of at least one real number in the intersec-tion of any sequence of nested, closed intervals.[7] Cantor shows how to extract a sequence of nested, closed intervals from any given enu-meration of real numbers—and from this, by the Bolzano-Weierstrass principle, he concludes the existence of $r \in \mathbb{R}$, which is not in the given enumeration. Notice that the 1874 proof is not the famous one by diagonalization. In a sense, this first proof is more interesting, since it exploits basic topological features of $\mathbb{R}$ and does not depend on a formal trick like the later, better-known, proof does. To sum up, one theorem in the 1874 paper exploits basic algebraic facts about the alge-braic numbers, while the other depends on characteristic methods of the new *arithmetized* analysis developed by Weierstrass.

The same emphasis on "purely arithmetical" methods can be found again in the surprising 1878 result that $\mathbb{R}$ and $\mathbb{R}^n$ have the same cardi-nality, i.e., can be put in one-to-one correspondence. In his paper, Can-tor (1878) even avoided more abstract set-theoretic language, to employ exclusively—and somewhat contortedly—the language of variables usual in analysis.[8] His methods relied on very simple manipulations of numbers in decimal representation, on the use of continued frac-tion representations, and on simple ideas about one-to-one functions. Cantor's concern with "arithmetically rigorous" proof procedures, with reducing things to "purely arithmetical form," is clearly expressed in

---

[7] This is sometimes called Cantor's principle, in acknowledgement that he published it in 1872, but it is perfectly well established that Weierstrass dealt with it carefully in his lectures (including the ones Cantor followed, 1863/64) and also that Weierstrass himself referred to Bolzano's famous 1817 paper which again contains it.

[8] See Ferreirós 1999, pp. 190, 193.

the letter of June 20, 1877, where he discusses this matter with Dedekind before publication (see Ewald 1996, volume 2, 853).

The meaning of "purely arithmetical" in Cantor's papers and letters of the 1870s accords very well with the manner of speech employed by Weierstrass and other members of the Berlin school (and with Dedekind's too). In a lecture course of 1874, Weierstrass stated that "the main difficulties in analysis" come "precisely from a vague and insufficiently comprehensive presentation of the basic arithmetical concepts and operations."[9] Cantor expressed the same idea when he remarked in 1882:

> Most of the difficulties of principle that are found in mathematics have their origins, it seems to me, in misjudging the possibility of a purely arithmetical theory of magnitudes and manifolds [*rein arithmetischen Grössen- und Mannigfaltigkeitslehre*]. (Cantor 1879/84, 156).[10]

Of course, we find here an addition of great interest to us: Cantor not only wants an "arithmetical theory" of magnitudes (i.e., of real and complex numbers, which was the goal of Weierstrass, with exactly the same terminology), but also such a theory of sets or "manifolds"—which is the point we are discussing.

Certain infinitary processes were taken to be purely arithmetical, such as the basic ones underlying analysis: infinite sequences, infinite series, and functional correspondences. Obviously, without this it would never have been possible to speak of an *arithmetization* of analysis. (Similarly, Dedekind says often that his theory of the real numbers is "purely arithmetical"—despite the fact that it relies on infinite sets, the cuts.[11]) One consequence of this standpoint is important: the system of real numbers was considered to be as much *arithmetical* and *pure* as the natural

---

[9] From the Hettner transcription (Weierstrass 1974, 78) of lectures given in 1874. Quoted also in Dugac 1973, 77.

[10] There he also emphasizes the "standpoint of a purely *arithmetical analysis*." Already, years before, in 1874, he had criticized others (Klein!) for not understanding the core problems of arithmetical analysis (letter to Dedekind, Jan. 1874).

[11] See the preface to Dedekind (1872), and the first preface of (1888), where he even speaks of the "purely arithmetical phenomenon of the cut." Cf. also the famous cry for autonomy and purity: "I demand that arithmetic shall be developed out of itself" (Ewald 1996, 771; see also 793–794).

numbers themselves. In fact, the same applied to the complex numbers for this community of mathematicians; with or without connection to the ideal of arithmetization, such a conviction was rather common: in the 1810s Gauss had already claimed "full citizenship" for the complex numbers, on equal standing with the naturals and the reals;[12] and Riemann agreed wholeheartedly, as he expressed in the first part of his dissertation on trigonometric series (Riemann 1854/1868).

For such historically conditioned reasons, the step of regarding the natural numbers as a completed whole, an actually infinite set, led quite immediately to a similar standpoint with regard to the real (and complex) numbers. The invention of $\mathbb{N}$ had to be accompanied by the invention of $\mathbb{R}$; naturally, this was also supported by the pointillist conception of geometric lines: the *line* is a mathematical object, just like the points that make it up.

As a matter of fact, with the set of naturals, many other sets were introduced: the integers, the rationals, the irrationals $\mathbb{R}\backslash\mathbb{Q}$, the set $\mathbb{A}$ of algebraic numbers, the set of transcendental numbers $\mathbb{R}\backslash\mathbb{A}$ (or $\mathbb{C}\backslash\mathbb{A}$),[13] the complex numbers $\mathbb{C}$. A good example of all this is Dedekind's work on algebraic number theory, published in 1871, where he developed his theory of ideals against the basic background image of the whole lattice of number fields that lie between the rationals $\mathbb{Q}$ and the field of complex numbers $\mathbb{C}$ (see Dedekind 1871).

And, once again, invention came accompanied by discovery. As I have reviewed, it was by applying traditional methods, already well established in mathematics as of 1860, that Cantor and Dedekind established the countability of the algebraic numbers $\mathbb{A}$ and the uncountability of the reals $\mathbb{R}$. To be sure, the problem Cantor posed was very new and involved a new concept (introduced explicitly in 1878: do $\mathbb{N}$, $\mathbb{A}$, and $\mathbb{R}$ have the same *power*, the same *cardinality*?), but the methods that allowed him to solve it formed a part of previous knowledge and practices. To emphasize the point, let me add this: there is good reason to think that the results would have been established even if Cantor had never lived.

---

[12] See p. 107n. His number theory for complex integers $a + bi$, with a, $b \in \mathbb{Z}$, strongly underscored the point.

[13] \ represents the set-theoretic difference: $\mathbb{R}\backslash\mathbb{A}$ is the set that remains from the reals when you take out the algebraic numbers, that is to say, the transcendental numbers. Obviously, $\mathbb{R}\backslash\mathbb{A} \cup \mathbb{A} = \mathbb{R}$.

Developments could have been much slower, and we may doubt whether the transfinite ordinals would have been introduced, but I do not see reasons to have similar doubts concerning the results in Cantor (1874).

It should now be clear to readers why I insist on the idea that the complex network of intertwined knowledge and practices causes links that restrict the admissible, leading to objective results, and guiding both the formation of new concepts and the adoption of new principles. *Even the introduction of radically new concepts and standpoints in mathematics is strongly constrained.* From our viewpoint, the legendary objectivity of mathematics is, more than anything, a strongly constrained form of intersubjectivity. The constraints range from roots in basic human cognition, at the elementary level, to intra-mathematical connections, and extend to include links with the sciences, with technology, and even with the arts.

9.3. OBJECTIVITY AND HYPOTHESES, II: THE CASE OF $\wp(\mathbb{N})$

Defining the real numbers on the basis of the ordered field $\mathbb{Q}$, Cantor relied on taking as given the totality of fundamental sequences of rational numbers, while Dedekind relied on the totality of cuts on the rationals. In the first case one needs to admit arbitrary (possibly nondefinable) sequences; in the second, one needs arbitrary (possibly nondefinable) cuts. But this postulation remained fuzzy in their work of 1872, a matter of hypothesizing that "all possible" sequences of rationals or subsets of $\mathbb{Q}$ are given, perhaps as elements of a new set. Indeed, the Dedekind cuts are all elements of $\wp(\mathbb{Q})$; although they all have a characteristic property that may be called *downward-closed*,[14] only some Dedekind cuts can be defined explicitly by properties in the language of rational-number arithmetic.

It must be relevant that neither Dedekind nor Cantor went on to formulate explicitly the Powerset principle, at least not in print. The

---

[14] I restrict my attention here to lower cuts, since it is well known that the upper cut is definable as $\mathbb{Q}\backslash C$, the complement of C. A cut C in $\mathbb{Q}$ is *downward-closed* in this sense: C is always such that $q \in C$ and $p < q$ imply $p \in C$.

idea was left somewhat vague, in the form of a practically crucial but implicit, unstated guiding principle, and it was the next generation of set theorists (especially Russell and Zermelo) who made it explicit.[15] At any rate, the theories of the real numbers presupposed full powersets; indeed, the axiom's necessity for understanding the continuum as a point set offers the strongest case in its favor. (There are other points in the works of Cantor and Dedekind related to Powerset, or at least to arbitrary subsets: Cantor's studies of point sets proceeded on the assumption of arbitrary sets of reals, and the same applies to Dedekind's work on ideals (infinite sets of complex integers), where he routinely assumed being given all possible ideals on any ring of integers. Cantor's work on ordinals and well-ordering seems also worth mentioning here, for, given any set, Cantor regards as given all possible well-orderings on it.)[16]

Retrospectively, and going further, the roots of this assumption (I mean the hypothetical principle Powerset in its standard, quasi-combinatorial interpretation) can be found in traditional views about decimal expansions. After the introduction of decimal expansions, it became clear that the different lengths within the unit interval (indeed, within any interval of unit length) can be associated with *all possible* decimal expansions after the colon, i.e., all possible assignments of ciphers to the finite ordinal numbers.

I have expressed myself with some care, writing "associated" intentionally, to avoid the modern idea of an *identification* of real numbers with infinite decimal expansions. From a methodological standpoint, this is an important distinction. Notice the following: it is one thing to admit that, *given* a point on a line, an infinite decimal expansion (possibly nonperiodic, i.e., irrational) is determined or can be produced; and it is quite another thing to *define* the real numbers in the interval (0,1) as all the possible infinite decimal expansions

---

[15] Russell (1903) and Zermelo (1904) and (1908). Cantor made the principle explicit in a letter to Hilbert (formulated as "The multiplicity of *all the subsets* of an available set is an available set," in an obscure context however), only to express doubts in the next letter (see Meschkowski and Nilson 1991, 396–398 and Ferreirós 1999, 447–448).

[16] Finally, some reconstructions of his celebrated categoricity argument in Dedekind (1888) assume that it relies on Powerset (Reck and Awodey 2002), although the matter is contentious (see, e.g., Parsons 2009).

$$0, c_1 c_2 c_3 \ldots c_i \ldots,$$

where the c_i are ciphers in decimal notation (0 to 9). The first principle was traditional, admitted since about 1000 (in Europe since about 1600),[17] but the second approach is typically modern and infinitarian. Within the first approach, questions of existence in mathematics can still be conceived in agreement with Euclid's *Elements*—points given by explicit diagrammatic constructions; see Chapter 5—and decimal expansions can be regarded as numerical approximations to such entities.

The second viewpoint is detached from such constructive and/or intuitive geometric underpinnings, and it necessitates the introduction of an infinitarian standpoint and abstract principles of existence. This modern approach originated around 1700 and was only made fully explicit in the second half of the nineteenth century. It emerged within the context of the new methodologies of pure mathematics, modern analysis, and algebra, and their arithmetization.

Let me employ the (Dirichlet) notion of arbitrary functions to make more explicit and clear this modern conception of the connection between real numbers and decimal expansions.[18] The real numbers in a unit interval can be understood in terms of sequences of the ten ciphers, and thus in terms of *all* functions

$$f \colon \mathbb{N} \to \{0, 1, 2, 3, 4, 5, 6, 7, 8, 9\}.{}^{19}$$

This standpoint suggests that the secret of the continuum will be found in arbitrary functions, in the arbitrary sequences of ciphers, in the idea of *all possible* assignments.

Of course, instead of the ten digits of decimal notation, one can just as well employ any finite number of digits—in particular, ternary or

---

[17] As we saw in Chapter 8 section 1, decimal expansions were used as early as the tenth century by al-Uqlidisi, by the Persian mathematician al-Kashi about 1400, and influentially presented by Simon Stevin around 1600.

[18] It is well known that Dirichlet introduced this notion in 1829 and 1837. See, e.g., Ferreirós 1999, 148.

[19] Obviously, one has to introduce a convention in order to exclude the possibility that two different functions correspond to the same number (corresponding to the identification 0.1000. . . = 0.0999. . .). From a more abstract point of view, one may avoid this distinction, leaving behind the topology of $\mathbb{R}$, which leads to the fundamental Baire space of descriptive set theory.

binary expansions. And if instead of $f: \mathbb{N} \to \{0, 1, 2, 3, 4, 5, 6, 7, 8, 9\}$ we consider the idea of binary notation,

$$f: \mathbb{N} \to \{0, 1\},$$

it becomes obvious that one can study the continuum via two-valued functions on $\mathbb{N}$.

These functions in turn can be seen to code subsets of $\mathbb{N}$, each one of them being the characteristic function of a subset S (say that $f(n) = 1$ means $n \in S$). Thus, we come to conclude that the "secret" of the continuum is nothing but the secret of the powerset $\wp(\mathbb{N})$. (Cantor saw clearly at least the first assertion, and understood fully clearly how to generalize to any set S whatever simply by considering $f: S \to \{0, 1\}$.[20]) The question was, then, that of studying "all possible" subsets of $\mathbb{N}$, or, in general, of S, however large the cardinality of S; that is to say, the question was that of *understanding arbitrary subsets*.

Incidentally, let me add at this point that Cantor's fundamental sequences can be regarded as elements of the set $\{f: \mathbb{N} \to \mathbb{Q}\}$ of all functions $f(x)$ that assign rational values to natural numbers. This, however, is a viewpoint that was not proposed by Cantor in his related work.

A historical annotation is in order. The rethinking of the real numbers by means of Dirichletian arbitrary functions is not a rational reconstruction of my own imagination; and not merely because the corresponding realization must have been lurking in the background, in particular among experts in arithmetic methods, such as Cantor and Weierstrass, well acquainted with continued fractions, power series, and, of course, digital expansions (decimal, binary, ternary). A well-known paper by Cantor [1892] and a manuscript of Dedekind [1891] are proof that the founding fathers of set theory reconceived the real number system exactly along these lines. Dedekind's short piece, published posthumously, is noteworthy because it introduces Baire space (in the sense of descriptive set theory, not the usual sense of topology) eight years before René Baire's dissertation [1899]. Working directly upon the structure $<N, 1, \sigma>$ of natural numbers defined in Dedekind

---

[20] Cantor 1892. His considerations, however, were complicated by doubts about the acceptability of $\wp(S)$ given S; see the references in a previous footnote. Especially noteworthy is the letter to Hilbert of Oct. 12, 1898, where Cantor ends up calling the Powerset principle "illusory" (Meschkowski and Nilson 1991, 396–398).

[1888], he considers the "continuous" set of all mappings $\varphi: \mathbb{N} \to \mathbb{N}$. That constitutes a substantial move of liberation from traditional ideas about the number system and its so-called "genetic construction," toward a more purely set-theoretic approach to the continuum.[21]

As one can see, ideas about arbitrary functions, arbitrary real numbers, and arbitrary sets were a crucial background to the emergence of set theory—from Dedekind and Cantor, to Baire and Lebesgue, and Zermelo and Gödel, and beyond. Under the influence of such ideas, the motivating principle of accepting totalities of arbitrary subsets became a central element of set-theoretic practice. This crucial principle would be the main bone of contention for critics of set theory in the decade of 1900, despite the fact that it was not made explicit by either Dedekind or Cantor. The Axiom of Choice, suggested by Erhard Schmidt and put to use by Zermelo (1904), made the principle surface more clearly.

9.4. ARBITRARY SETS AND CHOICE

Gödel's iterative conception presupposes the notion of arbitrary set, the quasi-combinatorial conception of sets. This is natural, because the iterative conception is meant to be a more or less intuitive account of sets in the tradition of Cantor and Dedekind, and the idea of arbitrary sets was a crucial motivation therein. Indeed, the tension between definable sets and arbitrary sets has been one of the central motives in the whole development of set theory. It would be easy to exhibit relevant passages from the works of almost any of the main authors—Dedekind, Cantor, Borel, Baire, Lebesgue, Zermelo, Hausdorff, Luzin, Gödel, Cohen, and so on—in the history of set theory.

Consider some sequence of numbers, the Fibonacci numbers to take a very simple example:

$$1, 1, 2, 3, 5, 8, 13, 21, \ldots ;$$

this is a *definable* sequence, generated by the simple rule $a_n = a_{n-1} + a_{n-2}$. The sequence of squares, 1, 4, 9, 16, . . . , is also definable and indeed in more than one way (the *nth* term a_n is not only the square of the *nth*

---

[21] A historical paper on this topic is in preparation.

natural number, but also the $(n-1)th$ term plus the $(n-1)th$ odd number). Math textbooks are filled with more complex examples of definable sequences. But in mathematics one often argues by saying something like "consider a monotonically increasing sequence of numbers . . .," where *any* such sequence is meant, and we reason about an *arbitrary* sequence. The same happens with regard to sets and functions.

In fact, when we talk about subsets of $\mathbb{N}$, we do not mean only definable subsets of $\mathbb{N}$, but any subset, definable or not. This is the idea of *arbitrary* sets of natural numbers, and the quasi-combinatorial viewpoint consists in admitting that all possible subsets (definable or not) of any given set are given. With S comes $\wp(S)$, and this is interpreted to be not just the collection of all subsets of S determined somehow,[22] but the collection of arbitrary subsets of S, the set of "all possible" subsets. In other words, the *quasi-combinatorial standpoint* involves complete rejection of the view that sets have to be defined, specified, or described by formal conditions or formulas or explicit processes.

Notice that it is one thing to reason about the assumption of an arbitrary function $f : \mathbb{R} \rightarrow \mathbb{R}$ having this or that characteristic (say, continuity or Lebesgue measurability), as Dirichlet started doing in 1830, and quite another to postulate the *existence of a complete totality* of such arbitrary real-valued functions, as Dedekind and Cantor were doing in the 1880s. Critics have seen here the original sin of set-theoretic mathematics.

The Axiom of Choice, AC, is the most direct embodiment of this quasi-combinatorial standpoint. Suppose we have partitioned $\mathbb{R}$ into denumerably many disjoint subsets $A_0, A_1, A_2, \ldots, A_n$, for all $n \in \mathbb{N}$; it is intuitively "evident," Zermelo (1908b) would say, that there is a set $S \subseteq \mathbb{R}$ of the form $S = \{ a_0, a_1, a_2, \ldots a_n, \ldots \}$ having one and only one element in common with each of the parts, $a_n \in A_n$. The Axiom of Choice simply extends this idea to the general case of an arbitrary family of sets. It is self-evident, I would reply to Zermelo, *once* you have understood *and* accepted the quasi-combinatorial viewpoint. The fact that you can understand it while rejecting it is simply proved by the statements of

---

[22] For example, by first-order formulas in the language of set theory with certain parameters (which may give rise to constructible subsets, or to HOD sets), or perhaps by those given in a certain model of the system we are working with.

many critics of AC, first-rate mathematicians such as Baire, Lebesgue, Weyl, and so on (some are mentioned below).

A set of numbers is conceived as if it were "the result of infinitely many independent acts," deciding for each number whether it should be included or excluded (Bernays 1935, 259). It does not matter how exactly we determine or circumscribe the set, for instance, it does not matter whether we violate the "vicious circle principle" established by Poincaré and Russell;[23] for,

> After all, such an object is not created through such a "determination"; rather, every object can be determined in a wide variety of ways. (Zermelo 1908b, 191)

From the point of view of set theory, concrete definitions or descriptions of specific sets (or functions, or sequences) "are only ways to pick out an object which exists [or is conceived to exist, JF] independently of, and prior to, the construction" (Bernays 1935, 260).

But, of course, some mathematicians insisted on the need to define or describe *explicitly* every infinite set introduced in a mathematical argument. In 1905, Borel emphasized that Zermelo's Well-ordering theorem established the equivalence of two problems: well-ordering a set M and being able to determine a distinguished element for each of its subsets $M' \subseteq M$. But the second he regarded as an open problem, to be investigated in each particular case, and one of the "most arduous" problems in the case of the continuum (Borel 1905, 194). He went on to propose a restriction to the *countable* version of AC,[24] saying that there are serious objections "to any argument in which one assumes an *arbitrary choice* made a nondenumerable infinity of times; such arguments are outside the domain of mathematics" (op. cit. 195).

Notice that Borel was a central figure in the development of modern theories of integration, and a pioneer in descriptive set theory. It was precisely out of such worries that the development of descriptive set theory began, and it became one of the most important branches of set

---

[23] The vicious circle principle forbids impredicative definitions. See Poincaré 1908, Whitehead and Russell 1910, and Weyl 1918.

[24] This is sometimes denoted by AC_ω; it says that there is a choice set for any countable family of sets.

theory in the twentieth century. Descriptive set theory restricts attention to sets of points (in $\mathbb{R}$ or $\mathbb{R}^n$)[25] that can be determined by quite simple procedures—taking complements, countable unions, and projections—from basic sets, such as open or closed sets. Mathematicians of a constructivist bent, like Baire, Lebesgue, Borel, and Luzin, were prominent among the first generations of descriptive set theorists who solved the main question for the hierarchy of Borel sets and for the analytic sets, while at the same time making clear the enormous difficulties confronting the study of other projective sets.

Some even went further than Borel, like Hermann Weyl, who required explicit definition or description for all infinite sets, regardless of their cardinality:

> No one can describe an infinite set other than by indicating properties characteristic of the elements of the set; no one can establish a coordination between infinitely many things without offering a *law*, . . . The notion that an infinite set is an "aggregate" [*Versammlung*] brought together by infinitely many individual arbitrary acts of selection . . . is nonsensical; "inexhaustibility" is essential to the infinite. (Weyl 1918, 15; 1987, 23 revised).

The situation is different with finite sets, which can be determined by explicit naming of its elements, or else by characteristic properties that apply only to the elements of the set; with infinite sets only the second way is possible (op. cit., 13). This, of course, is at odds with the Dedekind-Cantor, quasi-combinatorial or quasi-finitist conception, which treats infinite sets *as if* they were like finite ones.

Notice that the Powerset Axiom is typically interpreted in the light of quasi-combinatorialism (this is what I have been calling its standard interpretation) even though the axiom by itself does nothing to capture this conception or make it explicit. This is the main difference between AC and Powerset, for AC in fact implements one implication of quasi-combinatorialism (as I have argued in Ferreirós 2011). Notice too that

---

[25] In the second half of the twentieth century it became common to work in the setting of Polish spaces, not metric spaces, and to give prominence to the Baire set. For further details on this important topic, see Moschovakis 1974, Kanovei 1985, Kanamori 1995, and Hausdorff's *Gesammelte Werke*, vol. 3. Moschovakis 1980 is a standard reference that contains a lot of historical information.

this conceptual meaning of "all subsets," the quasi-combinatorial one, is also presupposed in the so-called standard interpretation of second-order logic (SOL). In this consists the famous vicious circle between SOL and set theory, which precludes any deep reliance on full SOL (or equivalently quasi-combinatorial systems, such as plural quantification) in the foundations of set theory.

Let me insist: the standard idea of $\wp(S)$ gains univocal meaning only at the conceptual level, seemingly at least,[26] and not at the formal logical level. This is actually one of the most important cases where the complementarity of thought and formula that I have claimed can be found (Chapter 4). No known set-theoretic axiom has pinned down formally what this standard interpretation of $\wp(S)$ is meant to be, which is why the powerset operation remains "amorphous" (Jensen 1995, 395). There is a sizable number of experts in the field who share that view, for instance, Shelah.[27] Baire, as we shall see below, was a very early exponent; even Gödel, at one point in his career, was inclined to think that the idea of arbitrary set was a "vague notion."[28]

We shall return to this topic in the next chapter. For the time being, it is enough to have shown (in this and the previous section) that the contentious Axiom of Choice, like the Powerset Axiom and indeed the notion of arbitrary functions, has strong roots in the theory of real numbers.

9.5. WHAT ABOUT CANTOR'S ORDINAL NUMBERS?

This is probably the next natural question, and apparently a more complex matter. We shall see, however, that it comes down mainly to an illustration of the issues previously discussed. The introduction of the transfinite ordinals by Cantor, in 1883, was arguably a turning point

---

[26] On this topic see the interesting work of Jané (2005).

[27] Shelah (2003, 211) explains that, although he is inclined to platonism, one cannot disregard the concrete evidence of independence results. As a result, "You may think 'does CH, i.e., $2^{\aleph_0} = \aleph_1$ hold?' is like 'Can a typical American be Catholic?'"

[28] *Collected Works*, vol. 2 (1990), p. 27: "The proposition [V = L] added as a new axiom seems to give a natural completion of the axioms of set theory, in so far as it determines the vague notion of an arbitrary infinite set in a definite way."

that transformed set theory into quite a new thing. Up until then, mathematicians were studying only sets of numbers or sets of points in Euclidean spaces such as $\mathbb{R}^3$, or at most sequences of real functions. The "absolutely unlimited" series of the transfinite ordinals, which Cantor described in a somewhat visionary way in 1883, provided quite a different domain for set-theoretic thinking.[29] Indeed, Cantor had sooner or later to find the set-theoretic paradoxes, for while considering sets of (say) complex numbers does not lead to logical inconsistencies, the opposite holds true for considerations about sets of ordinals.

"*Foundations for a general theory of manifolds*" [sets], known as the *Grundlagen,* is a long essay in which Cantor introduces the transfinite ordinals by means of what he calls the two "principles of generation," which lead to numbers such as

$$\omega, \omega + 1, \omega+2, \ldots \omega^2, \omega^2 + 1, \ldots \omega^3, \ldots \omega^n, \ldots \omega^\omega, \ldots \omega^{\omega^\omega}, \ldots^{30}$$

And so on and so forth, in an "absolutely infinite" way, because the second principle is so powerful that it "gives us the capacity to *break through every boundary* in the conceptual formation of numbers" (Cantor 1883, 883 and 908). The second generative principle is responsible for the "creation" of ω_0 and ω_1 (denoted Ω by Cantor in 1883) and all other limit ordinals. He also presents the idea of the "*principle of limitation*," by means of which this apparently limitless [*Grenzenlose*] succession of numbers will find "*provisional* termination." The idea is that all ordinals that represent sets with the same cardinality form a *number class*, e.g., all the countable ordinals form the "second number class." (In axiomatic set theory, by the von Neumann definition, an ordinal is the set of its predecessors. In this case one can say, simply, that all ordinals that have the same cardinal form a number class.) Cantor set the principle of limitation as "*a requirement which is set on all the numbers that have to be*

---

[29] This crucial point has been emphasized by Tait (2000) and Jané (2005). See also Ferreirós 1999, 267*ff* and 446–447.

[30] All of the ordinal numbers just written can be realized in well-orderings of the natural numbers; they are countable. The first principle *generates* the immediate successor, $\alpha+1$, to any ordinal α. The second principle, given any "determinate succession" of ordinals "of which no greatest exists," *generates* a new ordinal "which is thought of as the *limit* of those numbers, that is, it is defined as the number immediately greater than all of them" (Cantor 1883, 907–908). The second principle is a bone of contention for interpreters of Cantor's work, since it raises questions about the *Grundlagen* and the paradoxes.

formed next," namely by requiring the following: "proceed to the creation of a new [ordinal] with the help of the other two principles *only* if the totality of all numbers that precede it has the power [cardinality] of a previously defined number class, which was *available* already in its entire extent" (op. cit. 911, my translation).[31]

The totality of all natural numbers, which we assume are given, defines the first infinite cardinal, the countable power $\aleph_0$. Now, Cantor's principles license the "creation" of all the countable ordinals: ω, $\omega + 1, \omega + 2, \ldots \omega^2, \omega^2 + 1, \ldots, \omega^3, \ldots, \omega^n, \ldots, \omega^\omega, \ldots$. Next, Cantor assumes that one has the totality of all countable ordinals α "available in its full extension," and in *Grundlagen* he proves that the second number class is nondenumerable—moreover, that its cardinality is immediately greater than the cardinality $\aleph_0$ of the set of finite ordinals. Thus, the set of *all* countable ordinals defines in a natural way the second infinite cardinality, $\aleph_1$. And now the process continues iteratively. The three principles license a new series of "creations," that of the ordinals of cardinality $\aleph_1$, which will give their totality, the third number class— and this is a set of cardinality $\aleph_2$. And so on and so forth, in a process by which the ordinals are created, and their number classes establish the prototypes of the alephs. This involves no vicious circle, but rather what one may call a *virtuous helix*.

The introduction of transfinite numbers was a more speculative development than other aspects of Cantor's work in set theory, e.g., his basic results on cardinalities of number sets, or his work on point set topology. As such, it was resisted by many mathematicians, though not all of them were as negative as Hermite ("disastrous") or Kronecker ("humbug"). Today, we read Cantor's work with the conviction that it is a capital contribution to mathematical thought, but of course things could never seem so simple to his contemporaries, even to unprejudiced readers. Members of the younger generation like Mittag-Leffler and Poincaré thought the new ideas important and employed the ordinals of the second number class in their own work on point sets.

---

[31] He continues: "Following this path, by observing these three principles, one can attain with the greatest security and evidence ever newer number classes, and with them all the different, successive, ascending powers that are realized in corporeal and mental Nature; and the new numbers obtained in this way are then always of the very same concrete definiteness and objective reality as the earlier ones."

In Poincaré's opinion, the *Grundlagen* was a "beautiful memoir" but French readers would be "absolutely *refractory*" to Cantor's mathematical and philosophical investigations, particularly toward the ordinal numbers of the third and higher classes, which would seem repugnant to them as "forms without content" [*forme sans matière*].[32] Let me explain why this is a perceptive comment.

An interesting feature of Cantor's first presentation of the transfinite ordinals is that he deals with them in isolation from any other set-theoretic idea, presenting them as an autonomous "extension or continuation of the series of [natural numbers] beyond the infinite." Indeed his presentation of the two "principles of generation" was disconnected from any notion of set theory or concept of advanced mathematics, as it was known in the nineteenth century. The transfinite numbers were fully autonomous and unheard of—but at the same time natural and indispensable (Cantor 1883, 882). According to authors such as Jané and Tait, this feature of Cantor's early theory of ordinals, its full autonomy and, especially, its *open-endedness* made it indispensable to him and irreplaceable even in his later years. As underscored by those authors, it is also of high interest in connection with the cumulative hierarchy.

Cantor needed the ordinals of the second number class (countable ordinals) to formulate and prove important results such as the Cantor-Bendixson theorem.[33] But as his own results made clear, the process of obtaining "derived sets" that he was studying would terminate for some countable ordinal α (in that either $P^\alpha = \varnothing$ or P^α would be a perfect set, equal to its own derived set). Moreover, all the kinds of numbers that he was able to present through explicit notations corresponded to countable ordinals. Why consider other ordinals beyond the countable ones? Would this not be a contentless formalism? Later, authors such as Borel

[32] Dugac 1984, p. 278: "Il me semble que ce qui rendrait la lecture de la traduction de ce beau mémoire très pénible aux Français qui ne sont pas familiers avec la culture allemande, c'est moins la partie philosophique qu'on serait toujours libre de passer, que le défaut d'exemples un peu concrets. Ainsi ces nombres de la 2ème et surtout de la 3ème classe ont un peu l'air d'une forme sans matière, ce qui répugne à l'esprit français." See Ferreirós (2004), section 2 for references and further details on the reception.

[33] For any set $P \subseteq \mathbb{R}^n$ such that its derived P' is uncountable, one obtains $P' = S \cup T$, where S is a perfect set such that $P^\alpha = S$ for some countable ordinal α, and T is a set at most countable, such that $P^\alpha \cap T = \varnothing$.

and Baire, who like Cantor needed the countable ordinals for work in analysis, came to think precisely that way. Notice also—introducing a remark from twentieth century set theory—that, in an axiomatic system without the Powerset Axiom (ZFC–), one cannot prove the existence of the first nondenumerable ordinal, usually denoted by ω_1. In order to prove that ω_1 exists, we need Powerset.

Yet Cantor's speculation was sustained by his results about $\mathbb{R}$ and about sets of functions. Knowledge that $\mathbb{R}$ is a nondenumerable infinity gave reason to accept the possibility of going beyond any countable ordinal. Thus, if we assume the possibility of well-ordering the set $\mathbb{R}$, naturally there will be an uncountable ordinal corresponding to that well-ordering. (In the *Grundlagen*, Cantor regarded the well-ordering principle as a "law of thought," a logical law (1883, 886), a position that he later abandoned; but this could hardly seem more than a petitio principii to critical readers.) Consider it differently: If a set such as $\mathbb{R}$ was acceptable, it seemed also possible (Cantor would have said necessary) to accept the set of *all* countable ordinals—and then the second and third generative principles would imply that there is a first uncountable ordinal, ω_1. Furthermore, in endnote 10 to *Grundlagen* Cantor remarked that the set F of all functions $f: \mathbb{R} \rightarrow \mathbb{R}$ ought to be of cardinality $\aleph_2$ (he was taking for granted the Continuum Hypothesis, card($\mathbb{R}$) = $\aleph_1$; indeed he was extending it to the Generalized Continuum Hypothesis). This indicates that he may have had already a proof that F is not bijectable with $\mathbb{R}$, perhaps the same proof he published a few years later[34]—or at the very least he conjectured that result.

At any rate, since Cantor conceived of the sets $\mathbb{R}$ and F as actually infinite, fully determinate objects, it was natural for him to regard results such as the nondenumerability of $\mathbb{R}$ (and the non-bijectability of $\mathbb{R}$ and F) as supporting the introduction of ordinals ω_1 and ω_2 (whose cardinals are, respectively, $\aleph_1$ and $\aleph_2$).

Let me add the following. The first time that Cantor proved satisfactorily and in detail the existence of higher and higher cardinalities leading to an "absolute infinity" of them and substantiating the

---

[34] This, of course, must remain unclear; if he had the proof, why did he not publish it at the time? He only published a proof in Cantor (1892), nine years later; see also the letters to Hilbert written in 1897.

intuition he had from 1883, was in a paper he read eight years later (Cantor 1892). This contains a proof that the set of functions $f : \mathbb{N} \to \{0, 1\}$ is uncountable, which is actually the usual proof of uncountability for $\mathbb{R}$; and a proof that, for any S regardless of its cardinality, the set of all $f : S \to \{0, 1\}$ is of still higher cardinality. Thus, the existence of higher and higher cardinalities depends on the assumption that, whenever S is a set-object, so is the set of all functions $f : S \to \{0, 1\}$; this is another formulation of the Powerset Axiom. As one can see, consideration of functions intertwined with sets was highly important in the early explorations of the world of sets.[35] Cantor understood the relation between the contents of his (1892) paper and powersets, in particular that his second result translates into this: the powerset $\wp(S)$ is of cardinality greater than S itself.

In fact, the same results that Cantor regarded as so many triumphs of mathematical reason (or gifts of God's benevolence) could be taken by others to constitute arguments against some of the assumptions that they, innocently, might have been willing to grant initially. If there is any reason to have second thoughts about the set $\mathbb{R}$ being equally acceptable as the simple infinity of $\mathbb{N}$, the implications of accepting the actual infinity of real-valued functions F might strongly fuel such doubts. The assumption that the set F is given provides reason to accept an ordinal such as ω_2 and the cardinal $\aleph_2$ (or even bigger cardinals if we disavow the GCH), and this (by analogy) leads to the procession of "forms without content" constituted by $\aleph_3, \aleph_4, \ldots \aleph_\omega$, and so on. One ends up, as Cantor did, thinking that to every transfinite ordinal α there is a corresponding cardinal $\aleph_\alpha$, a truly exorbitant picture; thus, to the transfinite number γ there corresponds the γth cardinality, to ω, $\aleph_\omega$, and to ω^ω, $\aleph_{\omega^\omega}$.[36]

At this point, a critical thinker can easily begin to explore the possibility that F may not be admissible; around 1870, independently of Cantor's results, the Berlin leaders Kronecker and Weierstrass already had strong doubts about the notion of F.[37] This doubt was also harbored by

---

[35] See again section 8.5, and consider that Dedekind's theory of sets had *set* and *map* as primitive notions; if forced to choose one, there is evidence that he would choose map.

[36] This dizzying perspective is explained in a rich endnote added to the *Grundlagen*; little wonder that Cantor viewed the realm of transfinite numbers as lying between the finite and the Absolute of God.

[37] See Ferreirós 1999.

the French analysts and pioneers of descriptive set theory, Borel, Baire, and Lebesgue; Poincaré, too, joined them in being skeptical toward Cantorian set theory. Their idea was to restrict investigations to explicitly defined functions, to functions "described" or "named," as they said rather rudely—whence the label *descriptive* set theory. The countable ordinals of Cantor's "second number class" were found necessary for their investigations into the Baire hierarchy of functions and the Borel sets, but Baire and Borel rejected the notion of a completed totality of all countable ordinals. Brouwer was just noting down and summarizing their point of view when he wrote (in this early text, by "intuitionist" he means what would later be called semi-intuitionism):

> Let us now consider the concept: "denumerably infinite ordinal number." From the fact that this concept has a clear and well-defined meaning for both formalist and intuitionist, the former infers the right to create the "set of all denumerably infinite ordinal numbers," the power of which he calls aleph-one, a right not recognized by the intuitionist. Because it is possible to argue to the satisfaction of both formalist and intuitionist, first, that denumerably infinite sets of denumerably infinite ordinal numbers can be built up in various ways, and second, that for every such set it is possible to assign a denumerably infinite ordinal number, *not* belonging to this set, the formalist concludes: "aleph-one is greater than aleph-null," a proposition that has no meaning for the intuitionist. (Brouwer 1913, 91)

It seems to me quite important to stress the coherence and reasonableness of this standpoint, as it often happens that set theorists reason as if that were not the case. After all, this constructivist position admits the simple infinity of $\mathbb{N}$, but rejects the notion of arbitrary infinite sets; and there is nothing incoherent in that. I have seen great experts in set theory present the simplistic dichotomy that one must be either a finitist or a defender of large cardinals. That is not the case.

However, if one has accepted the postulation of $\mathbb{R}$ as a set-object—as concrete as $\mathbb{N}$ may be—and one understands (as Cantor, Baire, and others did) that $\mathbb{R}$ is related to $\mathbb{N}$ essentially as the set of all functions $f : \mathbb{N} \to \{0, 1\}$, that is to say, as $\wp(\mathbb{N})$, then it would only seem natural to accept the move from $\mathbb{R}$ to all functions $f : \mathbb{R} \to \{0, 1\}$, i.e., to $\wp(\mathbb{R})$. And one can also show that the set of all real-valued functions F does not introduce any further essential complexity, set-theoretically speaking, above and beyond the set of all $f : \mathbb{R} \to \{0, 1\}$. Lebesgue apparently was ready to accept the set $\mathbb{R}$, but had qualms about the set we have called F. This, in my view, is a very unstable position, not sufficiently crisp and coherent. If you want to push the argument coherently, then the natural conclusion is that the domain of real numbers $\mathbb{R}$ does *not* form a set of the same kind as $\mathbb{N}$. But this is a radical position that very few authors were willing to take—even if among them we find first-rate mathematicians like Kronecker, Baire, Brouwer, Weyl, and Bishop.[38]

Let me quote in detail a relevant passage from the well-known five letters exchanged between the French analysts in 1905, "in the great debate resulting from Zermelo's note" proving that AC implies the Well-ordering theorem. René Baire, who in my opinion took a coherent approach—more so than Lebesgue—wrote to Hadamard:

> The expression *a given set* is used continually. Does it make sense? Not always, in my opinion. As soon as one speaks of the infinite (even the denumerable, and it is here that I am tempted to be more radical than Borel), the comparison, *conscious or unconscious*, with a bag of marbles passed from hand to hand must disappear completely. We are then, I believe, in the realm of the *virtual* [dans le *virtuel*]. That is to say, we establish conventions that ultimately permit us, when an object is defined *by a new convention*, to assert certain properties of this object. But to hold that one can go farther than this does not seem legitimate to me. In particular, when a set is given (we agree to say, for example, that we are given the set of sequences of positive integers), *I consider it false to regard the subsets of this set as given*. I refuse, *a fortiori*, to attach any meaning to the act of supposing that a choice has been made in every subset of a set.[39]

This passage contains a clear and explicit repudiation of the Powerset Axiom in standard reading, but it was hardly taken into account at the time (except by Brouwer and Weyl).

---

[38] See Kronecker's material in Ewald 1996, Baire's letters in appendix to Moore 1982, the Brouwer material in van Heijenoort 1967, Weyl 1918 and 1949, and also Mancosu 1998.

[39] Translation taken from Appendix 1 of Moore 1982, 313, with a slight change.

The fact that Baire remained isolated in his clear-sighted analysis of the situation constitutes, in my view, a good argument for the perspective on mathematical objectivity that I am proposing. The idea that the irrational numbers are as basic and "real" as the natural or the rationals, the indispensability of a notion of the continuum, and the admissibility of the pointillist analysis of the continuum were so deeply entrenched in the "classical mathematics" of 1900 that a position like Baire's seemed too hard to take into account—even too painful. But, since a middle-ground perspective like Lebesgue's is unstable and ultimately indefensible, the well-established notion of the number-continuum backed up the classical approach to set theory of Cantor and Dedekind.

Cantor had been aware of this, and expressed it in 1887 by writing that the transfinite numbers stand or fall with the irrationals:

> One can say unconditionally: the transfinite numbers *stand or fall* with the finite irrational numbers; they are like each other in their innermost being; for the former like the latter are definite delimited forms or modifications (*aphorismena*) of the actual infinite.[40]

On the other hand, Baire ended his letter saying: "For me, progress in this matter would consist in delimiting the domain of the definable." Two methodological orientations had been clearly delineated, corresponding to different philosophical presuppositions. They would give rise to two different lines of mathematical practice.

9.6. OBJECTIVITY AND THE CONTINUUM PROBLEM

The Continuum Problem, in elementary form, is the question of what different cardinalities are realized in subsets of the set $\mathbb{R}$ of real numbers, the number-continuum. Cantor was able to prove, under the usual assumptions of set theory, that there are at least two different cardinals;

---

[40] *Mitteilungen zur Lehre vom Transfiniten* (1887), in Cantor's *Abhandlungen* p. 395–96: "Man kann unbedingt sagen: die transfiniten Zahlen *stehen oder fallen* mit den endlichen Irrationalzahlen; sie gleichen einander ihrem innersten Wesen nach; denn jene wie diese sind bestimmt abgegrentzte Gestaltungen oder Modifikationen ($\alpha\varphi\omega\rho\iota\sigma\mu\varepsilon\nu\alpha$) des aktualen Unendlichen."

by 1878 he had proven that the set of all irrational numbers $\mathbb{R}\backslash\mathbb{Q}$ has the power of the continuum (there is a one-to-one correspondence between $\mathbb{R}\backslash\mathbb{Q}$ and $\mathbb{R}$), while the set of algebraic real numbers is countable. Cantor's Continuum Hypothesis, CH, suggests that the answer is the simplest possible: there are only two infinite cardinals for $S \subseteq \mathbb{R}$, namely the first cardinal, $\aleph_0$, and the cardinal of the continuum, $2^{\aleph_0}$, which therefore is $= \aleph_1$. There is no unanimity among mathematicians, even among experts in set theory, as to whether this is an objective question or not. I hope readers who have followed the previous discussion will be in a better position to understand how this is possible.

Notice that $2^{\aleph_0}$ is *defined* to be the cardinality of a set of functions—all $f\colon \mathbb{N} \to \{0, 1\}$—while $\aleph_1$ is defined as the cardinality of the set of all uncountable ordinals. It is noteworthy, in light of the previous discussion, that both sides of the famous equation $2^{\aleph_0} = \aleph_1$ depend for their existence on the Powerset Axiom! It seems particularly important to emphasize this point, as even experts tend to think that $\aleph_1$ is neatly defined, and the problem lies in determining $2^{\aleph_0}$, measuring it against the scale of ordinals and number classes. However, if the powerset "operation" is somehow vague or "amorphous" (see above),[41] this will affect both sides of the equation. In the worst scenario, we might be trying to determine whether two vaguely defined objects are the same or not, which might be a perfectly arbitrary question. If so, Cantor's CH is very good as a "solution," as it would establish the simplest convention.

The problem lies deeper than the mere realization that CH is independent from the axiom system ZFC. Gödel proved in 1939 that the system ZFC+CH is consistent if and only if ZF is consistent (see Gödel 1940), using the method of inner models in its simplest form of the constructible hierarchy of sets, L. Cohen proved in 1963 that the system ZFC + ¬CH is likewise consistent, using the pathbreaking method of forcing. Many mathematicians have, in a formalist vein, relied on these two deep metatheoretic results to provide the answer I have mentioned: there is no longer a question about CH once we know that this axiom is acceptable but independent from ZFC—as if an independence result

---

[41] See also the interesting discussion of "power domains" in Jané 2005. I write "operation" in quotes when applied to powerset, because it is totally different from most things we call operations: normal operations are constructively specified.

could be the end of the story. Let me propose a comparison to make things more easily accessible to readers who are not experts.

Hilbert's axiom of Completeness, like Dedekind's axiom of Continuity [*Stetigkeit*], or the Bolzano-Weierstrass axiom,[42] are different propositions that can be used to state that the set of real numbers $\mathbb{R}$ is "gapless," as one usually says, i.e., continuous or complete. But, is the set of all real numbers complete or not? Well, it turns out that the axiom of Completeness is independent of the remaining axioms for the $\mathbb{R}$-structure. So, by analogy with the previous line of argument, this mere realization is the answer: there is no objective fact of the matter. (Similarly, to the question of whether the set of all natural numbers is really characterized by mathematical induction, we may answer that this axiom is independent of the remaining axioms.) Such answers are natural for a strict formalist, but strict formalism cannot make sense of real mathematical practice. Even the most cursory consideration of mathematical practice, in connection with these issues, raises the question: Why is the truth or falsity of CH considered as an open problem, while all mathematicians take $\mathbb{N}$ to be characterized by induction, and almost all take $\mathbb{R}$ to be complete?

So we are back to the issue of whether the Continuum Problem is an objective question or not. Taking into consideration the results obtained along the twentieth century in the field of set theory, there are at least two possible perspectives on the topic, depending on whether one adopts a realist view (in the sense of strong, ontological platonism) or a conceptualist view. The naïve student of mathematics will tend to consider these two perspectives as more or less equivalent, i.e., will think that in order to develop mathematical knowledge it is immaterial whether you believe it describes an independent reality or not. The deep lesson that foundational studies of the metatheory of ZFC teaches us is that this common sense opinion is wrong. The naïve person will think that CH must be either true or false, since he or she tends to believe that the principles of set theory provide a "complete" determination of

---

[42] This says that every sequence of nested, closed intervals of real numbers (i.e., such that $[a_{i+1}, b_{i+1}] \subset [a_i, b_i]$) has a nonempty intersection. In other words, for every such sequence of nested, closed intervals, there is at least one real number r that belong to all of them, $r \in [a_i, b_i]$ for all i.

the theory in some informal sense (which would correspond to categoricity or quasi-categoricity at the formal level). But we have to deny the assumption that the conceptual framework of set theory offers a full determination of the intended domain, in light of twentieth century results, most importantly in the light of *independence* results.[43]

Thus, the two possible perspectives have different implications on our reflection on the status of the continuum problem. To simplify my exposition, we can assume that all the parties involved accept (with Gödel and Woodin, and many others) that new axioms will become central to a future, more complete system of set theory. The two views are as follows:

- the realist thinks that our grasp of the set-theoretic universe is still quite deficient, but this can be remedied; the background realism offers a powerful reason to believe (like Gödel fifty years ago, and Woodin nowadays) that new axioms that settle CH as true or false can be objectively established;[44]

- the conceptualist—uncommitted to realism—should say that set theory does not present us a fully determinate setting, but describes partially (hence vaguely to some extent) a maximal realm of mathematical possibilities.

It is particularly what has to do with arbitrary subsets, and hence with the standard interpretation of the "operation" $\wp(x)$, that is unclear, not fully determined, and that thus contaminates axiomatic set theory with vagueness.[45] Let me insist (section 3) that from this point of view AC is a concrete, positive contribution to specifying what is meant by *quasi-combinatorial maximality*. The problem is that we would need further specifications, i.e., increased control over the postulation of arbitrary sets—and this seems to be out of reach.

---

[43] See Shelah 2003. Of course this is not a mathematical result: one can still hope that someday someone will find the missing conceptual determinations. Koellner (2013) employs the terminology of pluralism vs. non-pluralism for the two viewpoints above—hiding somewhat the realist underpinnings of the latter.

[44] Notice, by the way, that as time goes by, the kinds of axioms expected to play that role become increasingly complicated: large cardinal axioms for G., axioms for strong logics in the case of W.

[45] I have argued this in more detail in Ferreirós (2011).

It is instructive to make a comparison with another deep set-theoretic problem, which, however, is of a lower level of complexity than CH (even though its formulation may not give that impression) and on which there has been impressive technical progress over the last fifty years. We explained how Borel and Baire opted, against the Dedekind-Cantor orthodoxy, to seek "progress in this matter" by "delimiting the domain of the definable." This was the start of descriptive set theory, which had already begun with the Baire hierarchy of functions,[46] studied in 1905 by Lebesgue in a pioneering paper (where he proved that each Baire class contains functions not in any smaller class, and that there exist functions not in the Baire hierarchy). The development of descriptive set theory continued in the following decades with important contributions due to Hausdorff, Luzin, Suslin, Sierpinski, Gödel, and others. However, this development reached the limits of ZFC relatively soon, with the introduction of the hierarchy of projective sets (a rather natural prolongation of the hierarchy of Borel sets).

Although the first projective level Σ_1^1 (analytic sets) was amenable to treatment, the difficulties were enormous beyond that point. Luzin, the foremost expert before 1940, was thus led to make his famous remark:

> one does not know, and one will never know, whether the projection of the complement of an analytic set of two dimensions ([a Σ_2^1 set] supposed uncountable) has the cardinality of the continuum, whether it is not a set «that [ha]s not [the Baire property]», nor whether it is measurable.[47]

The assertiveness of this statement reminds one of Hilbert's famous talk on problems (1900, introduction), but its negative conclusion contrasts very strongly with the optimistic confidence expressed by

---

[46] This was obtained from continuous functions by transfinite iteration (over the ordinals of the first and second number classes) of the operation of forming pointwise limits of function sequences; the Baire functions coincide with the Borel-measurable functions and correspond neatly to the Borel sets.

[47] Luzin 1925, 1818: "on ne sait pas et l'on ne saura jamais si la projection même d'un complémentaire analytique à deux dimensions (supposée non dénombrable) a la puissance du continu, si elle n'est pas des ensembles « qui ne sont pas Z», ni même si elle est mesurable. Ces difficultés et la légèreté même avec laquelle on trouve le transfini exclu nous amènent à examiner plus attentivement la légitimité de ces ensembles, ainsi que l'existence des ensembles non mesurables B, d'ailleurs."

Hilbert—reasserted around 1930 with the famous saying *"Wir müssen wissen, wir werden wissen."*[48]

Forcing methods later clarified the situation when they established that the questions Luzin wanted to answer are in fact independent from the ZFC system. For example, one cannot prove in ZFC that all uncountable Σ_2^1 sets have a perfect subset; hence it is undecidable whether they have the cardinality of the continuum.[49] It is noteworthy, and deserves to be more widely known, that it was relatively easy to reach the limits of the axiomatic set theory ZFC by transcending the Borel sets with the addition of the operation of projection. As this corresponds to geometric projection, the idea was that this is a perfectly well-understood operation, which from simple sets should only produce other simple sets.

Faced with this situation, Gödel famously proposed in 1947 (see his 1964 paper) what many call the "Gödel program" of completing the theory of sets by means of large cardinal axioms. The domain of descriptive set theory has precisely been an area of success for this program, already in the 1960s. Solovay (1969) established that the existence of a measurable cardinal implies that uncountable Σ_2^1 sets have the perfect subset property, i.e., have the cardinal of the continuum. Other authors then proved that, under the same hypothesis, all Σ_2^1 sets have the property of Baire and are Lebesgue measurable. These are typical kinds of problems and results in descriptive set theory. Meanwhile, the situation with CH was much worse, since it was established that no large cardinal axiom can settle CH (forcing arguments produce models that comply with CH or with ¬CH, at will).[50]

We cannot enter here into the more recent (1970s and 1980s) resolution of questions about the projective hierarchy by means of a still stronger hypothesis, the axiom PD of *projective determinacy*; suffice it to remark that the nature of the evidence in support of PD remains problematic.[51] Like other strong assumptions in set theory, from measurable

---

[48] "We must know, we shall know." This was even inscribed as an epitaph in Hilbert's tomb.

[49] As stated by Moschovakis (1974), 253, this result was implicit in Cohen (1963). In fact, Gödel (1940) established, for the constructible universe $V = L$, that there is already a Π_1^1 (co-analytic) set that does not have a perfect subset, and there are non-Lebesgue measurable Σ_2^1 sets. See Moschovakis 1980, or Kechris 1996, for textbook treatments.

[50] This was shown for ZFC + measurable cardinal by Levy and Solovay (1967), and the argument extends to all known types of large cardinals.

[51] PD implies that every uncountable projective set (of any level Σ_n^1) has a perfect subset, is Lebesgue measurable, and has the property of Baire. The best discussion of the nature

cardinals upward, PD has no clear basis in the iterative conception or in generally accepted reflection principles. On the other hand, PD has been derived from a large large-cardinal assumption, the existence of infinitely many Woodin cardinals (Martin and Steel 1989).

The main evidence favoring PD comes from the very fact that it provides a neat and elegant theory of the projective sets, but the skeptics fear that this risks becoming circular. The reasoning goes: it is desirable to have a theory extending to all projective sets the patterns of simplicity that are found at the basic levels of the Borel sets and the analytic (Σ^1_1) sets; and we have found this principle that provides what is desired—so be it. One of the strongest advocates of PD is apparently admitting this when he writes:

> The old self-evidence requirement on axioms is too subjective, and more importantly, too limiting. In the future, what "forces itself upon us as true" is more likely to be a theory as a whole, and the process is more likely to be gradual. We may very well never reach the level of confidence in the new theory that we have in, say, Peano Arithmetic. Nevertheless, new axioms may emerge, and be rationally justified. The self-evidence requirement would block this kind of progress toward a stronger foundation. (Steel in Feferman et al. 2000, 422)

The worry, put very simply, comes from experience that in these complex matters of infinite sets, things easily turn out not to be what they seemed—intuition and expectation are never to be trusted; Cantor's famous "*Je le vois, mais je ne le crois pas*" ("I see it but I don't believe it") is the norm. Complete abandonment of intrinsic evidence in favor of extrinsic arguments of the kind advocated by Steel, his holistic theory-oriented processes of justification aimed ultimately at the satisfaction of our expectations, aesthetic judgments of the simplicity of Σ^1_n sets, could degenerate into a form of "wishful thinking," if you allow me the use of this expression. No doubt, deeper studies are needed into questions of evidence in higher set theory.

The realist, of course, will hold the belief that the Continuum Problem does have a unique solution; but the conceptualist will have to

---

of the evidence underlying this assumption is probably that of Martin 1998; see also Steel's paper in Feferman *et al.* 2000, and the recent exposition in Koellner 2013.

suspend judgment, leaving open the possibility that the deficiencies in axiomatic set theory have no remedy. It is a perfectly conceivable scenario that we can safely (consistently) introduce hypotheses about arbitrary sets in mathematics, yet cannot bring the ensuing theoretical system under full mathematical control. We are back in the labyrinth of the continuum.

What should be clear by now, and remains undeniable, is that the accumulation of hypothetical assumptions becomes rather thick in the domains of higher set theory. If there was disagreement with Cantor's work, one should only expect to find more of it, and stronger, in the context of assumptions about large cardinals, determinacy, and the like. Even the critical realist may accept that human mathematicians might never be able to remedy the insufficiencies of axiomatic set theory, even though she believes that CH is either true or false. The key difference is that the conceptualist finds no objective basis for this last assertion. This, of course, is unsettling, but the reader knows that post-Gödelian epistemology of mathematics must be unsettling to some extent.

We often dream of things such as the structure of natural numbers or the structure of real numbers being *univocal, categorical*, and perfectly determinate. That may be an assumption regulating mathematical practice due to the role of nonformal ingredients (i.e., of thoughts added to the formalisms; cf. Chapter 4), but it is not a reality of mathematical theories: it is not a reality of the formal science of mathematics, nor of the theories developed in mathematical practice. This "dream" or hope is usually interpreted, from a refined perspective, as the assumption that the structure of the reals is fully determinate *relative to a background set theory* (i.e., that there is a unique intended model). But this implies that the structure of the real numbers will inherit any indeterminacy or vagueness that may affect set theory. Employ Descartes's evil demon, if you will, to make the situation more vivid: you and I may believe that the $\mathbb{R}$-structure is categorical, but the mischievous demon confuses both of us, so that your $\mathbb{R}$-structure is not isomorphic to mine.

10

The Problem of Conceptual
Understanding

The philosophy of mathematics we are advocating puts the agent—
the mathematician, the math student, the math user—center stage,
and as a result it emphasizes not only practices, but also the thought-
content of mathematics as complementary to the symbols and formalisms that sustain the development of mathematics. As discussed in
previous chapters, it thus becomes natural to consider questions of
meaning and understanding in relation to practice and use. In this last
chapter, we shall consider how those issues interplay with the vexed
question of the acceptability of postulational ("classical") mathematics,
a question usually formulated in terms of consistency.

As the reader probably knows, the evolution of foundational-
justificational matters in the domain of mathematics during the twentieth century was somewhat of a tragicomedy. The wonderful increase in
rigor and system obtained by the arithmetization and set-theoretization
of classical mathematics, by the establishment of satisfactory modern
axiom systems for all theories, and by the emergence of modern mathematical logic, was followed by the shock of Gödel's incompleteness
results—from which apparently we have not fully recovered, even after
80 years. The system ZFC, augmented with the Continuum Hypothesis, has been proven to be consistent *relative to* the system ZF (without
CH and the axiom of Choice),[1] but on the other hand it was proven
that CH and many other questions of higher set theory are undecidable
from ZFC. The phenomena of independence and undecidability are

---

[1] This means that a contradiction in ZFC + CH can be translated into a contradiction in
ZF. See Gödel 1940. The independence of CH from ZFC was the result that won Cohen the
Fields medal.

the norm in foundational studies of set theory, and absolute consistency is nowhere within reach. This has led some authors to defend radical views, such as that mathematics does not have "nor need" foundations, or that, in the philosophy of mathematics, "nothing works".[2] Even better is the witty pronouncement attributed to André Weil: "God exists since mathematics is consistent, and the Devil exists since we cannot prove it."

My aim in this chapter, which has an experimental character, is to offer a fresh look at this matter. Instead of the technically crisp—but rather intractable—question of consistency, we shall focus our attention on a more human concern: understanding. This is a noteworthy dimension of mathematical practice, which often drives the developments of the discipline, the agent's search for new proofs, new concepts, new theories, and so on. Let me quote a relatively old paper of Manders:

> Any dispassionate look at mathematical sciences should teach that the "mathematical way of knowing" seeks as much to render things understandable as it does to establish theorems or avoid error. The process of establishing deductive relations is subsidiary to the larger goal of rendering understandable. . . . Matters affecting understanding are held to be irreducibly psychological, and therefore incapable of serious philosophical treatment. That inference is as blatantly fallacious as the claim that, because our grasp of the physical world necessarily involves experience, mathematics cannot possibly be the key to a theory of that world. (Manders 1989, 562)

If the reader would ask me whether I think that axiomatic set theory ZFC is consistent, the answer would be, "yes, I do." This belief of mine seems to coincide with that of the majority of experts in set theory, and probably with a majority of mathematicians—though *certainly* not all.[3]

---

[2] Putnam (1967) and Putnam (1994).

[3] To offer a rather extreme example: Princeton professor Edward Nelson has long been trying to establish the inconsistency of Dedekind-Peano Arithmetic; he shocked the community of mathematicians in 2011 with an announcement (later retracted) that he had succeeded. (For a highly interesting blog entry on the topic, see https://plus.google.com/117663015413546257905/posts/HH5X7728k34 and look for T. Tao's comment.) Also, Fields medalist Vladimir Voevodsky caused a stir when in 2010 he gave a talk entitled "What if Current Foundations of Mathematics are Inconsistent?" (this is available online) as part of a celebration of the 80th anniversary of the founding of the Institute for Advanced Study in Princeton. His doubts and ruminations have subsequently led to a research program

The question is—where do such convictions come from? And the typical answer is—from empirical sources: the experience of more than a century's probing of set-theoretic ideas, plus the solidity (again "empirical," but now based on three centuries of science) of theories such as mathematical analysis, that can be based on set theory. However, this rushed and ready answer does not quite satisfy me. There are many ways to go about trying to obtain a deeper understanding of this issue, and they could easily fill a book three times the length of this one. Given the limited space available, I will offer an incomplete account that is centered on considerations concerning the set-theoretic universe and the cumulative picture of sets.

Notice that such issues are quite independent of the problem of whether axiomatic set theory is the best system to employ as a basis for current mathematics. Our concentration on mainstream set-theoretic math, here and now, in no way implies a lack of recognition of alternative frameworks for current mathematics, such as constructivist, predicativist, or category-theoretic approaches. In fact, if the reader would ask me about this matter, I would certainly speak in favor of pluralism, and specifically against set-theoretic reductionism.

The first three sections of this chapter offer a brief presentation of the iterative picture of sets and its presuppositions, a new look into it from the standpoint of the web of practices, and a more general discussion of the issue of conceptual understanding in mathematics, as exemplified by the theory ZFC. Finally, in sections 4 and 5 we consider arguments based on the idea of the real-number continuum as a source of justification for the axioms of set theory. For this is, after all, what I regard as the ultimate source of the conviction that set theory is perfectly coherent and will never be proved inconsistent.

10.1. THE UNIVERSE OF SETS

In the second half of the twentieth century, it became usual practice to present the universe of sets on the basis of the iterative conception. Much has been written about the iterative conception as explanation

---

on what is called homotopy type theory (HoTT; see http://homotopytypetheory.org/), very active right now.

and motivation for the set-theoretic axioms, from its first appearance with Gödel (1933; 1947) to very recent times (e.g., Tait 2005). It purports to be a consistent conceptual framework for motivating the usual axioms of Zermelo-Fraenkel set theory, ZFC, and more ("small" large cardinal axioms). The topic is very large, including as it does different versions of the iterative conception (some maximal, some minimal), issues of justification of large cardinal axioms, questions about reflection principles,[4] and so on. These matters have been widely discussed by philosophers and logicians, especially since the 1970s, and I shall not try to cover the issue again.[5]

For our present purposes it is more than enough to deal with basic features of this conception. In this connection, the reader may be interested in the work of Tait (2005, Chapter 6), which operates on a relatively small basis, showing how this suffices for justifying the axioms of set theory and "small" large cardinals. Here, we shall just be interested in the role the conception has played in enhancing understanding of set theory, offering a conceptual picture of this subtle theory that can be easily shared, and convincing mathematicians of its inner consistency. These are aspects of the intersubjective elaboration of knowledge that form the soil from which mathematical objectivity grows.

The full universe of sets is ready to emerge as soon as we have at hand (or in mind) infinite sets, arbitrary subsets, and powersets, thanks to the idea of iterating the process of formation of sets of sets. Let me remind you that objecthood is—here, but also in the historical origins of the theory—a defining trait of sets, so that admission of S as a set immediately gives sense to the question of whether $S \in T$ for another set T. Dedekind, for instance (1888, 344), declared emphatically that a set is as much an entity (*ein Ding*) as any other "object of our thinking" may be; and this is enough for the iteration of sets of sets.[6] As we have seen, actually infinite sets were at hand in 1872, both with Cantor and Dedekind—not only in their theories of $\mathbb{R}$, but also in Dedekind's algebraic number theory (1871) and Cantor's work (1872) on pointsets, especially derived sets. And the same is true for arbitrary subsets, on

---

[4] See Koellner 2009.

[5] See, among others, Tait 1998, Boolos 1971, Wang 1983, Jané 2005, and Arrigoni 2007.

[6] This only made explicit views that he had entertained since the early 1870s.

exactly the same basis; with the work of Dedekind and Cantor, the principle of their existence and the idea of powersets were made necessary, even though it took three more decades for them to be made explicit.

The surprising conclusion is that, at the time when set-theoretic considerations made their first public appearance, the scene was almost ready for the universe of sets V to emerge. But one should be a bit more cautious: for V to emerge, a certain concentration on sets for their own sake is needed, and that cannot yet be found in the 1870s. Considering this, a safe bet for the "birth" of full-blown set theory might be the year 1883, when Cantor published his *Grundlagen* containing the first presentation of the never-ending ladder of the transfinite ordinals.

Gödel was inspired by the work of the early set theorists (such as the reconstruction of the number systems, due to Dedekind, Cantor, Peano, and others), by the role set theory plays in modern mathematics, and by ideas taken from Russell's theory of types. In 1947, he emphasized that the paradoxes, although a very serious problem for general logic, are really no problem for Cantorian set theory if properly conceived. There is a concept of set that "has never led to any antinomy whatsoever," namely the conception that "a set is anything obtainable from the integers (or any other well-defined objects) by iterated application of the operation «set of»" (Gödel 1947, 180).

It is crucial to know that Gödel understands this concept of *set of* as inextricably linked not only with the Axiom of Separation (informally, any property determines a subset of a given set), but also with the Axiom of Choice that gives access to arbitrary subsets.[7] Thus if, like Gödel, we begin with a domain $V_0 = \mathbb{N}$, in the next step we have all possible sets of natural numbers (definable or not), so we have $V_1 = \wp(\mathbb{N})$ interpreted standardly in the light of postulating arbitrary sets. The usual picture of Gödel's iterative conception *presupposes* the notion of arbitrary set, the quasi-combinatorial conception of sets (section 8.5).

A new step will take us to V_2, comprising all possible sets whose elements are numbers or number sets, and so on. In this way, at the very

---

[7] In (1944, 139) he wrote that "nothing can express better the meaning of the term 'class' than the axiom of classes [Separation, as made explicit on p. 131] and the axiom of choice," and, in (1947, 177 note), he insisted that AC "is exactly as evident as the other axioms for sets in the sense of arbitrary multitudes." See also the p. 255 note, which dates to 1964.

least, we conceive of levels V_n for any natural number n, an idea that Gödel took from Russell's type theory. Notice that even this basic version of the iterative conception makes room for all of the sets encountered in normal mathematical practice, and much more, offering arguments for accepting higher and higher cardinals. At the base level we have $\aleph_0$ objects, and (assuming the Continuum Hypothesis) at the next level $\aleph_1$, then $\aleph_2$ elements, and so on.

But Gödel also warned that under "iterated application" he meant to include *transfinite* iteration: the totality of sets obtained by finite iteration form again a set $\bigcup_{n \in \mathbb{N}} \{V_n\}$, and we obtain a new basis V_ω for further application of the operation *set of*. In fact, Gödel seems to have taken inspiration from both technical results in the metatheory of ZFC and developments of type theory. The above is essentially the idea of *simple* type theory (a great simplification of Russell's theory due to Chwistek, Ramsey, and others) extended to transfinite types, making the types cumulative.[8]

The iterative conception of sets contrasts with the earlier logicist conception due to Riemann, Dedekind, Frege, and others, which Gödel called *dichotomic* because it starts with the "set" of everything[9] and proceeds to more particular sets via properties that create dichotomies. (The idea is just this: imagine a property "being a number," or, in symbols, $N(x)$; property $N(x)$ dichotomizes everything into the complementary sets of numbers and non-numbers.) The iterative conception starts somewhere, from a given domain of well-defined objects, while the logicist conception—as contemplated by Dedekind and Frege—started in a void of pure logic, without existential assumptions, seeming to offer some kind of *creatio ex nihilo*. Perhaps there is some deep lesson in the fact that this attempt of a logocentric creation of the world of mathematics led irredeemably to antinomies.

Set theorists prefer to start from one single object, the empty set, taking advantage of the assumed *transfinite extension* of the process of iteration; this makes the whole process purely set-theoretic. Then, more

---

[8] I have discussed this point historically in Ferreirós 1999, Chapter XI.

[9] As the reader knows, this in the light of ZFC is not a set but a proper class: it can be accepted as a multiplicity, but not a single object (in Cantor's terminology, it must be thought as a Many which is not a One). Also, by V we mean a proper class.

concisely and mathematically, the iterative conception gets us a hierarchy of domains defined over the series of transfinite ordinals:

   I. $V_0 = \varnothing$
  II. $V_{\beta+1} = \wp(V_\beta)$
 III. $V_\lambda = \cup_{\alpha < \lambda} V_\alpha$ for all $\alpha < \lambda$, whenever λ is a limit ordinal.

Notice that each of the domains is identified with a set; had we not
done that, we could still identify domain V_α with a certain set, namely
one that belongs to $V_{\alpha+1}$.[10] The proper class V formed by all of these
domains V_α is called the *universe of sets*; it is also commonly called the
von Neumann hierarchy or the cumulative hierarchy.[11]

The expression "the universe" of sets assumes that it identifies a
unique "meta-thing," if one can use this word. This can be construed as
a slight abuse of language, meaning "the" universe of sets *in a model* and
thus merely a relative "meta-thing." It can also be construed realistically,
assuming that there is indeed one single Big Universe—a view that is
attractive to mathematicians and philosophers akin to ontological platonism; V would develop into a hierarchy that has levels α for each and
every one of the ordinal numbers that there "really is." Not surprisingly,
I find more attractive the alternative perspective presented in (1930) by
Zermelo, namely the view that one can always conceive of further extensions of the universe, in which the former proper class V now becomes
a set inside a bigger class V'. Indeed, according to Zermelo the wrong
idea that V is uniquely determined would be the source of the antinomies.[12] But for our purposes here we may avoid this issue.

There has been a long debate over what the iterative conception
can and cannot justify viz. motivate. This is not the place to review the
debate or enter into a very careful discussion, but I would like to make
clear my own views. First and foremost, the iterative conception has

[10] But here there is a common misunderstanding. The starting point is just one object
(the empty set), not the pure void of nothingness; we may as well start with the number 1, or
with a geometric point.

[11] Zermelo (1930), following von Neumann, showed how, given any model of ZFC, the
Axiom of Foundation allows one to "develop" the model into a cumulative hierarchy. This
was one of the technical origins of the iterative conception, together with type theory.

[12] In other words, the expression "*the* universe of sets" would be grammatically incorrect.
In each application of set theory we may think of a universe of sets V, but there is not a single
such thing, for, as we have already said, one can always conceive of further extensions.

certain *presuppositions,* and it seems obvious that it cannot justify what is presupposes! These assumptions range from the mild one of the empty set to two decisive postulations:

A.   Determined adoption of an *infinitarian standpoint,* admitting the existence of actual infinities.

This postulate can take at least two forms, depending on whether one starts by admitting actually infinite sets (such as the classical $\mathbb{N}$ or $\mathbb{R}$) or the transfinite iteration of the operation *set of.* The weakest version is to admit simple infinity ($\mathbb{N}$) without transfinite iteration; this is simple type theory as Gödel or Tarski used it around 1930 (see Ferreirós 2001). Strong versions admit transfinite iteration up to very large ordinals, whose postulation is in practice made explicit by adopting strong reflection principles or large cardinal axioms.

Interestingly and importantly, technical results in the metatheory of ZFC show that the extent of the iteration is unrelated to whether $\mathbb{R}$, classically interpreted, is or is not in the cumulative hierarchy. This is linked to the fact that large cardinal axioms do not settle the truth or falsehood of the Continuum Hypothesis. But, of course, $\mathbb{R}$ is meant to be in the hierarchy, according to the second key assumption:

B.   Adoption of the *quasi-combinatorial viewpoint,* i.e., admitting that all possible *arbitrary* subsets of any given set are likewise given.

Only this second assumption gives (a sense of) univocal meaning to the powerset operation, $\wp(S)$, in the sense of the standard interpretation of set theory. It gains univocal meaning, seemingly at least, at the conceptual level—but not at the formal level (see Chapter 4). As I explain elsewhere, no known set-theoretic axiom has pinned down formally what this standard interpretation of $\wp(S)$ is meant to be.[13]

Notice that this conceptual meaning of "all subsets" is also presupposed in so-called standard second-order logic (SOL)—and in this consists the famous vicious circle between full SOL and set theory. This problem is easily resolved by ceasing to call "standard" the full

---

[13] Ferreirós 2011 and Chapter 9 above. See also Jané 2005.

(set-theoretic) semantic interpretation of second-order logic. One only needs to avoid a simplistic conflation of two different things—roughly speaking: property-theoretic reasoning and the theory of arbitrary subsets. SOL is a perfectly acceptable logical system as long as one does not try to make it do serious mathematical work of the kind of fixing the meaning of $\wp(S)$ and determining the truth value of CH.[14]

The iterative conception cannot justify its own presuppositions. It does not justify the existence of $\varnothing$ (this may seem trivial)[15] and it does not justify the admission of actual infinity in either the form of the classical structure $\mathbb{N}$ or of the ordinal ω and its successors.[16] Concerning powersets, the iterative picture can be used to explain that given a set A of cumulative rank α the powerset $\wp(A)$ will be of rank $\alpha + 1$. But it certainly cannot justify the standard interpretation of $\wp(A)$ as comprising all arbitrary subsets of A, which it presupposes. To sum up, this Gödelian conception has nothing to say—in terms of justification—about the Axiom of Infinity, and little to say about the Axiom of Powersets.

On the other hand, there are many axioms that the conception motivates very well. This happens with Separation, for, given a set S, it will have a certain rank, α, and all its subsets will be there precisely at rank α, so certainly any subset obtained by Separation "is" there already. It happens with Union, for the union of a family of sets will be there in the same rank as the member of the family with highest rank. And also with Choice, for if a family of sets is there at rank α, corresponding choice sets must be there at a lower rank—this depends crucially on hypothesis B. I agree with Gödel (1944, 139; 1947, 177): no set-theoretic axiom is more natural than the Axiom of Choice, but this is because when doing set theory we adopt wholeheartedly the quasi-combinatorial

---

[14] We have touched upon this matter already in Chapter 7. Interpretations of SOL along lines that are free of commitment to "full" set theory have long been in use, e.g., among proof theorists. From a rather traditional, set-theoretic point of view, the standard semantics for SOL should be something like the general (Henkin) semantics. On this topic see Jané 1993 and 2005b, and the references therein.

[15] One might as well begin with $V_0 = \{a\}$; then $\varnothing$ would be an object in V_1 (for $\varnothing$ is the subset that corresponds to a contradictory property, or to the property $x \neq a$).

[16] Here ω need not be identified with the von Neumann set in the usual way; it may be regarded as an autonomous entity, perhaps defined in Cantor's way through generating principles. See Jané 2005 and Tait 2000.

standpoint. The iterative conception is not needed to justify AC: it is a consequence of assumption B alone.

As the reader can see, I side with those who have insisted on the idea that the iterative picture cannot do the whole work of motivating set theory, that it needs to be supplemented.[17] But more than discussing this side of the story, I am interested in offering a new look at the iterative conception from our practice-oriented standpoint.

10.2. A "WEB-OF-PRACTICES" LOOK AT THE CUMULATIVE PICTURE

When he proposed seven axioms for set theory in (1908), Zermelo said that he was unable to offer a simple definition of sets like those offered by Cantor (in 1895) or by the logicists. Thirty years later, a coherent, albeit quite complex, "narrative account" of the unfolding of the world of sets was available: Gödel's iterative conception. The fact is that the iterative picture of the universe V of sets comes up often, not only in basic presentations of set theory, but also in sophisticated exchanges among experts concerning technical results in set theory. The drawing of "cone representations"—as I shall call them, see Fig. 6—is very common in specialized courses on set theory, which aim at a first serious introduction of students to this topic. It is also frequent in connection with talk about models of set theory, their extensions, their inner models, or the elementary embeddings of a model into another.

This is just another example of the omnipresent use of visual representations in mathematical practice, not just at the elementary level (important examples being the real line, the complex plane), but also in nonelementary math (consider topological structures, Riemann surfaces, etc.). In the case of set theory, as of many others, such visual representations seem to emerge more frequently in verbal communication between mathematicians, and much less in print. As in the other cases, they deserve to be considered by philosophers of mathematical practice.

---

[17] See, e.g., Lavine (1994), 147*ff.*

The case of "cone representations" in set theory is particularly interesting when considered from the present standpoint of an interplay of practices. As discussed in previous chapters, my analysis moves away from the systematic reductionism that is often encountered with experts in foundations, to emphasize cognitive and practical interplay. Considered from this point of view, one should not forget that the study of set theory and in particular the iterative picture come into play at a time when the agents already have some knowledge of the real numbers, basic geometry, and basic analysis. What does this have to offer us? In my view, the iterative picture plays an important role in promoting conceptual understanding of set theory, i.e., in helping agents put together the different pieces into a coherent overall conceptual scaffolding (a "*Fachwerk von Begriffe*," as Hilbert liked to say).

Our basic question is thus: what resources does this kind of representation mobilize? Notice first that, in spite of talk about the universe of sets being "developed" or "constructed," the basic thrust of this line of approach is purely mathematical. It employs ideas and resources taken from more basic levels of mathematical knowledge and practice, to suggest a coherent picture of this new and sophisticated set-theoretic world. What resources, precisely? The cone representation relates back to arithmetic notions of iteration, to geometric ideas, and, also, to analytic refinements of the geometry.

Notions of iteration are arguably the most primordial idea in all of mathematics: our first encounter with them is in arithmetic, and they relate to mathematical induction; we found them also in definable sequences (Chapter 8, section 5). In the present case, a certain operation (the step from V_β to $V_{\beta+1}$) is iterated and the universe V is defined recursively. Notice also that the picture draws on basic ideas of the new set-theoretic practice, for it uses the notion of *transfinite* iteration, i.e., the step to limit ranks V_λ is assumed and V is defined by transfinite recursion.[18] This is not so difficult to present and to grasp, at least at the most basic level: the example of V_ω functions as a paradigm ("imagine now that all of the sets found in finite ranks, taken together,

---

[18] This kind of impurity—the mix of previous concepts with some of the newer ideas—may be typical: it may be found in Cantor, and one might also compare it with the case of the reals.

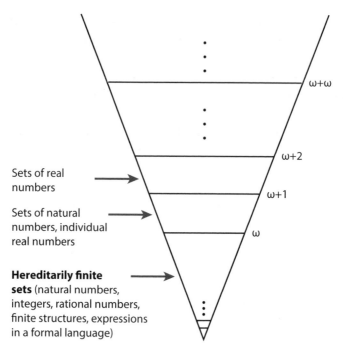

Figure 6 The "cone representation" of the world of sets from a hand-drawing by mathematician John Steel.[19] (Courtesy J. Steel)

form a new domain . . ."), all other cases V_λ, where λ is a limit ordinal, being understood by analogy with the first. In all cases, what happens is "simply" that one forms the union of all sets in previous stages: rank V_λ is defined to be the union of all previous ranks ($= \cup V_\alpha$ for all $\alpha < \lambda$). At the same time, a simple geometric conception comes in to further justify these steps.

Indeed, another obvious feature is that the iterative conception is linked with a very simple geometric picture, the drawing of a cone (or its plane section, a triangle) with an axis. Along the axis we envision the ordinal numbers, while the parallel cone sections going through each one of the ordinals represent the different ranks V_α. Not much emphasis is put on the admissibility of writing ω_0 at a finite distance from the base, or ω_1 at a finite distance from ω_0, when there is "actually" an

---

[19] See also Kunen 1983 and Friedman 2002, among many others.

infinity in between. Here, I suggest, previous experience with the basics of analysis is likely to play an important role: if, mixing representations, we apply the conception of a real line to the vertical axis of the ordinals, it becomes easy to see how one can mark a sequence of points corresponding to 0, 1, 2, . . . that converge to ω_0 (think of a monotonically increasing convergent sequence). This justifies placing ω_0 at a finite distance from the base, the paradigmatic case, and now the idea is simply to iterate the same process.

On the basis of such considerations, which are rather elementary ideas of a geometric-analytic character, "it is not difficult to see"—as an instructor may well say—that the above geometric picture with its ordinal marks along the axis is, indeed, coherent and admissible.

Rank V_λ is defined as the union of all previous ranks ($= \cup V_\alpha$ for all $\alpha < \lambda$) whenever λ is a limit ordinal, regardless of its cardinality—this notion of course is a very sophisticated one; it belongs to advanced mathematics. But the student is likely to be ready for it, due to previous exposure to mathematical analysis. The cone representation draws very heavily on infinitarian ideas, but students of mathematics are indoctrinated with such ideas from the first day of university courses. The iterative picture is presented to mathematicians or students of mathematics who have already accepted the infinitarian standpoint—if only in the guise of the real numbers (which are marked by infinity no matter how we conceive of them; see Chapter 8).

Before presenting the picture, the learners are carefully instructed to conceive of subsets of infinite sets by analogy with the subsets of a finite set. ("Exercise: write all of the subsets of set $T = \{a, b, c, d, e\}$, not forgetting T itself and $\varnothing$." Immediately, students are invited to view the subsets of infinite sets analogously.[20]) This makes them acquainted with the key thought of the quasi-combinatorial perspective, the admission of arbitrary subsets. Whether the student is fully convinced at this point or not, perhaps it does not matter so much. As a matter of fact, the good pedagogical practice followed by most instructors is to downplay the conceptual and foundational difficulties implicit in the material,

---

[20] Of course, some explanations may be given of the kind: "if we are dealing with $\mathbb{N}$, of course, due to the (psychological? practical?) limitations we have, one cannot write explicitly all the subsets, but *in set theory we regard them as given*, each and every single one of them, both the finite subsets and the infinite subsets."

presenting all of it as natural and almost necessary. The point for us is that, at this stage, students have available the two key assumptions A and B that lie at the basis of the iterative conception (section 1). On this basis, the cone representation is set in place, mobilizing basic ideas of arithmetic, geometry and analysis; and the emerging picture motivates very well many axioms of set theory, integrating the different elements of set theory into a coherent whole.

The argument for drawing ω_1 at a finite distance from ω_0 is never done explicitly, as far as I know, and here of course great difficulties would be faced. We have already discussed aspects of this matter in Chapter 9. Iterating the argument for ω_0, one can easily justify the extension of the "cone representation" to quite large countable ordinals.[21] The problem, however, is that the set of all countable ordinals cannot be thought of as the union of countably many disjoint subsections (each of them countable); it is, rather, an uncountable union of such subsections. And we face once again the difficulty that a well-ordering of an uncountable set cannot be given explicitly. As one can see, from this point of view we also find a basis for the objections that Borel, Baire, Poincaré, and others made to set theory.

In effect, as already mentioned, the cone representation does much more than that. For Zermelo, the cumulative picture motivated the idea of an *open-ended universe* of sets, which can always be extended further. That, in turn, suggested an analysis and conceptual solution to the paradoxes, "apparent" contradictions which—in his view—rest merely on "a confusion between *set theory itself*, which is not determined categorically by its axioms, and the particular *models* that represent it" (Zermelo 1930, 47). Set theory is not categorical, its models are always extensible,[22] the universe is open-ended—and this is an "*advantage*" (op. cit. 45): what in one model appears as an "ultrafinite non-set or

---

[21] Cantor and others in his time employed the following idea: take a sequence of (countably many) disjoint intervals on the line converging to a point (the lengths of the intervals will be diminishing); consider an infinite sequence of points in each of the intervals. This can be adapted to the "cone representation" and its ordinals.

[22] Considering a *second-order* version of ZFC, Zermelo was led to a quasi-categoricity result according to which models of set theory are characterized by their "height," which must be an inaccessible cardinal. See Hallett's introduction to the paper in Ewald (1996), vol. 2. To emphasize once more: relying on second-order ZFC is in effect a kind of naïve set theory, a relapse into an uncritical approach to powersets.

super-set" is already in the next higher model a fully valid "set" with its cardinal number and order type. To the unlimited series of ordinals corresponds an equally unlimited series of essentially distinct models of set theory, "in each one of which the whole classical theory expresses itself." And so the antinomies of set theory, well understood, lead the science of mathematics "not to a narrowing and impoverishment, but rather to an unfolding and enrichment that is still impossible to gather with our gaze" (Zermelo 1930, 47).

The later proposals of Gödel (1947, 181) that the operation "set of" can always be iterated further with the help of stronger axioms of infinity (large cardinal axioms), and that the set-theoretic universe cannot be characterized mathematically—which justifies strong reflection principles—can be regarded also from a similar standpoint. That, however, was not exactly Gödel's point of view, since his opinion was that "to say that the universe of all sets is an unfinishable totality does not mean objective indeterminedness, but merely a subjective inability to finish it."[23] This seems to be close to Cantor's point of view: all-embracing totalities are objectively given and determined, but the human mind cannot grasp them; the Absolute is given in itself, though it cannot be given to us.

It is easy to see that such points of view will not be attractive to everyone. The concept of an open-ended (not to mention a transcendent, incomprehensible) mathematical universe can easily make nervous many a practically oriented mathematician. Thus, it is little wonder that many mathematicians have preferred to trim the set-theoretic system, bringing it down to a minimum as long as this suffices for their theoretical developments and problem-solving practices. Let me give three examples: the Bourbaki group oriented their choice of set-theoretic axioms in a down-to-earth, pragmatic way, and thus Dieudonné, a core member of the group, proposed adhering to the ZFC system (with just the countable version of the axiom of Choice) as a basis for "classical analysis" and for "the most spectacular progresses obtained in mathematics after 1950."[24] For another point of view, which takes

---

[23] Gödel, according to Wang (1997), 260.

[24] See his preface to Dugac (1976, 10–11): "Thus one understands better the reticences of Dedekind, Poincaré, or Lebesgue vis-à-vis the Cantorian school." In general, the Bourbaki tended to despise logic and consider a waste of time the more speculative questions of higher set theory, unrelated to more explicitly mathematical concerns.

more seriously some of the problems that are simply disregarded by Dieudonné, Devlin (1984) has been one of the proponents of the axiom of constructibility, $V = L$, as a legitimate way of rounding up the axiomatic system and fixing many of the open problems (Quine too, unsurprisingly, viewed this as natural). And, more radically, aside from proposals that simply give away with set theory as a foundation, there exists a large body of work on constructive and intuitionist set theories CZF and IZF.[25]

10.3. CONCEPTUAL UNDERSTANDING

Conceptual understanding of a theory is no doubt a complex thing, but even the understanding of a single result can be multilayered. Here, we are focusing on the role of a certain kind of representation—the spatial picture of the world of sets as a cumulative hierarchy, a well-ordered system of ranks—but this by itself is obviously not enough. To avoid the impression that understanding is being reduced here to the availability of a visual picture, let me briefly sketch some of the factors that, in my view, are linked with the phenomenon of mathematical understanding. Then, I will say a few more words on the particular case of set theory.

Issues of understanding arise at many different levels, and there is of course the level of learning a theory, i.e., getting to know it and learning to use it. From the standpoint of our investigation in this work, it may seem better to concentrate on the *advancement* of understanding, but it is worthwhile dwelling on the seemingly simpler question of understanding by the learner (among other reasons, because it is not obvious). This aspect of practice can be seen reflected in problems and examples in textbooks, designed and chosen (more or less consciously) to promote understanding and clarification. It involves not only solving problems but—importantly—avoiding mistakes; learning about the rationale of certain things (e.g., a definition or even an axiom); grasping the coherence of results in the global configuration of theories; and

---

[25] See Crosilla 2009.

more. As Giaquinto (2005) comments, a successful justification of an axiom or axiom system or definition does not prove its truth, but rather warrants our adopting it.

In a case such as that of the set theory ZFC, one finds relevant work of understanding by the learner in comprehending its role in systematizing math: deriving the number systems; introducing in a uniform way algebraic structures and topological or metric spaces and functions and function spaces; and so on. This is the goal of relevant parts of introductory courses in set theory, but it is also served by the use of set theory (if only as a language, with little theoretical development on top of it) in specialized textbooks and courses, e.g., of algebra or analysis or topology. Implicitly or explicitly, this provides motivation and—significantly—extrinsic evidence for the axiom system, for as long as one accepts classical analysis, the fact that it can be derived from ZFC lends support to this system.

Such forms of understanding by comprehension of the role in a systematic presentation of branches of mathematics, I believe, are not limited to a foundational system such as ZFC. Comparable elements form a relevant part of the learning about the system of real numbers $\mathbb{R} = <R, 0, 1, <, +, \cdot>$ when we get to know their role in connection with problems of measuring or with processes studied in classical analysis, or in providing coordinates for points in a given space and making possible a systematic interconnection of algebra with geometry. And the same could be said of topological spaces, or algebraic structures, or, even more, of category-theoretic notions. Much of the learning of any central mathematical concept has to do with learning to use it in the context of some particular subdiscipline or branch of mathematics, and there are paradigmatic cases of such uses. Cartesian coordinates have been a paradigm (in the sense of an exemplary case studied generation after generation) in the study of real numbers, just as the investigation of equations via Galois theory provides paradigmatic instances for the understanding of algebraic structures of group and field, and how they are interconnected.

But representations also play a role in such contexts. Mathematical explanations often involve visuo-spatial representations, and understanding a mathematical fact, as opposed to merely knowing it, may depend on visuo-spatial cognition (Giaquinto 2005, 86). At this level, as in relation to the aspects mentioned in previous paragraphs, one can and should expect to find *individual differences*. When talking about ways in which individuals attain understanding, different cognitive styles are obviously going to emerge: the geometric-spatial ingredient that may be centrally relevant to some, might be despised by others as mere illustration;[26] the careful step-by-step derivation of a long chain of results, in Landau style, may be criticized by some mathematicians as totally boring and unenlightening.[27] Hilbert himself wrote that, notwithstanding the abstract tendency of modern geometry and its importance, "it is still as true as it ever was that *intuitive* understanding plays a major role in geometry. And such concrete intuition is of great value not only for the research worker, but also for everyone who wishes to study and appreciate the results of research in geometry" (Hilbert and Cohn-Vossen 1932, Preface).

All aspects of understanding that appear in the context of learning already-established mathematics are pointers to elements of understanding that may be important in the development of math, that is, in the *advancement* of understanding. Let me draft a provisional list, which the reader should merely take as an initial step in the discussion of this matter:

1. Problems are posed and ways of solving them have to be found, which can involve conceptual difficulties—the need to find the proper *formulation* (basic concepts, basic propositions) that can make things explicit in the right way, enabling proper methods and avoiding mistakes.

---

[26] Hadamard 1954 (with a preface by Johnson-Laird). Here, Hadamard takes over the distinction between "logical" and "intuitive" mathematicians proposed by Klein and Poincaré, those who think in images and synthetically vs. those who think in symbols through step-by-step inference. Hadamard, and the leading scientists he asked, used introspection to describe mathematical thought processes. Like Einstein, he described his own mathematical thinking as largely wordless, often accompanied by mental images that represent the entire solution to a problem, only in a second phase linking to logical concepts. I should note that introspection is heavily discredited today.

[27] Landau's books and lectures were famous for their dry, laconic style, and the unintuitive (but simple and rigorous) way of presenting concepts; short and precise proofs, the classical structure of theorem/proof, theorem/proof, with no motivation. In the preface to Landau (1930), he writes: "Please forget everything you have learned in school; for you haven't learned it."

2. The definitions, axioms, and methods employed are sought and motivated by taking into account the overall coherence of results in the field, by reflecting on their interdependences—with an aim at simplicity and rigor and explanatory power.

Judging those virtues can be influenced by individual differences in expertise (areas and methods to which much time has been devoted) or in cognitive style of the kind mentioned before (e.g., preference for geometric-spatial representations vs. preference for analytic-symbolic presentation).[28]

3. Gains in understanding often involve the elaboration and use of spatial representations—think again of geometric representations of the algebra of complex numbers, of the Riemann surfaces in complex analysis, or of diagrams such as the cumulative picture of sets.

4. Improved understanding is obviously attained via deepening knowledge of the theory, which is often gained by concentrating on the study of special cases (descriptive set theory in our case). The Berlin school of Weierstrass and Kronecker insisted on the importance of special cases as a guide to the truly general, and Hilbert picked up on this view.

5. A typical by-product of such developments is reformulation of the theory, which may affect concepts, methods, and even axioms. A typical theory in math (or in any science) will suffer significant changes over relatively short periods of time.

6. Understanding may also be gained by critical examination of alternatives, e.g., to classical analysis, such as constructive analysis; or in our case radical alternatives to ZFC and their interrelations with that system (like NBG, Morse-Kelley set theory, or, more radically, type theories, Quine's system NF, and CZF).

---

[28] The word "intuition" should rather be avoided, unless it is carefully defined and refined. Common talk of intuition, even (or perhaps especially) among philosophers, is heavily polysemic and thus confusing.

7. Improved understanding is also obtained by examination of the "deductive network" of axioms in Hilbert's style (axiomatic analysis): studying what depends on what, what can be done without what, strengthenings and weakenings of principles.

8. Also important can be metamathematics, work in foundations and mathematical logic, which in the case of set theory has significantly reshaped understanding: consider independence results, ZFC models, large cardinals and their consistency strength, and so on.

Our guide in the exploration of all these aspects of understanding must be careful attention to past and present mathematical practice. As a rule, whenever you find elements of practice that are stabilized and consistently repeated, you should pay attention to them. This is precisely the case with the "cone representation" as a visuo-spatial representation of the cumulative picture of sets (or, as some may prefer, as a mental image of the same), which is a "fixed point" in set-theoretic practice at many levels—including the crucial metatheoretic level.

The case of set theory is interesting, too, because unlike other examples from the history of mathematics, the common conceptions of its basic notions had to be abandoned and one had an axiom system (Zermelo 1908) long before a crisp understanding of the basic concept of set behind the axiom system or a clear view of an intended model.[29] In the subsequent decades, experts faced problems of justification of the axiom system, including the question of finding a model that incorporates the intended subject matter. Thus, the von Neumann hierarchy and Gödel's conception of sets as the product of iterated applications of the "set of" operation were received as significant steps toward a deeper understanding of set theory. Moreover, in this development one finds the problem of reformulation, understanding, and justification of relatively complex bodies of theory. An example is the Cantorian theory of

---

[29] Perhaps there are similarities with non-Euclidean geometry, where the conviction that the axiom system was functional and consistent came before indubitable examples of models were available.

transfinite cardinals and ordinals, whose basics have not changed since 1895, but which has seen enormous transformations nonetheless.

Cantor (1883, 1895/97) had left quite a well-rounded body of theory, including the "discovery" of the alephs, their links to "classical" examples of sets, the arithmetic of the $\aleph$s, the basic theory of the ordinals, and the intriguing connection between ordinals and alephs by means of the "number classes."[30] A few core issues remained to be settled, especially the Cantor-Bernstein theorem, the Well-ordering theorem, and the problem of the continuum—the last two were highlighted by Hilbert (1900, problem 1). But, the solution to the first issue was found before 1900: Bernstein (and Dedekind) were able to prove the Cantor-Bernstein theorem; and few years later Zermelo established the Well-ordering result on the basis of AC, albeit this became a source of great controversy. With Zermelo's theorem, there was assurance that the cardinality of *any* set is an aleph.

Subsequently, worries about the basic concept of set induced by the paradoxes and by doubts concerning the Axiom of Choice destabilized the whole situation. It was necessary to find a proper formulation of the theory, to redefine the basic concepts and axioms, or else to motivate the abandonment of the whole theory. Thus an initial period, roughly from 1900 to 1940, saw a good number of reformulations of set theory, and the proposal of alternative mathematical foundations that abandoned the Cantor-Dedekind orthodoxy. Not everything from Cantor could be retained, e.g., it was basic to admit that one could not speak of all cardinals and say that they formed "a well-ordered set" (as Cantor was still doing in 1892). This led to an era of set theory inside ZFC (or inside type theory, or the Neumann-Bernays system, NBG), more and more stable from 1920 on, which saw a good number of triumphs; an example from the study of "describable" sets (descriptive set theory) are the proofs that the Borel and analytic sets comply with CH and are Lebesgue measurable.[31] Notice that the transition to an axiomatic setting initially forced Zermelo to abandon Cantor's ordinal and cardinal

---

[30] Class II includes all the countable ordinals, and the class as a whole is of power $\aleph_1$; class III is formed by the ordinals whose cardinality is $\aleph_1$, and the class as a whole is of cardinality $\aleph_2$. The pattern goes on, see section 9.5.

[31] For a historical account, see Kanamori 1995.

numbers (replaced by theories of well-ordering and of cardinals), so it also took some effort to obtain ways of reintroducing the transfinite numbers in the axiomatic setting.

However, this period was troubled by the general "atmosphere of insecurity" in foundational studies, by the resolute rejection of arbitrary sets among key players such as Weyl and Brouwer, and by doubts about logical principles and consistency. Moreover, the period ended in unsolvable questions, in particular due to the lack of progress on the basic question about the power of the continuum (truth or falsity of CH) and the desperation of descriptive set theorists when confronted with the study of co-analytic and other projective sets (remember Luzin's "one does not know, and one will never know" quoted in Chapter 9 section 6). Perhaps one can say that the Cantorian theory of transfinite numbers had a proper foundation by 1940, though the following period brought increased understanding of the problems and complexities it involved.

This second period was marked by a meta-theoretic turn initiated by Gödel that would increasingly affect set theory and color the experts' understanding of all relevant issues. Gödel introduced the first effective moves in the model theory of ZFC, and his study of the minimalistic system of the constructible hierarchy $V = L$ (which in his mind came accompanied by the cumulative picture) led to significant progress with the proof that AC and CH are consistent relative to the basic Zermelo-Fraenkel system without Choice (Gödel 1940). At the same time, there was renewed insecurity due to Gödel's discovery of the phenomenon of incompleteness of formal systems (undecidable sentences), but Gödel himself reacted positively and insightfully proposing large cardinal axioms and the *Gödel program*. After the Second World War, sophisticated study of reflection principles, measurable cardinals, and so on, was finally supplemented by Cohen's invention of the forcing technique in 1963. The leap forward that axiomatic set theorists were able to take within the next decade can only be called enormous: a plethora of independence results emerged to make the overall picture more complex; but also candidate axioms providing a satisfactory theory of the whole projective hierarchy (determinacy axioms, PD) eventually derived from large cardinals in the 1980s. Nevertheless, this period also witnessed another blow to hopes of resolving the continuum problem: the Levy-Solovay theorem, establishing that no large cardinal axiom, from the level of measurables onward, would bring a solution (Levy and Solovay 1967).

The greater and greater entanglement of metatheory and theory, led by Gödel and Cohen and Solovay, has continued to this day. In the absence of basic new insights into the properties of sets at low levels of the cumulative hierarchy, which may bring an answer to the continuum problem, set theorists nowadays tend to find inspiration for new developments in the properties of models of ZFC under forcing. What started as a theory of infinite arbitrary sets is nowadays, to a large extent, a theory of forcing variance or invariance (depending on whether one is a pluralist or a non-pluralist):[32] forcing axioms of different kinds, Woodin's initiatives, etc. The recent, closely followed attempts by Woodin to prove or disprove the CH by means of his Ω-logic, and his valiant search for "Ultimate L", fall under this rubric too (see Koellner 2013). The period also brought a satisfactory outcome in the empirical phenomenon that large cardinals appear to be linearly ordered—the scale of consistency strength—although the deeper reasons behind this are still to be clarified. As usual, the search for understanding continues, and it aims beyond the mere establishment of new results inside the known systems.

Let us now come back to the question of consistency of the Cantorian theory of ordinals and cardinals. It was Hilbert's hope that this could be proven by finitary metamathematical means, and Gödel made a highly relevant step forward with his proof of relative consistency for AC and CH. But Gödel also made clear that a consistency proof was not possible by finitary means, and a further 80 years of development of proof theory suggest that no reasonable means to prove consistency are within reach. Yet a great majority of mathematicians believe that the Cantorian theory and ZFC (or even ZFC augmented by large cardinals axioms)[33] are consistent. This is a fact of mathematical practice, and so the question arises, What are the sources of this belief in consistency?

---

[32] See the work of Hamkins (2012) for details about recent proposals of a set-theoretic "multiverse."

[33] How large exactly is a debated question. Inaccessible cardinals are generally admitted, but measurable cardinals have been much less clear. A great expert in set theory, Jack Silver allegedly was guided in his work (1970s) by the goal of establishing the inconsistency of the assumption of measurable cardinals. The fact that he failed in this has actually augmented the confidence of large-cardinal experts in their matter.

The answer is not so simple as to be given in a single sentence, but part of the answer lies in the conceptual understanding of set theory that has been gained in the last century. To simplify our discussion, consider just ZFC, a system that establishes the existence of $\aleph_1, \ldots \aleph_\omega$ and higher cardinals (but not of inaccessible cardinals, "small" as these are considered today). Strict metamathematical results reduce the consistency of ZFC + CH to that of ZF, whose axioms include Infinity, Powerset, and Replacement. This gives a boost to reasonable expectations of consistency, and it marked the end of the long tradition of treating the Axiom of Choice as a dangerous assumption. However, due to the impredicativity of the system ZF, advances toward a proof of consistency (even, say, relative to intuitionistic set theory)[34] have been impossible.

Meanwhile, of course, a century of theoretical developments have led to many alternative formulations of axiomatic set theory, to important results concerning their interrelations, and to highly sophisticated work on the model theory of ZFC and its extensions. And all of this provides what are often called "empirical" reasons for belief in the consistency of ZFC. Added to this is the apparent necessity of ZFC for classical analysis developed in the usual way—though, in effect, proof theorists have established that axiom systems much weaker than ZFC are sufficient to derive the theorems of classical analysis.[35]

That, I surmise, would be *insufficient* to promote the levels of confidence in set theory that we experience in the absence of clear conceptual perspectives on set theory, its basic principles, and the concepts underlying them. This is in the spirit of an analogous statement that I find quite remarkable, due to Vaughn Jones, a Fields Medallist known mostly for his work on von Neumann algebras and knot polynomials; he writes:

> Proofs are indispensable, but I would say they are necessary but not sufficient for mathematical truth, at least truth as perceived by the individual. (Jones 1998, 208)

[34] There is work by H. Friedman proving consistency relative to an intuitionistic version of ZF set theory—but this is only *formally* intuitionistic, insofar as the basic logic is; still, the system is not acceptable for strict constructivists. See Crosilla 2009.

[35] On this topic, see Feferman 1998 and Simpson 1999; the first considers predicative systems, the second subsystems of second-order arithmetic in the style of Friedman.

In effect, truth as perceived by the individual is what philosophers call belief; but most mathematicians would have no trouble saying that the ZFC axioms are true. Jones wants to say that intuitive arguments allowing us to better understand or explain the results are indispensable at the level of research mathematical practice. My point is that this has applied in the history of set theory, too. We may replace the unclear talk of "intuition" by considerations of conceptual interpretation—which often include spatial representations like the "cone picture" of the cumulative hierarchy of sets.

10.4. JUSTIFYING SET THEORY: ARGUMENTS BASED ON THE REAL-NUMBER CONTINUUM

Let us assume that the multilayered complex of ingredients reviewed in the previous sections have lent support to the iterative conception of "sets of," so that we have, on those grounds, arguments for the axiom systems ZFC. Still, as we emphasized, two assumptions remain insufficiently grounded: the two basic postulates underlying the cumulative picture, namely, the adoption of

A. an infinitarian standpoint, i.e., acceptance of actual infinities;

B. the quasi-combinatorial standpoint, i.e., admission of all arbitrary subsets of any given set.

In Chapter 8 we argued that, when the set-theoretic approach to mathematics got started, in 1872, acceptance of the actual infinite arose in the dual form of *simple* infinity (the natural numbers as a totality) and *arbitrary* infinity (the collection of all sets of natural numbers). This was precisely due to the centrality of analysis and the real numbers to math and to continuum-based representations of the physical world. The parallel between these dual principles and the postulates A and B should be obvious.

Underlying all this was the *pointillist conception* of the continuum, whose exploration was to Luzin (1930, 2) the main goal of set theory and a "question of the utmost importance." By 1900, analysis formed the main bulk of mathematics and all of physics was continuum-based; the notion of the continuum, interpreted through the lens of Dedekind-Cantor-Hilbert theories of the real numbers, was at the very center of mathematical knowledge. It was then that the most crucial existential hypotheses of modern mathematics were codified and accepted—especially in the form given to them by Zermelo in his axiom system. (Analysis and the real numbers are still today at the center of the university curriculum of mathematics, being a sine qua non of the subject, pure or applied.) We can now complete the picture by considering again how the pointillist conception of the continuum as an actually infinite set of points lies behind our adoption of set theory.

The strongest hypotheses of set theory can be justified on the basis of the classical continuum, but hardly without it. In particular, the two basic assumptions, A and B, that are necessary for the iterative conception of sets arise naturally out of the idea of the pointillist continuum. Postulate A is the resolute adoption of the actual infinite, and indeed, in order to conceive the continuum as built up from points, one is forced to admit actual infinities. To see this, it is not even necessary to invoke the structure of continuity (completeness) of $\mathbb{R}$: already the denseness structure of the continuum implies that any interval $[a, b]$, however small, will contain an infinity of points.[36] This entanglement with actual infinity was Aristotle's argument, in the *Physics*, to justify his rejection of the pointillist conception of the continuum.

Once a point set continuum such as, say, $[0, 1]$ is accepted, arguments similar to those used in Chapter 9 section 1 force us to accept definable subsets of this, such as the set of points in the unit interval whose coordinates are rational, which can be defined as $\mathbb{Q}_{[0,1]} = \mathbb{Q} \cap [0, 1]$. And the admission of $\mathbb{Q}_{[0,1]}$ or, for that matter, $\mathbb{Q}$ itself as an infinite set will lead to admitting $\mathbb{N}$ (in Chapter 9 we saw a different, more historical line of argument leading to the same conclusion). Moreover (Chapter 9 section 3), once we admit actually infinite sets like the classical $\mathbb{N}$ and $\mathbb{R}$, arguments based on properties of the classical structure of $\mathbb{R}$ force us into accepting that $\mathbb{R}$ cannot stand in one-to-one correspondence with $\mathbb{N}$; hence the existence of two different cardinalities.

I am led to conclude that, once we start moving along the path of a pointillist conception of the continuum articulated in terms of sets of

---

[36] Since $[a, b]$ is dense, there is a c such that $a < c < b$, and this can be iterated indefinitely, taking $[a, c]$ and $a < d < c$, and then $[a, d]$, etc.

points, Cantor's conclusion that there exist different "kinds" of infinities, with different "sizes," is inescapable.

Some authors have questioned the *naturalness* of this last conclusion, arguing that the idea of cardinality as based on equipollence (one-to-one correspondence) is only one theoretical option among others.[37] I would reply that Cantor's concept of cardinality is the natural one in the context of abstract sets, freed from extra structure of a metric or topological or algebraic kind. As a matter of historical fact, the distinction between metric and topological structures was already sufficient for Cantor to articulate it: his concept of cardinality (based on equipollence) emerged in the context of studies of point sets, when he was still assuming topological structures (e.g., limit points). A process of liberation from metric notions and gradual movement into topology was needed for set theory to become possible.[38] Cantor ended up conceiving pure set theory as dealing with sets without metric, topological, and even order structures. The concept of cardinality or "power," however, was not an end result of this process, but was obtained relatively early; it was instrumental, together with ideas about ordinals, in leading to the completion of the process.

Be that as it may, it should be clear that the pointillist continuum motivates adoption of an infinitarian standpoint, and thus leads to postulate A. Even the work of Bolzano can be used to substantiate this point, for although he did not arrive at a theory of sets like Cantor's, his efforts to rigorize and arithmetize the theory of real numbers forced him (from 1830) to accept a fully infinitarian standpoint. It is noteworthy that his work before 1820 had exactly the opposite orientation: he was explicitly rejecting actual infinity.[39]

Let us now consider B. As we saw in Chapter 9 section 3, classical ideas about real coordinates and, in particular, decimal expansions as tools for the determination of points in such a continuum build a

---

[37] See, e.g., Mancosu 2009, which reports on recent work exploring the part-whole principle (if a collection A is properly included in a collection B then the "size" of A should be less than the "size" of B).

[38] This, I believe, is an important philosophical and mathematical insight; for a historical exposition of the idea, see Ferreirós (1999), sections II.6 and IV.3. Cantor elaborated on the concept of cardinal or power in the 1870s; he presented it for the first time in (1878).

[39] See Bolzano (2004), ed. Russ. The case of Bolzano is interesting from the point of view of the part-whole principle (Mancosu 2009), as he privileged it against the one-to-one correspondence principle. I keep thinking that this was because Bolzano had not freed himself from the dominance of metric considerations (see Ferreirós 1999, 74–76).

powerful argument for the admission of *arbitrary* subsets—that given any set, all its possible arbitrary subsets are given. (One should be careful here with the methodological complexities involved, for it is possible to adopt a more or less "Aristotelian" conception of the continuum and treat the reals and decimal representations only as tools. But once the whole issue is rethought in the context of postulate A with $\mathbb{N}$, $\mathbb{Q}$, and $\mathbb{R}$ admitted as actual infinities—extensional objects, i.e., sets—one is led at the very least to admit arbitrary subsets of $\mathbb{N}$.) It is perfectly clear that decimal expansions of the real numbers are not restricted to those specified by formal conditions or formulas; and this leads also to rejecting the view that infinite subsets of $\mathbb{N}$ have to be defined (section 9.3)—and likewise for subsets of $\mathbb{Q}$ or sequences of rational numbers.[40]

Postulate B, that is, the full adoption of a quasi-combinatorial viewpoint, is a simple and natural generalization of that conclusion. What is true of $\mathbb{N}$ and $\mathbb{Q}$ should be true of $\mathbb{R}$ itself, especially given that nineteenth century mathematicians, in general, regarded the real (and complex) numbers as perfectly equal citizens, on par with the naturals. Thus, one admits arbitrary sets of real numbers, which poses the problem of the continuum (truth or falsity of CH) in all its complexity.

Even though the important and subtle conceptual and methodological points involved here were not made explicit, this much was presupposed in early work on point sets such as that developed by Hankel, Cantor, du Bois-Reymond, and others in the 1870s (see Ferreirós 1999, Chapter V). When they analyzed the properties of infinite point sets (meaning sets of points on the real line or in Euclidean space), they were implicitly assuming that the point set may be arbitrary, i.e., not given by an explicit definition. As we have explained, the quasi-combinatorial viewpoint is tantamount to a rejection of definability and constructivity as requirements, and it also involves an analogy between the finite and the infinite.[41]

---

[40] Conversely, if one were to insist on the view that mathematical notions must be defined explicitly on the elementary basis of a theory of natural numbers, this would lead to a rejection of the real-number continuum. The work of Kronecker, Weyl, and Brouwer is a perfect illustration of this point.

[41] The contrary standpoint, adopted, for example, by Weyl (1918), underscores the *inexhaustibility* of the infinite (in particular the continuum); hence its radical disanalogy from the finite.

We have seen that only assumption B gives meaning to the powerset operation, $\wp(S)$, in the sense of its standard interpretation in set theory. Thus, it gains univocal meaning at the conceptual level—seemingly at least—though not at the formal level; this is one of the most important instances of the complementarity of thought and formula (see Chapter 4). It should also be underscored that the Axiom of Choice is a consequence of postulate B, as Bernays (1935) remarked, for this postulate makes explicit that the subsets of a given set are by no means restricted to those definable in some theoretical framework or formal system; arbitrary subsets are fully countenanced in set theory, and this justifies the particular—and very simple—kind of case envisioned in the Axiom of Choice.[42]

With these considerations, based on postulating the pointillist continuum, we obtain motivation and even justification for the basic postulates A and B on which the iterative conception is based. Thus, we motivate the Axiom of Infinity, the Axiom of Choice, and, also, the standard reading of the Axiom of Powersets. Adding now the iterative picture of the universe V of sets, one can go on motivating the Axioms of Separation (though this is also justifiable on the basis of B), of Union, and of Foundation. (The case of Replacement is less clear, and may need extra consideration.) Thus the process of accounting for all, or almost all, of the axioms in the system ZFC can be completed.

Notice that, from the points of view of conceptual interpretation and visuo-spatial representation, this adds something new. We have seen that, to a large extent, the ZFC system is simply needed to make sense of the pointillist continuum. If there were any inconsistency in ZFC, this would, with high probability, translate into inconsistencies in the conception of the (real) line as an actually infinite set of points (linearly ordered, dense, and complete). The pointillist concept of the line has been known since antiquity, and has been the topic of innumerable philosophical and mathematical discussions. But its modern reformulation in set-theoretic fashion, obtained by Dedekind and Cantor, has been so satisfactory as to almost erase the traces of polemics.

---

[42] In fact, Zermelo (1908b) was already arguing along these lines.

Even if the conception of the line as point set were a mental construction, in the sense that nothing physical corresponds to it, should we
expect to find inconsistencies in it? Are we justified in believing that
this conception is perfectly admissible, and hence consistent? I take it
that the answer is yes, and this lends further support to the idea that
the ZFC system will not be found inconsistent any time in the future.

10.5. BY WAY OF CONCLUSION

This book has offered a long and winding argument for the cogency
and the well-justified nature of so-called classical mathematics. According to our view, the basic modern edifice (number systems, geometries,
algebra, analysis, basic set theory) cannot be regarded as a priori or as
a body of necessary truths; but we have presented it as a body of knowledge strongly interconnected, whose branches are linked systematically
among themselves and ultimately with elementary mathematics—
where, I argued, certainty claims may be upheld. In previous chapters
we argued for the tight links between the theory of real numbers and
many parts of the body of math, reviewing its well-known connections
with geometry. In the last two chapters we have discussed the justification of axiomatic set theory.

The argument has been that an axiom system such as ZFC can be
recognized to be a well-grounded infrastructure for classical math
in general, and the real number system in particular. If we take for
granted the need for a notion of the continuum, and also accept a certain conception of the continuum (the mainstream view since at least
1870), namely the pointillist conception, then we find justification for
the assumptions A of actual infinity and B of arbitrary subsets.[43] These
two assumptions are the basic presuppositions behind the celebrated
iterative conception of sets. Take the collection of points on a line, or
the collection of real numbers, to be a mathematical object, a set, just

---

[43] An alternative is to prefer an Aristotelian conception of the continuum, for which
Brouwer offered an insightful modern reconstruction inside intuitionistic analysis. Another
is to employ systems that adopt infinite sets but not arbitrary sets.

like the simpler set of natural numbers.[44] Then you have good justification for considering the cumulative structure of sets, at least in its more basic versions.[45] This kind of universe, interpreted in light of the mentioned assumptions, suffices for justifying basically all of the axioms (except Replacement). Furthermore, the conceptual understanding obtained through knowledge of set theory and of its "cone representation" (including technical results such a Gödel's relative consistency proofs) provides convincing arguments for the cogency of this theory—indeed, for believing it consistent, even in the absence of formal proofs.

Considering that the most delicate parts in the (soft) justification of the $\mathbb{R}$-structure and ZFC that we have been discussing boil down to issues about the real numbers, i.e., about mathematical analysis, the large body of knowledge having to do with pure analysis and with analysis-based science (mathematical physics in particular), and their long history during the last four centuries, become relevant. Continuum-based representations of physical phenomena such as the electromagnetic field (Maxwell) or the gravitational field (Einstein) may be wrong, of course, but there seems to be no reason for doubting their consistency. All of this adds to the reasons for belief in the consistency of axiom systems for $\mathbb{R}$ and for the class of sets V. Of course, that will be insufficient for the rigorist logician or the staunch formalist, but it is certainly enough for the great majority of practicing mathematicians and for many logicians.

What kind of justification is this? Ultimately that depends on the grounds on which we postulate the pointillist continuum, and on the kinds of claims that we aim to infer from the cogency of analysis and set theory. It is also a historical justification; as Weyl wrote: "Without the concepts, methods and results found and developed by previous generations, right down to Greek antiquity, one cannot understand either the aims or achievements of mathematics in the last 50 years [1900–1950]" (Weyl 1951, 523).

---

[44] A natural alternative would be to deny that $\mathbb{R}$ is like $\mathbb{N}$, to regard the domain of real numbers as open-ended, and thus to go in the direction of the predicativism of Weyl (1918) and Feferman (1998).

[45] For example, the first ω ranks built on top of the set $\mathbb{N}$, taking the naturals as urelements; or perhaps by admitting transfinite ranks without strong large cardinal assumptions. For developments along relatively minimal lines, see Tait 1998.

The best conceivable situation would be one in which we know for sure that the continuum is physically real or, at least, conceptually necessary for understanding physical phenomena. Kant believed that a continuum structure is given a priori by the constitution of the subject's forms of sensibility, space and time; Poincaré regarded the continuum as a necessity imposed on us by our experience of the world. Many others, with different nuances, have agreed with them, though not, for instance, Hilbert (1926). Notice that this still leaves room for the question, what is the correct (or the best, if any) conceptual analysis of the continuum structure? Again, the best situation would be one in which we know that the pointwise analysis of the continuum is fully coherent conceptually, in particular that it does not lead to contradictions. To summarize, the best conceivable situation would consist in the conjunction of two claims: *a*) physical or phenomenal necessity, and *b*) conceptual flawlessness.

Unfortunately, the human situation is frail, and one must remain skeptical. Although continuum structures have been assumed to be a basis for all the successful theories in the history of physics,[46] the truth is that doubts concerning their physical adequacy have been widespread since the dawn of the quantum world in the 1900s. It is not just that physicists have been forced to question the necessity thesis; there have been reasons to believe that the models obtained on a continuum basis are inadequate. In the case of QED and the standard model, the most convincing interpretations of the field theories (e.g., the ensemble interpretation) do not understand the field phenomena as describing real processes at the atomic level. Thus, several attempts have been made by physicists like Heisenberg and Penrose to establish discrete structures at the basis of new fundamental physical theories. Such attempts are surrounded by difficulties (to mention a deep but simple problem, how do we model the very well-confirmed wave phenomena in discrete space-time?), and the matter must be left open.

Thinking about this situation, I find it very noteworthy that, 150 years ago, Riemann formulated the antinomy between:[47]

---

[46] Not just the classical theories and relativity, but most noteworthily also quantum theory in its Schrödinger formulation, quantum electrodynamics (QED)—in sum, the theories behind the Standard Model of particle physics.

[47] See his philosophical fragments in Riemann 1892, 509–538, in particular the Antinomies on pp. 518–520. I have discussed aspects of them in Ferreirós 2006b.

Discrete elements of space and time	*Continuity*
[*Finite, representable*]	[*Infinite, . . . at the limits of representation*]

Although in all his work (purely mathematical or physico-mathematical) he sided with the continuum and the actual infinite, Riemann never concealed from himself the possibility that "the reality which lies at the basis of space [may] form a discrete set" (Riemann 1854/68, 268).[48] By calling this an antinomy, he seriously meant—like Kant—that human reason inevitably falls into contradictions in this field, being unable to decide which one of the alternatives is real. Riemann expressed here a profound thought, which is seemingly confirmed by subsequent events in the development of science and mathematics, and it might well be true that we cannot attain a final decision on this matter. Hilbert, who hated the idea of having to acknowledge an *ignorabimus*, would not be pleased.

In my view, prospects are much better for the thesis *b*) that the pointillist analysis of the continuum is conceptually flawless. But even here, no proof of logical consistency is in sight. So we must frame a different and, again, more fragile line of argument for that thesis.

In the end, the process of justification that we have tried to get underway moves in circles, at least insofar as we cannot claim that classical mathematics is *strictly* necessary for scientific understanding of natural phenomena. Ultimately the process is sending us back to the Western traditions of mathematical and physical thought—it is the representations favored by these traditions, not the phenomena themselves, that reinforce a mathematical background such as classical analysis and set theory. The argument, far from being logical or purely mathematical, moves beyond the boundaries of mathematics, mobilizing mathematical physics and the uncertainties of traditional assumptions in modelling natural phenomena, very specially, uncertainties in the assumption of continuous space and time and movement.

The issue resonates also with well-known debates about extrinsic justification of mathematical axioms or systems: Quine liked to say that

---

[48] "das dem Raume zu Grunde liegende Wirkliche eine discrete Mannigfaltigkeit bilden [mag]."

set-theoretic systems are less certain than the mathematical superstructures built on them (classical analysis in particular); before Quine, Hilbert, Weyl, and others had made similar remarks. But even his position has to be modified in two directions: first, by considering the Riemann antinomy of continuity vs. discreteness in physics, and second, by taking into account that a set-theoretic foundation is not strictly necessary for establishing the classical mathematics that has proven to be scientifically relevant.[49]

Let me add, before concluding, that there is no need to be monistic here, one should rather promote pluralism. There is reason to think that the set-theoretic universe ought to be regarded as a background setting for free development of mathematical structures and mathematical models—but to turn it into a goal in itself may be a mistake. So, let Woodin's maximalist program develop and try to advance describing the one true universe of set theory; but let other restrictive perspectives on set theory exist and develop—they will probably convince practitioners of algebra. Let also the category theorists denounce the absurd excess of structure in set theory and develop their top-down approaches to characterizing structures by morphisms and functors; and let us see what comes from new programs such as homotopy type theory (HoTT).

Incidentally, even if we were to opt for category theory as the foundation of mathematics, our previous discussion of reasons for the postulates A and B would still be relevant. This is because Infinity and the axiom of Choice are no less necessary to category theory than they are to ZFC.[50]

Mathematics is not given a priori, but is the "science that is sought" (to abuse words of Aristotle), constantly being reshaped and reconstructed.

---

[49] This deserves to be better known by philosophers of science. See Feferman 1998, and work on reverse math (Simpson 1999).

[50] See, e.g., the basic category-theoretic "axioms on functions" presented by Mac Lane (1985), 398*ff.*

References

Aristotle, *Metaphysics*. W. D. Ross, ed. 2 vols. Oxford: Clarendon Press, 1924.

——— *Physics*. Book III (Γ; 200b–208a) and Book VI (Z: 231a–241b). R.P. Hardie and R. K. Gaye, eds. *Physica* in *The Works of Aristotle*, vol. 2, W. D. Ross, ed. Oxford: Clarendon Press, 1930.

V. Arnol'd, 1997. On teaching mathematics. *Russian Math. Surveys* 53:1, 229–236.

T. Arrigoni, 2007. *What is Meant by V? Reflections on the universe of all sets*. Paderborn: Mentis.

M. van Atten, 2007. *Brouwer Meets Husserl: On the Phenomenology of Choice Sequences*. Dordrecht: Springer Netherlands.

J. Avigad. 2003. Number theory and elementary arithmetic. *Philosophia Mathematica* 11 (2003), 257–284.

——— 2008. Computers in mathematical inquiry, in Mancosu (2008), 302–316.

J. Avigad, E. Dean and J. Mumma, 2009. A Formal System for Euclid's Elements. *Review of Symbolic Logic*, 2, no. 4, 700—768.

R. Baire, 1899. *Sur les fonctions de variable réelles*. Ph.D. thesis. Milan, Bernardoni Rebeschini.

I. Barrow, 1685. *Lectiones Mathematicæ XXIII, in quibus principia matheseos generalia exponuntur* (1664, 1665, 1666). London: J. Playford.

C. Bartocci, 2012. *Una piramide di problemi. Storie di geometria da Gauss a Hilbert*. Milano, Raffaello Cortina Editore.

J. L. Bell, 1998. *A Primer of Infinitesimal Analysis*. Cambridge University Press.

——— 2005. *The Continuous and the Infinitesimal in Mathematics and Philosophy*. Milan: Polimetrica.

——— 2010. Continuity and Infinitesimals, In: Zalta, ed. *The Stanford Encyclopedia of Philosophy* (Fall 2010 Edition). http://plato.stanford.edu/archives/win2014/entries/continuity/

P. Benacerraf and H. Putnam, eds. 1983. *Philosophy of Mathematics: Selected Readings*, Cambridge University Press.

P. Benoit, K. Chemla and J. Ritter, 1992. *Histoire de fractions, fractions d'histoire*. Basel, Birkhäuser.

P. Bernays, 1935. Sur le platonisme dans les mathématiques. *L'Enseignement Mathématique*, vol. 34: 52–69. English in Benacerraf and Putnam 1983.

E. Bishop and D. Bridges, 1985. *Constructive Analysis*. New York: Springer.

J. Bolyai, 2004. *Non-Euclidean Geometry and the Nature of Space*. Ed., with commentary, by J. J. Gray. Cambridge: MIT Press.

P. Du Bois-Reymond, 1882. *Die allgemeine Functionentheorie*. Laupp, Tübingen.

B. Bolzano, 1817. *Rein analytischer Beweis des Lehrsatzes, dass zwischen je zwey Werthen, die ein entgegesetztes Resultat gewahren, wenigstens eine reelle Zahl liege*. Prague. German edition with notes by Jourdain in Ostwalds Klassiker der exakten Wissenschaften, No. 153. Leipzig. English by Russ in *Historia Mathematica* 7 (1980), 156–185.

———— 1851. *Paradoxien des Unendlichen*. Leipzig, Reclam. Reprints in Leipzig, Meiner, 1920, and Hamburg, 1975. English trans. London, Routledge, 1950.

———— 2004. *The Mathematical Works of Bernard Bolzano*, ed. *S.* Russ. Oxford University Press.

E. Bombieri, 2000. The Riemann Hypothesis. *Problems of the millennium.* Clay Mathematics Institute. www.claymath.org/sites/default/files/official_problem _description.pdf.

R. Bonola. 1906. *La Geometria non-euclidea. Esposizione storico-critica del suo sviluppo.* Bologna, N. Zanichelli editore.

G. Boolos, 1971. The iterative conception of set. *The Journal of Philosophy* **68**, 8 (Apr. 22), 215–231. Also in: Benacerraf and Putnam 1983, 486–502.

E. Borel 1905. Quelques remarques sur les principes de la théorie des ensembles. *Math. Annalen* 60, 194–195.

H. J. M. Bos. 2001. *Redefining Geometrical Exactness: Descartes' transformation of the early modern concept of construction.* New York, Springer.

R. J. Boscovich, 1754. *De continuitatis lege.* Roma, Typographia Generosi Salomoni.

U. Bottazzini, 1986. *The Higher Calculus: A history of real and complex analysis from Euler to Weierstrass.* Berlin, Springer-Verlag.

U. Bottazzini and A. Dahan-Dalmedico, eds. 2001. *Changing Images in Mathematics: From the French revolution to the new millennium.* London: Routledge.

U. Bottazzini and J. J. Gray, 2013. *Hidden Harmony—Geometric Fantasies. The Rise of Complex Function Theory.* Berlin, Springer.

N. Bourbaki, 1949. Foundations of mathematics for the working mathematician. *The Journal of Symbolic Logic* Vol. 14.1 (Mar. 1949): 1–8.

———— 1950. The architecture of mathematics. *The American Mathematical Monthly*, Vol. 57:4 (Apr. 1950), pp. 221–232. Reprinted in Ewald 1996, vol. 2.

L. E. J. Brouwer, 1905. *Leven, Kunst en Mystiek.* English translation by W. van Stigt in *Notre Dame Journal of Formal Logic*, 37(3): 389–429.

———— 1913. Intuitionism and Formalism. *Bulletin of the American Math. Soc.* 20, 81–96.

———— 1927. Über Definitionsbereiche von Funktionen, *Mathematische Annalen* 97: 60–75. English trans. of sec. 1–3 in van Heijenoort 1967, 457–463.

———— 1948. Consciousness, Philosophy, and Mathematics. Reprinted in Benacerraf and Putnam 1983, 90–96.

B. Butterworth, 1999. *The Mathematical Brain.* London: Macmillan.

F. Cajori, 1918. Origin of the Name "Mathematical Induction". *The American Mathematical Monthly*, Vol. 25, No. 5 (May, 1918), pp. 197–201.

J. I. D. Campbell, ed. 2005. *Handbook of Mathematical Cognition.* New York, Psychology Press.

G. Cantor. 1872. Über die Ausdehnung eines Satzes aus der Theorie der trigonometrischen Reihen, *Math. Annalen* **5**, 123–32. In: *Abhandlungen mathematischen und philosophischen Inhalts* (Springer 1932), 92–101.

———— 1874. Über eine Eigenschaft des Inbegriffes aller reellen algebraischen Zahlen, *Journal für reine und angew. Mathematik* **77**, 258–62. English trans. in Ewald 1996, vol. 2.

———— 1878. Ein Beitrag zur Mannigfaltigkeitslehre, *JrM* **84**, 242–58. In: *Abhandlungen mathematischen und philosophischen Inhalts* (Springer 1932), 119–133.

———— 1879/84. Über unendliche, lineare Punktmannichfaltigkeiten, *Math. Annalen* vols. 15, 17, 20, 21, and 23. In: *Abhandlungen mathematischen und philosophischen Inhalts* (Springer 1932), 139–244.

———— 1883. *Grundlagen einer allgemeinen Mannigfaltigkeitslehre.* Leipzig, 1883. English trans. in Ewald 1996, vol. 2.

———— 1892. Über eine elementare Frage der Mannigfaltigkeitslehre, *DMV* 1, 75–78. English trans. in Ewald 1996, vol. 2.

G. Cardano, 1545. *Ars Magna or the rules of algebra.* New York, Dover, 1993.

J. Cavaillès, 1938. *Methode axiomatique et formalisme,* Paris, Hermann.

C. Cellucci, 1998. *Le raggioni della logica.* Roma, Laterza.

K. Chemla, ed. 2012. *The History of Mathematical Proof in Ancient Traditions.* Cambridge Univ. Press.

K. Chemla and S. Pahaut, 1992. Remarques sur les Ouvrages Mathématiques de Gersonide, in G. Freudenthal (ed.), *Studies on Gersonides—A 14th Century Jewish Philosopher-Scientist,* Leiden: Brill, 1992, p. 149–191.

K. Chemla and Guo Shuchun, 2005. *Les neuf chapitres: Le classique mathématique de la Chine ancienne et ses commentaires.* Critical, bilingual edition. Paris, Dunod.

P. J. Cohen, 1963. The Independence of the Continuum Hypothesis, I and II. *Proceedings of the National Academy of Sciences of the U.S.A.,* Vol. 50:6 (Dec. 15, 1963), pp. 1143–1148; Vol. 51:1 (Jan. 15, 1964), pp. 105–110.

———— 2002. The discovery of forcing. *Rocky Mountain Jour. Math.* 32:4, 1071–1100.

M. Cole, J. Gay, and J. Glick, 1974. Some experimental studies in Kpelle quantitative behavior. In. J. W. Berry and P. R. Dasen (eds.), *Culture and Cognition: Readings in cross-cultural psychology.* London, Methuen.

L. Corry. 1996. *Modern Algebra and the Rise of Mathematical Structures.* Basel, Birkhäuser; 2nd edn. 2004.

———— 2004. *Hilbert and the Axiomatization of Physics (1898–1918): From "Grundlagen der Geometrie" to "Grundlagen der Physik",* Dordrecht, Kluwer.

———— 2008. The development of the idea of proof. In: T. Gowers (ed.), *Princeton Companion to Mathematics,* Princeton Univ. Press, 129–142.

———— 2010. On the History of Fermat's Last Theorem: Fresh views on an old tale, *Mathematische Semesterberichte* Vol. 57, Issue 1, 123–138.

L. Crosilla, 2009. Set Theory: Constructive and Intuitionistic ZF. In: *The Stanford Encyclopedia of Philosophy,* E. N. Zalta (ed.), <http://plato.stanford.edu/archives/spr2014/entries/set-theory-constructive/>.

M. J. Crowe, 1967. *A History of Vector Analysis: The evolution of the idea of a vectorial system.* University of Notre Dame Press.

C. Cullen, ed. 1996. *Astronomy and Mathematics in Ancient China: The Zhou bi suanjing.* Cambridge University Press.

———— 2002. Learning from Liu Hui? A Different Way to Do Mathematics. *Notices of the AMS.* Aug. 2002, 783–790.

D. Van Dalen, 1999/2005, *Mystic, Geometer, and Intuitionist,* 2 vols., Oxford: Clarendon Press.

H. Dales and G. Oliveri, eds. 1998. *Truth in Mathematics.* Oxford University Press.

P. Damerow, 2007. The material culture of calculation: A theoretical framework for a historical epistemology of the concept of number. In: U. Gellert, E. Jablonka (eds.) *Mathematisation and demathematisation: Social, philosophical and educational ramifications*, Rotterdam, Sense Publ., 2007.

J. W. Dawson, 2006. Why do mathematicians re-prove theorems?, *Philosophia Mathematica* 14:3, 269–286.

H. De Cruz, 2009. An enhanced argument for innate elementary geometric knowledge and its philosophical implications. In: B. van Kerkhove (ed.), *New perspectives on mathematical practices*, World Scientific, 2009, pp. 185–206.

R. Dedekind, 1871. Supplement X to *Vorlesungen über Zahlentheorie*, 2nd edn. Partial reprint in Dedekind's *Gesammelte Werke*, vol.3, 223–261.

―――― 1872. *Stetigkeit und irrationale Zahlen*. Braunschweig, Vieweg. English in Ewald 1996, vol. 2.

―――― 1888. *Was sind und was sollen die Zahlen?* Braunschweig, Vieweg. English in Ewald 1996, vol. 2.

―――― 1891. Stetiges System aller Abbildungen der natürlichen Zahlenreihe *N* in sich selbst, in Dedekind's *Gesammelte Werke*, vol. 2, 371–72.

S. Dehaene, 1997. *The Number Sense*. Oxford University Press. 2nd edn. 2011.

S. Dehaene et al. 2006. Core Knowledge of Geometry in an Amazonian Indigene Group. *Science* Vol. 311 (20 Jan 2006), 381–384.

R. Descartes, 1637. *The Geometry*. New York, Dover, 1954 (with facsimile of 1st edn.).

K. Devlin, 1984. *Constructibility*, Berlin, Springer-Verlag.

―――― 2000. *The Maths Gene*. London, Phoenix.

P. J. G. Lejeune-Dirichlet, 1829. Sur la convergence des séries trigonométriques qui servent à représenter une fonction arbitraire entre des limites données, *Journal für reine und angew. Math.* 4, 157–69. In Dirichlet's *Werke*, vol.1, 283–306.

―――― 1837. Beweis des Satzes, dass jede unbegrenzte arithmetische Progression, . . . unendlich viele Primzahlen enthält. *Akademie der Wissenschaften zu Berlin*, 1837. In Dirichlet's *Werke*, vol. 1, 313–342.

R. M. W. Dixon, 1980. *The Languages of Australia*. Cambridge University Press (Cambridge Language Surveys).

F. R. Drake, 1974. *Set Theory: An introduction to large cardinals*. Amsterdam, North-Holland.

E.J. Dubuc, C. Sanchez de la Vega, 2000. On the Galois Theory of Grothendieck. *Boletín Academ. Nac. de Ciencias de Córdoba*, 65 (2000). (Accessed Dec. 2013 in http://arxiv.org/abs/math/0009145v1)

P. Dugac, 1973. Eleménts d'analyse de Karl Weierstrass. *Archive for Hist. of Exact Sciences* 10, 41–176.

―――― 1976. *Richard Dedekind et les fondements des mathématiques (avec de nombreux textes inédits)*, Paris, Vrin.

―――― 1984. Georg Cantor et Henri Poincaré, *Bolletino di Storia delle Scienze Mathematiche* 4, 65–96.

K. Dunlop, 2009. Why Euclid's Geometry Brooked No Doubt: J. H. Lambert on Certainty and the Existence of Models. *Synthese* 167 (1): 33–65.

J. Dupré, 1993. *The Disorder of Things*. Harvard University Press.

A. Dürer, 1525. *Underweysung der Messung mit dem Zirckel und Richtscheyt (Instruction in measuring with compass and straightedge)*. Nuremberg; facsimile edn. by A Jaeggli and C Papesch, Zurich, 1966.

H.-D. Ebbinghaus *et al.* 1991. *Numbers.* Berlin, Springer Verlag.

H. M. Edwards, 1977. *Fermat's Last Theorem.* New York: Springer-Verlag.

P. Ehrlich, ed. 1994. *Real Numbers, Generalizations of the Reals, and Theories of Continua.* Dordrecht, Kluwer.

—— 2006. The rise of non-Archimedean mathematics and the roots of a misconception. I. The emergence of non-Archimedean systems of magnitudes. *Archive for History of Exact Sciences* 60, 1–121.

A. Einstein, 1933. *On the Method of Theoretical Physics.* Clarendon Press, Oxford; also in: *Ideas and Opinions* (New York: Bonanza Books, 1954).

F. Engel and P. Stäckel, eds. 1895. *Urkunden zur Geschichte der Nichteuklidischen Geometrie.* Leipzig.

Euclid, *Elementaria. Obra completa.* Madrid: Editorial Gredos. ISBN 978–84–249–1463–9. *Volumen I: Libros I–IV.* 1991. *Volumen II: Libros V–IX.* 1994. *Volumen III: Libros X–XIII.* 1996.

—— 1908. *The thirteen books of Euclid's Elements.* Cambridge University Press, ed. T. L. Heath. See also: *Euclid's Elements—All thirteen books in one volume,* Based on Heath's translation, Green Lion Press, 2002.

L. Euler, 1737, De fractionibus continuis dissertatio. Comm. Acad. Sci. Petrop. 9 1737 (1744), 98–137, In: Opera Omnia, series 1, vol. 14, 187–216; translation by M. F. Wyman and B. F. Wyman, *Math. Systems Theory* 18 (1985), 295–328.

—— 1748. *Introductio in Analysin infinitorum.* Lausanne: M.-M. Bousquet, 1748. In: *Opera Omnia,* series 1, vol. 8. English translation in Springer, 1988.

—— 1771. *Vollständige Anleitung zur Algebra,* Berlin. In: *Opera Omnia,* vol. 1, Leipzig, Teubner, 1911. English transl. Cambridge University Press; 3rd edition, 2009.

W. Ewald, ed. 1996, *From Kant to Hilbert: A source book.* 2 vols. Oxford University Press.

S. Feferman. 1965. Some applications of the notions of forcing and generic sets, *Fundamenta Mathematica,* vol. 56, pp. 325–345.

—— 1991. Reflecting on Incompleteness. *The Journal of Symbolic Logic,* Vol. 56, No. 1, 1–49.

—— 1998. *In the Light of Logic.* Oxford University Press.

—— 2009. Conceptions of the continuum, *Intellectica* 51 (2009), 169–189.

S. Feferman and G. Hellman, 1995. Predicative foundations of arithmetic. *Journal of Philosophical Logic* 24 (1995): 1–17.

S. Feferman, H. Friedman, P. Maddy and J. Steel, 2000. Does Mathematics Need New Axioms? *The Bulletin of Symbolic Logic* 6:4 (Dec. 2000).

L. Feigenson, S. Dehaene and E. S. Spelke, 2004. Core systems of number. *Trends in Cognitive Sciences,* 8 (10), 307–314.

J. Ferreirós, 1999. *Labyrinth of Thought: A history of set theory.* Basel, Birkhäuser; 2nd edn. 2007.

—— 2001. The Road to Modern Logic—An interpretation. *The Bulletin of Symbolic Logic* 7, 441–484.

—— 2004. The Motives Behind Cantor's Set Theory: Physical, biological and philosophical questions. *Science in Context* 17, n° 1/2 (2004), 1–35.

—— 2005. Certezas e hipótesis: Perspectivas históricas y naturalistas sobre la matemática. In: *Filosofía de las ciencias matemáticas, naturales y sociales,* ed. A. Estany (Trotta, Madrid, 2005), pp. 45–73.

———— 2006. Ο θεος αριθμητιζει: The rise of pure mathematics as arithmetic with Gauss. In: *The Shaping of Arithmetic*, ed. C. Goldstein, N. Schappacher, J. Schwermer (Springer, Berlin, 2006), pp. 206–240.

———— 2006b. Riemann's *Habilitationsvortrag* at the Crossroads of Mathematics, Physics and Philosophy. In Ferreirós and Gray 2006, 67–96.

———— 2009. The Dynamics of Experimentation and its Role within a Philosophy of Scientific Practice. In: *Observation and Experimentation in Science: New methodological perspectives*, ed. W. González. La Coruña, Net · biblos, 2011.

———— 2010. *In* or *About* Mathematics? Concerning some versions of Mathematical Naturalism. In: *Philosophical Perspectives on Mathematical Practice*, ed. B. van Kerkhove, J. de Vuyst, J.P. van Bendegem (London: College Publications), 117–154.

———— 2011. On Arbitrary Sets and ZFC. *The Bulletin of Symbolic Logic*, Vol. 17:3, 361–393.

———— 2013. On the very notion of applied mathematics. *Oberwolfach Reports* 12 (2013): 70–74.

———— forthcoming. On the semantics of higher-order logic. 1st CLE Colloquium for Philosophy and History of Formal Sciences, ed. F. Bertato. Campinas, Cadernos CLE/Manuscrito.

J. Ferreirós and J. J. Gray, eds. 2006. *The Architecture of Modern Mathematics: Essays in history and philosophy*. Oxford University Press. [Introduction, pp. 1–43, is available online.]

R. Firth, 1973. *Symbols: Public and Private*. Cornell University Press.

D. Flament, 2003. *Histoire des nombres complexes: Entre algèbre et géométrie*. Paris, CNRS.

D. H. Fowler, 1999. *The Mathematics of Plato's Academy*. Oxford, Clarendon Press. 1st edn. 1987.

A. Fraenkel 1928. *Einleitung in die Mengenlehre*. 3rd revised edn. Berlin, Springer.

M. C. Frank, D. L. Everett et al. 2008. Number as a cognitive technology: Evidence from Pirahã language and cognition. *Cognition* 108, 819–824.

G. Frege, 1884. *Grundlagen der Arithmetik*. Breslau, Koebner. Translated as *The Foundations of Arithmetic* by J.L. Austin, Oxford: Blackwell, 1974.

———— 1893/1903. *Grundgesetze der Arithmetik*, Jena, Pohle, 2 vols. Partial translation of vol. I, *The Basic Laws of Arithmetic*, by M. Furth, Berkeley, Univ. of California Press, 1964.

———— 1976. *Wissenschaftlicher Briefwechsel*. Felix Meiner Verlag, Hamburg.

S. Friedman, 2002. Cantor's Set Theory from a Modern Point of View. *Jahresbericht der Deutschen Mathematiker-Vereinigung* 104, Heft 4, pp. 165–170.

C. F. Gauss 1801. *Disquisitiones arithmeticae*. Leipzig, G. Fleischer. In: *Werke*, vol. 1. English translations: Yale University Press, 1965; Springer Verlag, 1986.

———— 1831. Anzeige der Theoria residuorum biquadraticorum, commentatio secunda [1832], *Göttingische gelehrte Anzeigen*. In: *Werke*, vol. 2, 169–178. English trans. in Ewald 1996, vol. 1.

———— 1849. Beiträge zur Theorie der algebraischen Gleichungen. In: *Werke*, vol. 3, 71–102.

R. Gelman and B. Butterworth 2005. Number and Language: How are they related? *TRENDS in Cognitive Sciences* Vol. 9:1 (Jan 2005), 6–10.

A. George and D. J. Velleman, 1998. Two Conceptions of Natural Number. In G. Dales and G. Oliveri, eds., *Truth in Mathematics* (Oxford University Press, 1998), 311–327.

M. Giaquinto, 2005. Mathematical activity, in: K. Jørgensen and P. Mancosu, eds., *Visualization, explanation and reasoning styles in mathematics.* Dordrecht, Kluwer.

—— 2007. *Visual Thinking in Mathematics: An epistemological study.* Oxford Univ. Press.

—— 2008. Visualizing in mathematics. In: Mancosu (2008), chap. 1, 22–42.

K. Gödel, 1933. The present situation in the foundations of mathematics (manuscript). In: *Collected Works*, vol. 3, pp. 45–53.

—— 1940. *The Consistency of the Axiom of Choice and of the Generalized Continuum Hypothesis with the Axioms of Set Theory,* Princeton University Press. In: *Collected Works*, vol. 2, 33–101.

—— 1944. Russell's mathematical logic, in P. A. Schilpp (ed.), *The philosophy of Bertrand Russell.* In: *Collected Works*, vol. 2, 119–141.

—— 1947. What is Cantor's Continuum Hypothesis? *The American Mathematical Monthly.* Revised with additions in 1964: Benacerraf and Putnam (1983). Also in: *Collected Works*, vol. 2, 176–187, 254–270.

—— 1986, *Collected Works*, ed. by S. Feferman et al. Oxford Univ Press, 1986, 1990, 1995. Vol. 1: Publications 1929–1936. Vol. 2: Publications 1938–1974. Vol. 3: Unpublished essays and lectures.

D. Goldfeld. 2003. The elementary proof of the prime number theorem: An historical perspective, In: *Number Theory, New York Seminar 2003,* edited by D. Chudnovsky, G. Chudnovsky, M.B. Nathanson, Springer Verlag, p. 179–192.

C. Goldstein, 2010. Les fractions décimales: Un art d'ingénieur?, preprint. http://halshs.archives-ouvertes.fr/hal-00734932/

C. Goldstein *et al.* 1996. *L'Europe mathématique—Mathematical Europe.* Paris, Editions MSH.

H. G. Grassmann, 1844. *Die Lineale Ausdehnungslehre,* Leipzig, Teubner. In *Gesammelte Schriften*, vol. 1, 1–319.

—— 1861. *Lehrbuch der Arithmetik für höhere Lehranstalten,* Berlin, Enslin.

I. Grattan-Guinness, ed. 1980. *From the Calculus to Set Theory, 1630–1910.* London, Duckworth.

—— ed. 1994. *Companion Encyclopaedia of the History and Philosophy of the Mathematical Sciences.* Vol. 1. London, Routledge.

—— ed. 2005. *Landmark Writings in Western Mathematics 1640–1940.* Elsevier Science.

J. J. Gray, 1989. *Ideas of Space: Euclidean, Non-Euclidean and Relativistic.* Oxford, Clarendon.

—— 2008. *Plato's Ghost: The modernist transformation of mathematics.* Princeton University Press.

—— 2012. *Henri Poincaré: A scientific biography.* Princeton University Press.

E. Grosholz, forthcoming. Reducibility and Meaning: Logic and number theory. (Talk at Midwest PhilMath Workshop › MWPMW 14, 2013). Explanation and Meaning in Modern Number Theory. (Talk at the conference Mathematical Cultures 2, London, 2013).

B. Grünbaum, 2006. What symmetry groups are present in the Alhambra? *Notices of the AMS,* 53(6): 670–673.

J. Hadamard, 1954. *The Mathematician's Mind: The Psychology of Invention in the Mathematical Field.* With preface by P. N. Johnson-Laird: Princeton Univ Press, 1996.

M. Hallett, 1984. *Cantorian Set Theory and limitation of size*. Oxford, Clarendon.

W. R. Hamilton, 1837. Essay on Algebra as the Science of Pure Time. *Transactions of the Royal Irish Academy* 17, 293–422. In: *Mathematical Papers* vol. 3, 3–96.

J. D. Hamkins, 2012. The set-theoretic multiverse. *Review of Symbolic Logic*, vol. 5, 416–449.

O. Harari, 2003. The Concept of Existence and the Role of Constructions in Euclid's Elements. *Archive for History of Exact Sciences* 57, 1–23.

——— 2004. *Knowledge and Demonstration: Aristotle's Posterior Analytics*. New Synthese Historical Library, Kluwer.

R. Hartshorne. 2000. *Geometry: Euclid and beyond*. New York, Springer.

F. Hausdorff, 1914. *Grundzüge der Mengenlehre*. Leipzig, Veit. Reprint New York, Chelsea, 1949. In: *Gesammelte Werke*, Springer, vol. 2, 2002.

——— 1927. *Mengenlehre*, Berlin, W. de Gruyter. In: *Gesammelte Werke*, vol. 3 (2002). 2nd revised edn. (1937) translated as *Set Theory*, New York, Chelsea, 1957.

T. W. Hawkins, 1981. The Berlin School of Mathematics, in H. Bos, H. Mehrtens and I. Schneider (eds.), *Social History of Nineteenth Century Mathematics* (Basel, Birkhäuser 1981), 233–245.

A. Heeffer, 2012. Epistemic Justification and Operational Symbolism. *Foundations of Science* 19:1 (2014), 89–113.

J. van Heijenoort, ed. 1967. *From Frege to Gödel: A source book*. Cambridge, Harvard Univesity Press.

J. L. Heilbron. 2000. *Geometry Civilized. History, Culture, and Technique*. Oxford University Press.

B. Heintz, 2000. *Die Innenwelt der Mathematik. Zur Kultur und Praxis einer beweisenden Disziplin*. Springer Verlag, Wien.

H. Helmholtz 1887. Zählen und Messen, erkenntnistheoretisch betrachtet. *Philosophische Aufsätze für E. Zeller*, Leipzig, Fues. In: *Schriften zur Erkenntnistheorie* (Berlin, Springer, 1921), 70–108. English trans. in Ewald 1996, vol. 2.

L. Hermer and E. Spelke 1994. A geometric process for spatial reorientation in young children. *Nature* 370, 57–59.

C. Herrenschmidt, 2007. *Les Trois Écritures. Langue, nombre, code*. Gallimard, Paris.

A. Heyting, 1956. *Intuitionism, an introduction*. Amsterdam: North-Holland.

D. Hilbert 1897. Die Theorie der algebraischen Zahlkörper, *Deutsche Math. Vereinigung* 4, 175–546. In: *Gesammelte Abhandlungen* vol.1, 63–363.

——— 1900. Sur les problèmes futurs des mathématiques [Mathematische Probleme]. *NG*, 253–97. In: *Gesammelte Abhandlungen* vol. 3, 290–339. English translation in *Bulletin of AMS* 8 (1902), 437–79. Partial translation in Ewald 1996, vol. 2.

——— 1903. *Grundlagen der Geometrie*. (1st edn. 1899, 7th 1930) Leipzig, Teubner. Translation as *Foundations of Geometry*, Open Court, 1902.

——— 1926. Über das Unendliche. *Math. Annalen* 95, 161–90. English translation in van Heijenoort 1967, 367–92.

——— 1930. Naturerkennen und Logik [Radio speech]. Excerpt in http://math.sfsu.edu/smith/Documents/HilbertRadio/HilbertRadio.pdf http://math.sfsu.edu/smith/Documents/HilbertRadio/HilbertRadio.mp3

D. Hilbert and Cohn-Vossen, 1932. *Anschauliche Geometrie*. New York, Dover. English translation New York: Chelsea, 1952.

J. Høyrup, 1990. Algebra and naive geometry. An investigation of some basic aspects of Old Babylonian mathematical thought, *Altorientalische Forschungen* 17 (1990) 27–69; 262–354.

———— 1994. *In Measure, Number, and Weight. Studies in Mathematics and Culture.* New York, State University of NY Press.

———— 2002. *Lengths, Widths, Surfaces: A portrait of Old Babylonian Algebra and its kin.* Berlin, Springer.

———— 2011. Written Mathematical Traditions in Ancient Mesopotamia: Knowledge, ignorance, and reasonable guesses. Preprint, talk at *Traditions of Written Knowledge in Ancient Egypt and Mesopotamia* (Frankfurt am Main).

J. Hurford. 1987. *Language and Number: The emergence of a cognitive system.* Basil Blackwell, Oxford.

E. Hutchins, 1995. *Cognition in the Wild.* MIT Press.

V. Izard, P. Pica, E.S. Spelke and S. Dehaene, 2011. Geometry as a universal mental construction. In: *Space, Time and Number in the Brain*, ed. by S. Dehaene, E. Brannon. Academia Press, 2011. (*Attention and Performance* Vol. 24, pp. 319–332)

C. G. J. Jacobi, 1881. *Gesammelte Werke*, vol. 1, Berlin, Reimer.

H. N. Jahnke, ed. 2003. *A History of Analysis.* American Mathematical Society / London Mathematical Society.

I. Jané, 1993. A Critical Appraisal of Second-Order Logic. *History and Philosophy of Logic* 14 (1): 67–86.

———— 2001. Reflections on Skolem's Relativity of Set-Theoretical Concepts. *Philosophia Mathematica* 9: 31–55. Reprinted in the *Philosopher's Annual XXIV* (2002).

———— 2005. The iterative conception of sets from a Cantorian perspective. In: *Logic, Methodology and Philosophy of Science. Proceedings of the Twelfth International Congress*, P. Hájek *et al.*, eds., 373–393. King's College Publications, 2005.

———— 2005b. Higher-order logic reconsidered. In: *The Oxford Handbook of Philosophy of Mathematics and Logic* (S. Shapiro, ed.). Oxford University Press: 781–810.

R. Jensen, 1995. Inner Models and Large Cardinals. *The Bulletin of Symbolic Logic*, vol. 1, pp. 393–407.

D. M. Jesseph, 2000. *Squaring the Circle: The War Between Hobbes and Wallis.* Chicago Univ. Press.

V. Jones, 1998. A credo of sorts. In: G. Oliveri and H. G. Dales, eds. *Truth in Mathematics* (Oxford University Press, 1998), 203–214.

A. Kanamori, 1995. The emergence of descriptive set theory. In: J. Hintikka, ed., *From Dedekind to Gödel*, Dordrecht, Kluwer.

———— 1994. *The Higher Infinite*, Berlin, Springer. 2nd edn 2002.

V. G. Kanovei, 1985. The development of the descriptive theory of sets under the influence of the work of Luzin. *Russ. Math. Surveys* 40:3, 135–180.

I. Kant, *Kritik der reinen Vernunft.* English in: P. Guyer and A. Wood, eds. *The Cambridge Edition of the Works of Immanuel Kant*, Cambridge University Press, 1998.

———— *Prolegomena.* In: Allison, H. and Heath, P., eds. *Theoretical Philosophy after 1781, The Cambridge Edition of the Works of Immanuel Kant*, Cambridge University Press, 2002.

———— *Correspondence*, translated and edited by Arnulf Zweig. *The Cambridge Edition of the Works of Immanuel Kant*, Cambridge: Cambridge University Press, 1999.

V. J. Katz, 1993. *A History of Mathematics: An Introduction*, Harper Collins, 2nd edn., 1998.

R.W. Kaye, 1991. *Models of Peano Arithmetic.* Oxford University Press.

A. S. Kechris, 1996. *Classical Descriptive Set Theory.* New York, Springer-Verlag.

B. van Kerkhove, J. de Vuyst and J. P. van Bendegem, eds. 2010. *Philosophical Perspectives on Mathematical Practice.* London, College Publications.

P. Kitcher, 1984. *The Nature of Mathematical Knowledge.* Oxford Univ. Press.

———— 1988. Mathematical naturalism. In: *History and philosophy of modern mathematics*, ed. by W. Aspray and P. Kitcher, 293–325. Minneapolis: University of Minnesota Press.

F. Klein, 1872/93. Vergleichende Betrachtungen über neuere geometrische Forschungen [*Erlangen Program*]. Revised in: *Mathematische Annalen*, 43 (1893): 63–100.

M. Kline, 1972. *Mathematical Thought, from Ancient to Modern Times.* 3 vols. New York, Oxford University Press.

W. Knorr, 1975. *The Evolution of the Euclidean Elements*, Dordrecht, Reidel.

P. Koellner, 2009. On Reflection Principles. *Annals of Pure and Applied Logic* 157:2, 206–219.

———— 2013. Large Cardinals and Determinacy, *The Stanford Encyclopedia of Philosophy* (Summer 2013), E. N. Zalta ed., http://plato.stanford.edu/archives/sum2013 /entries/large-cardinals-determinacy/.

R. Krömer, 2007. *Tool and Object. A history and philosophy of Category Theory.* (Science Networks Historical Studies 32), Basel: Birkhäuser.

L. Kronecker, 1887. Über den Zahlbegriff. *Journal für die reine und angewandte Mathematik* 101, 337–55. Also in Kronecker's *Werke* (Leipzig, Teubner, 1895/1930), vol.3/1, 249–274. English trans. in Ewald (1996), vol. 2.

T. S. Kuhn, 1970. *The Structure of Scientific Revolutions.* Chicago Univ. Press. 1st edn. 1962.

K. Kunen, 1983. *Set Theory: An introduction to independence proofs.* Amsterdam: North-Holland.

J.-L. Lagrange, 1788. *Mechanique Analytique*, Paris: Desaint.

———— 1797. *Théorie des fonctions analytiques*, Paris: Imprimerie de la Republique.

I. Lakatos, 1967. A Renaissance of Empiricism in the Recent Philosophy of Mathematics? In: *Mathematics, Science and Epistemology. Philosophical Papers Vol. 2*, eds. J. Worrall and G. Currie, Cambridge Univ Press, 1978, 24–42.

———— 1976. *Proofs and Refutations.* Cambridge University Press.

J. H. Lambert, 1770. Vorläufige Kenntnisse für die, so die Quadratur und Rectification des Circuls suchen. In: *Beyträge zum Gebrauche der Mathematik und deren Anwendung*, vol. II, Berlin, 1770. References to F. Rudio (1892), 133–156.

———— 1786. Theorie der Parallellinien. *Leipziger Magazin für reine und angewandte Mathematik*, Band I (1786), 137–164 and 325–358. Reprinted in Engel and Stäckel (1895), 152–208.

D. F. Lancy, 1983. *Cross-cultural Studies in Cognition and Mathematics.* New York: Academic Press.

E. Landau, 1930. *Foundations of Analysis*, Chelsea Pub Co. 1951.

D. Laugwitz, 1999. *Bernhard Riemann 1826–1866. Wendepunkte in der Auffassung der Mathematik*, Basel, Birkhäuser. English translation.

S. Laurence and E. Margolis 2007. Linguistic Determinism and the Innate Basis of Number. In P. Carruthers, S. Laurence, and S. Stich (eds.) *The Innate Mind: Foundations and the Future*, pp. 139–169. Oxford University Press, 2007.

A. Lautman, 2006. *Les mathématiques, les idées et le réel physique*. Paris, J. Vrin.

S. Lavine, 1994. *Understanding the Infinite*. Harvard University Press.

W. Lawvere and Stephen H. Schanuel, 2005. *Conceptual Mathematics: A First Introduction to Categories*. Cambridge Univ. Press.

G. W. Leibniz, 1702. Specimen novum Analyseos . . . *Acta eruditorum* 1702. In: *Mathematische Schriften* vol. V (Hildesheim, Georg Olms, 1971).

F. Lemmermeyer, 2000. *Reciprocity Laws from Euler to Eisenstein*. Berlin: Springer.

A. Levy and R. Solovay, 1967. Measurable cardinals and the continuum hypothesis, *Israel Journal of Math.* 5, 234–248.

O. Linnebo, 2011. Platonism in the Philosophy of Mathematics. *Stanford Encyclopedia of Philosophy*, http://plato.stanford.edu/entries/platonism-mathematics/

J. E. Littlewood, 1962. The Riemann hypothesis, in: *The scientist speculates: An anthology of partly baked ideas*, edited by Good, M. and M. Smith, 1962, New York: Basic books.

P. Luckey, 1951. *Die Rechenkunst bei Ğamšīd b. Masʿūd al-Kāšī*, Steiner, Wiesbaden.

———— 1953. *Der Lehrbrief über den Kreisumfang (ar-risala al-Muhitiya) von Gamsid b. Masʿud al-Kaši übersetzt und erläutert* (ed. Albred Siggel). Berlin.

J. Lützen, 2005. *Mechanistic Images in Geometric Form: Heinrich Hertz's principles of mechanics*. Oxford University Press.

N. Luzin, 1925. Les propriétés des ensembles projectifs. Note présentée par M. Émile Borel. *Comptes Rendus Acad. Sciences Paris*, 180, Ier Semestre, 1817–1819.

———— 1930. *Leçons sur les Ensembles Analytiques et leurs Applications*. Paris: Gauthier-Villars.

D. Macbeth, 2010. Diagrammatic Reasoning in Euclid's *Elements*. In: *Philosophical Perspectives on Mathematical Practice*, ed. B. Van Kerkhove, J. De Vuyst, J. P. Van Bendegem. London: College Publications.

S. Mac Lane, 1985. *Mathematics, Form and Function*. Berlin, Springer.

P. Maddy, 1988. Believing the axioms, parts I and II. *Journal of Symbolic Logic*. [For a copy with corrections, http://www.lps.uci.edu/lps_bios/pjmaddy]

———— 1990. *Realism in Mathematics*. Oxford University Press.

———— 1997. *Naturalism in Mathematics*. Oxford University Press.

M. S. Mahoney, 1990. Barrow's mathematics: Between ancients and moderns. In: M. Feingold, ed. *Before Newton: The Life and Times of Isaac Barrow*, pp. 179–249, Cambridge Univ. Press, Cambridge, 1990.

A. Malet, 2006. Renaissance notions of number and magnitude. *Historia Mathematica* 33, 63–81.

———— 2007. Just before Viète: Numbers, polynomials, demonstrations, and variables in Simon Stevin's Arithmétique. In: *Liber amicorum Jean Dhombres*, Louvain-la-Neuve, Centre de recherche en histoire des sciences.

P. Mancosu, ed. 1998. *From Brouwer to Hilbert*. Oxford University Press.

———— 2005. Harvard 1940–41: Tarski, Carnap and Quine on a finitistic language of mathematics for science. *History and Philosophy of Logic* 26, 4: 327–357.

———— ed. 2008. *The Philosophy of Mathematical Practice*. Oxford University Press.

———— 2009. Measuring the size of infinite collections of natural numbers: Was Cantor's theory of infinite number inevitable? *Review of Symbolic Logic*, 2, 612–646.

K. Manders, 1989. Domain extension and the philosophy of mathematics. *Journal of Philosophy* 86, pp. 553–62.

———— 2008. Diagram-Based Geometric Practice. Chapter 3 in Mancosu, 2008, pp. 65–79.

———— The Euclidean Diagram (1995). Chapter 4 in Mancosu, 2008, pp. 80–133.

E. Maor, 2007. *The Pythagorean Theorem: A 4,000-Year History*. Princeton University Press.

J. P. Marquis. 2009. *From a Geometrical Point of View: A study in the history and philosophy of Category Theory*, Berlin: Springer.

D. A. Martin, 1998. Mathematical evidence. In: Dales and Oliveri 1998, pp. 215–231.

D. A. Martin and J. R. Steel, 1989. A Proof of Projective Determinacy. *Journal of the American Mathematical Society* 2 (1): 71–125.

B. Mazur, 1997. Conjecture. *Synthese* 111 (2): 197–210.

K. Menninger, 1969. *Number Words and Number Symbols. A cultural history of numbers*. MIT Press. Reprinted in Dover, 1992.

M. Merleau-Ponty, 1962. *Phenomenology of Perception*. London: Routledge.

H. Meschkowski, 1983. *Georg Cantor. Leben, Werk und Wirkung*. Mannheim, Bibliographisches Institut.

H. Meschkowski and W. Nilson, eds. 1991. *Cantor—Briefe*. Berlin, Springer.

G. H. Moore, 1982. *Zermelo's Axiom of Choice*. Berlin, Springer.

L. Moreno-Armella and G. Waldegg, 2000. An epistemological history of number and variation. In: V. Katz *et al.* (eds.) *Using History to Teach Mathematics: An International Perspective*, Math. Assoc. America, p. 183–190.

Y. N. Moschovakis, 1974. New Methods and Results in Descriptive Set Theory. In: *Proceedings of the International Congress of Mathematicians Vancouver, 1974*. Canadian Mathematical Congress, 1975.

———— 1980. *Descriptive Set Theory*. North-Holland. 1980. 2nd edn. 2005.

———— 2006. *Notes on Set Theory*. New York: Springer, 2nd edn.

J. Mumma, 2008. Ensuring Generality in Euclid's Diagrammatic Arguments. In: G. Stapelton, J. Howse and J. Lee (eds.), *Diagrammatic Representation and Inference*, New York, Springer.

———— 2010. Proofs, Pictures and Euclid. *Synthese* 175 (2), 255–287.

E. Nagel, 1935. «Impossible Numbers»: A Chapter in the History of Modern Logic. In *Teleology Revisited and Other Essays*, Columbia University Press, 1979, 166–194.

J. Needham 1959. *Science and Civilisation in China*. Vol. 3. Cambridge University Press.

R. Netz, 1999. *The Shaping of Deduction in Greek Mathematics: A Study in Cognitive History*. Cambridge University Press.

I. Newton, 1687. *Philosophiae naturalis Principia mathematica*. London. *The Principia: A New Translation*, I. B. Cohen and Anne Whitman, Berkeley: University of California Press, 1999.

———— 1720. *Arithmetica Universalis*. Cambridge. English translation, London: printed for J. Senex, W. and J. Innys, J. Osborne and T. Longman, 1728.

———— 1691. *De Quadratura Curvarum*, published in 1704 as an Appendix to the *Opticks*. In: *The Mathematical Papers*, ed. Whiteside, vol. VIII, Cambridge University Press, 1981.

———— ed. 2003. *The History of Early Mathematics—Ways of Re-Writing*, special issue of *Science in Context* (2003).

C. O'Callaghan, 2007. Perception. In: *Cambridge Handbook to Cognitive Science*, eds. K. Frankish and W. Ramsey. Cambridge Univ. Press.

M. Panza. 2011. Rethinking geometrical exactness. *Historia Mathematica* 38:1, 42–95.

C. Parsons, 1992. The impredicativity of induction. In: *Proof, logic and formalization*, ed. M. Detlefsen, Routledge, 1992, pp. 139–61.

———— 2009. *Mathematical Thought and its Objects*. Cambridge University Press.

C. S. Peirce, 1955. *Philosophical Writings of Peirce*, ed. Justus Buchler. Dover Publications.

———— 1974. *Collected papers of Charles Sanders Peirce*. Cambridge (Mass): The Belknap Press of Harvard Univ Press. Vol. IV: The simplest mathematics, edited by C. Hartshorne and P. Weiss.

B. Petri and N. Schappacher, 2006. On Arithmetization. In: *The Shaping of Arithmetic*, ed. C. Goldstein, N. Schappacher, J. Schwermer (Springer, Berlin, 2006), pp. 343–374.

H. Poincaré, 1902. *Science and Hypothesis*. Walter Scott Publishing Co., 1902. (Paris: Flammarion)

———— 1905. *The Value of Science*. New York: Dover. (Paris: Flammarion)

———— 1908. *Science and Method*. London/New York: Th. Nelson and sons. (Paris, Flammarion)

M. Potter, 2005. *Set Theory and its Philosophy: A critical introduction*. Oxford University Press.

Proclus *Diadochus*. 1992. *A Commentary on the First Book of Euclid's Elements*, ed. G.R. Morrow. Princeton University Press, 1970; reprinted 1992, with a new foreword by I. Mueller.

W. Purkert, 1986. Georg Cantor und die Antinomien der Mengenlehre. *Bulletin de la Société Mathématique de Belgique* 38: 313–27.

H. Putnam, 1965. The Analytic and the Synthetic, in his *Philosophical Papers*, vol. 2, Cambridge University Press, 1975.

———— 1967. Mathematics without Foundations. *Journal of Philosophy* 64 (1): 5–22.

———— 1994. Philosophy of Mathematics: Why Nothing Works. In: *Words and Life* (Harvard Univ. Press, 1994), pp. 499–512.

W. V. Quine, 1951. Two Dogmas of Empiricism, *The Philosophical Review* 60: 20–43. In: *From a Logical Point of View*, Harvard University Press, 1980.

R. Rashed, 1972. L'induction mathématique: al-Karaji, as-Samaw'al. *Archive for History of Exact Sciences*, 9 (1972).

———— 1994. *The Development of Arabic Mathematics: Between Arithmetic and Algebra*. Boston Studies in Phil. of Science, Vol. 156. Springer Verlag. (French original *Entre arithmétique et algèbre*, Paris, Les Belles Lettres, 1984.)

Y. Rav, 1999. Why do we prove theorems? *Philosophia Mathematica* 7:5–41.

E. Reck and S. Awodey, 2002. Completeness and Categoricity. Part I and II. *History and Philosophy of Logic*, 23 (2002), 1–30; 77–94.

B. Riemann, 1851. Grundlagen einer allgemeinen Theorie der Functionen complexer Grössen. In: Riemann (1892), 3–45.

———— 1854/1868. Über die Hypothesen, welche der Geometrie zu Grunde liegen (lecture, 1854). In: Riemann (1892), 272–287. English translation by Clifford, in Ewald (1996), vol. 2.

———— 1892. *Gesammelte mathematische Werke und wissenschaftlicher Nachlass.* Leipzig, Teubner, 1892.

E. Robson, 2008. *Mathematics in Ancient Iraq: A Social History.* Princeton University Press.

E. Robson and J. Stedall, eds. 2009. *Oxford Handbook of the History of Mathematics.* Oxford University Press.

B. A. Rosenfeld, 1988. *A History of non-Euclidean Geometry: Evolution of the concept of a geometrical space.* New York, Springer.

D. Rowe, 2003. Mathematical schools, communities and networks. In: *The Cambridge History of Science: Vol. 5, Modern Physical and Mathematical Sciences*, ed. M J Nye (Cambridge University Press, 2003), pp. 113–132.

F. Rudio, ed. 1892. *Archimedes, Huygens, Lambert, Legendre. Vier Abhandlungen über die Kreismessung.* Leipzig, Teubner.

D. Ruelle, 2000. Conversations on mathematics with a visitor from outer space. In: V. Arnold *et al.* (eds.), *Mathematics: Frontiers and Perspectives* (American Mathematical Society, 2000).

B. Russell, 1903. *The Principles of Mathematics.* Cambridge Univ. Press.

A. S. Saidan, 1978. *The Arithmetic of al-Uqlidisi*, Dordrecht, Reidel.

Ken Saito, 2011. Diagrams from Greek and Arabic manuscripts. Digital documents in http://www.greekmath.org/.

J.-M. Salanskis, H. Sinaceur, eds. 1992. *Le labyrinthe du continu* (Colloque de Cerisy). Springer-Verlag, Paris.

M. Serfati, 2005. *La révolution symbolique. La constitution de l'écriture symbolique mathématique.* Paris, Pétra.

S. Shapiro, 1991. *Foundations without Foundationalism: A case for second-order logic.* Oxford, Clarendon Press.

———— 1997. *Philosophy of Mathematics: Structure and Ontology.* Oxford University Press.

S. Shelah, 1991. The Future of Set Theory. In: *Set Theory of the Reals* (Ramat Gan: Bar-Ilan Univ.), 1–12.

———— 2003. Logical Dreams. *Bulletin of the American Mathematical Society*, 40: 203–228.

W. Sierpinski, 1916. Sur le rôle de l'axiome de M. Zermelo dans l'Analyse Moderne, *Comptes Rendus*, Acad. Sciences Paris, 688–691.

———— 1918. L'axiome de M. Zermelo et son rôle dans la théorie des ensembles et l'analyse. *Bulletin international de l'Académie des sciences de Cracovie*, pp. 97–152. (Classe des sciences mathématiques et naturelles, Krakau).

S. G. Simpson, 1999. *Subsystems of Second-Order Arithmetic.* Berlin, Springer/ASL.

T. Skolem, 1923. Einige Bemerkungen zur axiomatischen Begründung der Mengenlehre. *Dem femte skandinaviska matematikerkongressen*, Helsinki, Akademiska Bokhandeln. English translation in van Heijenoort 1967, 290–301.

———— 1934. Über die Nicht-charakterisierbarkeit der Zahlenreihe mittels endlich oder abzählbar unendlich vieler Aussagen mit ausschliesslich Zahlenvariablen, *Fundamenta Mathematicae* 23, 150–161.

J. D. Smith, 1992. The Remarkable Ibn al-Haytham, *The Mathematical Gazette* 76 (475): 189–198.

R. M. Solovay, 1969. On the cardinality of Σ^1_2 sets of reals. In: *Foundations of Mathematics* (Symposium Commemorating Kurt Gödel, 1966), Springer, New York, 1969, pp. 58–73.

S. Stevin, 1585. *De thiende*. Leiden. Trans. by R. Norton, *Disme, the Art of Tenths* (London, 1608); also in D. E. Smith, ed. *Source Book of Mathematics* (New York–London, 1929).

———— 1585b. *L'Arithmétique* (Leiden, 1585, 1625), which also contains French translation of *La Disme;* and *Pratique de Arithmétique*. Thèses mathématiques. See *The Principal Works of Simon Stevin*, eds. E. Crone, E. J. Dijksterhuis, R. J. Forbes, *et al.*, (Lisse: Swets and Zeitlinger, 1955–1966), Vol. II B, Mathematics.

J. Stillwell, 2002. *Mathematics and its History*. New York, Springer.

———— 2012. *Roads to Infinity*. A K Peters, 2012.

———— 2013. *The Real Numbers: An Introduction to Set Theory and Analysis*. New York, Springer.

W. Tait, 1998. Foundations of set theory. In: Dales and Oliveri 1998, pp. 273–290. Revised version as chap. 6 of Tait 2005.

———— 2000. Cantor's *Grundlagen* and the paradoxes of set theory. In: *Between Logic and Intuition: Essays in Honor of Charles Parsons,* ed. G. Sher and R. Tieszen. Cambridge Univ. Press, 269–290.

———— 2005. *The Provenance of Pure Reason. Essays on the Philosophy of Mathematics and on its History*. Oxford University Press.

J. Tappenden, 2006. The Riemannian Background to Frege's Philosophy. In: Ferreirós and Gray (2006), 97–132.

R. Torretti, 1978. *Philosophy of Geometry from Riemann to Poincaré*. Dordrecht, Reidel.

A. S. Troelstra and D. van Dalen, 1988. *Constructivism in Mathematics*. 2 vols., Amsterdam: North Holland.

G. C. J. Ulrich, 1836. *Lehrbuch der reine Mathematik*. Göttingen.

S. Unguru, 1975. On the need to rewrite the history of Greek mathematics. *Archive for Hist. of Exact Sciences* 15:1, 67–114.

J. Väänänen, 2001. Second order logic and foundations of mathematics, *Bulletin of Symbolic Logic*, Volume 7:4, Dec 2001.

———— 2012. Second order logic or set theory? *Bulletin of Symbolic Logic*, 18:1, 91–121.

N. J. Wade, 2000. An upright man, *Perception* 29 (2000): 253–57.

S. von Waltershausen, 1856. *Gauss zum Gedächtniss*. S. Hirzel, Leipzig. English translation in https://archive.org/details/gauss00waltgoog

H. Wang, 1983. The concept of set. In: Benacerraf and Putnam (1983), 530–70.

———— 1997. *A Logical Journey: From Gödel to Philosophy*. Bradford Books.

C. Weierstrass, 1874. *Einleitung in die Theorieen der analytischen Functionen* (Nach den Vorlesungen im SS 1874, transcription G. Hettner). Mathematisches Institut der Georg-August-Universität Göttingen, 1974.

H. Weyl, 1918. *Das Kontinuum*. Leipzig, Veit. Reprint: Chelsea, New York.

———— 1927. Comments on Hilbert's Second Lecture on the Foundations of Mathematics, in: van Heijenoort 1967, 482–84.

———— 1949. *Philosophy of Mathematics and Natural Science*. Princeton Univ. Press. (1st ed. 1928 in *Handbuch der Philosophie*, Oldenbourg)

———— 1951. A Half-Century of Mathematics. *The American Mathematical Monthly*, Vol. 58, No. 8 (Oct., 1951), pp. 523–553.

A. N. Whitehead and B. Russell, 1910. *Principia Mathematica*, vol. 1. Cambridge Univ. Press.

E. Wigner, 1960. The Unreasonable Effectiveness of Mathematics in the Natural Sciences. *Communications in Pure and Applied Mathematics* 13: 1–14.

M. Wilson, 2000. The Unreasonable Uncooperativeness of Mathematics in the Natural Sciences. *The Monist* 83 (2): 296–314.

L. Wittgenstein, 1976. *Wittgenstein's Lectures on the Foundations of the Mathematics, Cambridge 1939.* Hassocks: Harvester Press.

Zeno. Fragments. In: G. S. Kirk, J. E. Raven, M. Schofield (1984) *The Presocratic Philosophers: A Critical History with a Selection of Texts, 2nd ed.* Cambridge University Press. See *Zeno's Paradoxes*, Wesley C. Salmon, ed. (Indianapolis, 1970).

E. Zermelo, 1904. Proof that every set can be well-ordered. *Math. Annalen.* English in van Heijenoort 1967, 139–141.

———— 1908. Untersuchungen über die Grundlagen der Mengenlehre. *Math. Annalen* 65, 261–281. English in van Heijenoort 1967, 199–215.

———— 1908b. Neuer Beweis für die Möglichkeit einer Wohlordnung, *Math. Annalen* 65, 107–128. English in van Heijenoort 1967, 183–198.

———— 1930. Über Grenzzahlen und Mengenbereiche, *Fundamenta Mathematicae* 14, 29–47. English in Ewald 1996, vol. 2.

Index

ABBREVIATIONS

$\mathbb{C}$	the structure of the complex number system
CH	Continuum Hypothesis
FOL	first-order logic
GCH	generalized CH
$\mathbb{N}$	the set of natural numbers (or the structure)
PA	Peano Arithmetic
PD	Projective Determinacy
$\mathbb{Q}$	the set of rational numbers (or the structure)
$\mathbb{R}$	the structure of the real number system
SOL	second-order logic
V = L	axiom of Constructibility
ZFC	Zermelo-Fraenkel system with axiom of Choice

Abel, Niels Henrik (1802–1829), 87, 239

Abū al-Wafā, Muḥammad al-Būzjānī (940–998), 210n8

agents, 3, 11–12, 28, 59–65; average, not genius, 60, 79–80; vs. Brouwer's creating subject, 61–64; and historical actors, 80–83; and meaning, 100; vs. subject, 60–64, 75–77

algebra, 126–27, 208–10, 215–17; algebraic geometry, 179, 181; Cartesian math and, 48, 85, 103, 147–48; complex numbers and, 105–08, 110; modern, 86–87, 160, 227; pre-Arabic math not algebraic, 20, 211n8. *See also* group theory

analysis (mathematical), 47, 81, 104, 127, 150, 153, 169, 220, 223, 230, 245, 295, 297, 305–06; AC and, 179–80, 295; complex, 33n18, 107, 183, 299; constructivistic, 172, 176, 203, 205, 235, 245, 273; Fourier series and, 240, 249; functional, 150, 179; infinitesimal, 222, 248; Weierstrass-style, 58, 81–82, 88, 254–55

Apollonius of Perga (c. 262–c. 190 BCE), 132, 133, 138, 142

a priori, apriorism, 61, 81, 126, 132, 156, 169, 204, 208, 245, 247, 310, 312, 314

Archimedes of Syracuse (c. 287–c. 212 BCE), 15, 18, 30, 116, 118, 132, 138, 165, 166, 238

Aristotle of Stagira (384–322 BCE), 56, 112, 125, 132, 134, 139–40, 143, 155; on the continuum, 166, 168–69, 243–44, 306

arithmetic, 27, 39, 81–82, 87, chapter 7 *passim*; basic, 182–84, 189–90; and conceptual understanding, 101–02; consistency relative to intuitionistic arithmetic, 203; fraction, 40; fundamental theorem of, 191, 196; Gödel incompleteness, 49–50; intuitionistic, 202–03; models of PA, 198–200; Peano Arithmetic (PA), 156, 190. *See also* number theory, in particular p. 183

arithmetization of pure math, 26, 80–82, 104, 126–27, 149, 225, 227, 254–56, 259, 281, 307

Arnol'd, Vladimir I. (1937–2010), 149, 242

Aryabhata the elder (476–550 CE), 193n22, 210n8

Axiom of Choice (AC), 161, 164, 172, 178–80, 224, 233, 244n61, 261, 262–65, 285, 301, 309; and category theory, 314; countable AC, 179, 263; dependent choice and global AC, 179

axioms, 23, 52, 62–63, 99, 131, 147, 153, 155, 162–63, 177, 189, 224, 276, 279, 299; Cantor-Dedekind, 86; of Completeness, 161, 164, 275; Euclid's axioms, 63, 113, 125, 128–31, 166; extrinsic vs. intrinsic evidence for, 23, 153, 155, 279, 297; of mathematical Induction (MPI), 95, 190–93, 197–98, (restricted) 196–97; and models, 199–200; of Parallels, 126, 155, 156n6, 163, 167–68; Peano's axioms, 190. *See also under* real number

axioms of set theory: Infinity, 155, 164, 251–52; large cardinals, 231n38, 278, 284, 288, 303; Powersets, 98, 233, 257–58, 264–65, 269–70, 272; Projective determinacy (PD), 278–79, 302; V = L, Constructibility, 23, 296, 278n49, 302

axiom system, 23, 31–32, 92–93, 119, 133, 138, 158, 240–41, 295, 299–300, 304; categoricity of, xiv, 258n16, 280, 294; consistency of, 87, 171–73, 177n25, 203, 266, 300, 303–04; and existence, 171–73

Baire, René L. (1874–1932), 88, 212n14, 234, 260, 261, 263, 264, 265, 269, 271, 272, 273, 277, 278, 294; Baire space, 207n2, 230, 234, 259, 260; property, 278

Banach, Stefan (1892–1945), 150, 159, 179

Barrow, Isaac (1630–1677), 214–15

Bernays, Paul (1888–1978), 1, 2, 111, 156, 170, 172, 173, 203, 233n40, 251n3, 263, 301, 309

Bernoulli, Daniel (1700–1782), 221, 238, 240

Bernoulli, Jacob (1655–1705), 111, 193n23

Bhaskara II (1114–1185), 210

Bishop, Errett (1928–1983), 176, 203, 272

Bloor, David (1942–), 22n7, 69, 185

Bohr, Niels (1885–1962), 61n17, 204

Bolzano, Bernard (1781–1848), 229, 307

Bolzano-Weierstrass: axiom, 275; theorem, 58, 235

Boole, George (1815–1864), 91n4, 200

Borel, Émile (1871–1956), 263–64, 277; and AC, 180, 263

Boscovich or Bošković, Rogerius (1711–1787), 221

Bourbaki group, 33, 36, 148, 149, 163, 207n1, 242, 295

Brahmagupta (598–c.670 CE), 210

Brouwer, L. Egbertus J. (1881–1966), 33n19, 65, 155, 164, 171, 203, 235n47, 244–46, 271, 272, 302; on the continuum, 245–46; and the creating subject, 61–64; and logic, 205; on ordinals, 271; on technology, 245–46

Cantor-Dedekind: continuum, 169, 223, 223–27, 230, 243, 245, 309; orthodoxy, 205, 252, 261–62, 273, 277, 284; and powerset, 258, 262; and quasi-combinatorialism, 264, 285

Cantor, Georg (1845–1918), 32n17, 59, 82, 88, 180, 234, 236, 252, 253–57, 261, 264, 277, 279, 284, 294n21, 306–07; and the absolutely infinite, 266, 269, 270n36, 295; non-denumerability of $\mathbb{R}$, 9, 92; and ordinals, 265–70, 271, 285; and powersets, 257–58, 260, 270, 285; and real numbers, 227–28, 230, 231–33, 237; on transfinite numbers, 273, 300–02, 303. *See also* Continuum Hypothesis

Carnap, Rudolf (1891–1970), 14

Cartan, Élie (1869–1951), 18

Category theory, 47, 84, 88, 90, 99–100, 102, 161, 173, 225, 241, 283, 297, 314

Cauchy, Augustin-Louis (1789–1857), 33, 58, 105, 107, 108n24, 180, 183, 222, 223, 239; sequence, 92, 171, 228n31, 230–31, 260

Cavaillès, Jean (1903–1944), 1, 2, 11, 15, 160, 163

Cavalieri, Bonaventura (1598–1647), 224, 244

Cellucci, Carlo (1940–), 49

Chemla, Karine (1957–), 21, 31n15, 51, 120, 152n52

Clavius, Christophorus (1538–1612), 134n27, 142

cognitive science, 65–66, 76–80; cognitive abilities, 33, 59–61, 65–69, 71–79, 141–42, 151; perception, 75–77, 140, 144

Cohen, Paul J. (1934–2007), 261, 274, 278n49, 281n1, 302, 303

complementarity in math (formula and meaning), 91–92, chapter 4 *passim*, 265, 309

computers, 14, 21, 59–60, 67, 140, 151, 205, 234; computer-assisted proof, 25; human, 60 conceptual understanding, 93, 100–02, 103–04, 296–305; conceptual approach to math, 90, 102–03

conjectures: Continuum Hypothesis (CH), 177, 274; generalized CH, 269–70; Goldbach's, 177, 201; Poincaré's, 176; Riemann Hypothesis, 178, 180–81

continuum, 64, 86, 146, 243–44, 273; in ancient times, 166–68; diversity of conceptions, 223, 225; in early modern period, 212–15; intuitionistic, 244–46, 310n43; and justification of ZFC and modern math, 305–14; in the nineteenth century, 221–27; pointillist continuum (and Luzin's views), 168–69; Riemann and Hilbert on, 161. *See also under* Aristotle; Brouwer; Cantor-Dedekind; Euler; al-Kashi; Peirce; Poincaré; Riemann; Stevin

Continuum Hypothesis (CH), 177, 274

counting practice, 38, 41–42, 65–68, 71–72, 85, 113–14, 188; counting numbers, 189, 194–95, 197

D'Alembert, Jean Le Rond (1717–1783): 107, 222, 238, 240

Dedekind, J. W. Richard (1831–1916), 15, 82, 84n53, 90, 110, 180, 227, 237n51, 253, 256, 260–61; ideal theory, 74–75, 256; on the infinite, 205, 217n19, 231–35; on natural numbers, 73–74, 96–97, 183, 188, 193, 198; on real numbers, 169, 207, 210n7, 229–30, 235, 255, 257; on set theory, 251–53, 256–57, 262, 270n35, 284–85, 286, 301. *See also* Cantor-Dedekind

Dehaene, Stanislas, 65, 68, 151–52

Democritus of Abdera, (c. 460–c. 370 BCE), 242

Descartes, René (1596–1650), 11, 48, 55, 85, 110, 147, 194, 217, 280

Dewey, John (1859–1952), 11

Dieudonné, Jean (1906–1992), 17n1, 131, 148, 295, 296

Dirac, Paul A. M. (1902–1984), 43n33, 204

Dirichlet, Gustav P. Lejeune (1805–1859), 88, 179, 239, 240; on arbitrary functions, 259, 260, 262; on conceptual math, 90

drawing practices, 17, 20–21, 30, 41, 113–14, 118–23, 133, 146, 292

du Bois-Reymond, Paul (1831–1889), 36n22, 308

Dürer, Albrecht (1471–1528), 41

Einstein, Albert (1879–1955), 9, 61n17, 163n12, 244, 298n26, 311

Euclid of Alexandria (3rd century BCE), 18, 24, 52, 117–18, 122, 125, 141–43, 149, 154, 224, 238, 239, 240; editions of the *Elements*, 143, 145. *See also* geometry: Euclidean

Eudoxos of Cnido (4th century BCE), 39, 42, 165; Eudoxian model, 42

Euler, Leonhard (1707–1783), 18, 24, 35, 47, 52, 55, 111, 148, 151, 179, 218; on complex numbers, 105–07; and the continuum, 221

Feferman, Solomon (1928–), 113n2, 171, 176, 195n27, 198, 206, 231, 304n35, 311n44

Fermat's last theorem, 101, 178–79, 182–84. *See also* Wiles, Andrew J.

Fibonacci, Leonardo of Pisa (c. 1170–c. 1250), 231, 261

formalism (foundational view), 89–91, 110, 271, 274–75

Fourier, Joseph (1768–1830), 150, 239; Fourier series, 240, 249; on mathematics, 242, 245

Fowler, David (1937–2004), 132–33

frameworks (symbolic and conceptual), 45–55, 114; for complex numbers, 105; for Euclidean geometry, 123–24, 137; vs. formal systems, 49–50, 198; and objects, 174–75

Frege, Gottlob (1848–1925), 15, 36n23, 38n25, 71, 81, 155, 186n14, 200, 205; and formal systems, 50, 90–91, 93, 198; and logicism, 97n12, 286; and properties, 95–96

Friedman, Harvey (1948–), 87, 304n35

functions, 35–36, 47, 88, 107n22, 218–19, 238, 240; analytic, 33; arbitrary (Dirichlet), 104, 259–60, 262; Baire hierarchy, 271, 277; complex, 108, 110; continuous, 180; or mapping, 84, 188, 241; Riemann on, 90

Galilei, Galileo (1564–1642), 244; and the paradox of infinite magnitude, 125, 232, 252

Galois, Evariste (1811–1832), 87; Galois theory, 86–87, 88, 250, 297

Gauss, Carl Friedrich (1777–1855), 15, 25n10, 33, 80n46, 81–82, 139n32, 148,

Gauss (*continued*)
154, 170, 204, 227, 239, 253; on complex numbers, 105–10, 256; on number theory vs. basic arithmetic, 184, 189; on pure math, 115–16

geometry, 8, 20, 30, 39, 63, 81–82, 85–86, 147–52, 247, 298; Cartesian, 37, 39, 48, 85, 86, 103, 114, 147, 297; Euclidean, chapter 5 *passim* (contrasted with Hilbert-style, 119; definitions in, 143–46; diagrammatic proof in, 25, 37, 126, 132–35); non-Euclidean, 8, 129, 153, 167, 300n29; practical (*see* drawing practices)

Gersonides, Levi ben Gershon (1288–1344), 53, 193n23

Giaquinto, Marcus, 2, 10, 26, 74n38, 141, 151, 297

God, 270, 282; the mathematician, 43n33, 81, 148n49; and platonic ideas, 4

Gödel, Kurt (1906–1978), 45, 79, 87, 95, 155, 158n8, 261, 265, 274, 277, 280, 288, 302; on axioms, 9n10, 23, 155; Gödel's program, 276, 278, 302; incompleteness results, 50, 91n4, 280–81, 303; and the iterative conception of sets, 261, 283–84, 285–86, 289, 290, 295, 300; and metatheory of ZFC, 302–303; on relative consistency (for PA and ZFC), 203, 274, 281, 311

Gowers, Timothy, 57

Grassmann, Hermann G. (1809–1877), 91n4, 193n22, 228n30

Gray, J. Jeremy (1947–), 2, 163

Grothendieck, Alexander (1928–2014), 18, 33, 87, 181n31

group theory, 30, 56–57, 179n28; Galois group, 86–87

Hadamard, Jacques (1865–1963), 272, 298n26

Hamilton, William Rowan (1805–1865), 57, 86, 107n21, 108n25, 109, 216n18

Hankel, Hermann (1839–1873), 110, 308

Hardy, Godfrey H. (1877–1947), 184, 196, 239

Hausdorff, Felix (1868–1942), 261, 264n25, 277; Hausdorff space, 207n1

Heisenberg, Werner (1901–1976), 312

Helmholtz, Hermann (1821–1894), 35n21, 113

Hermite, Charles (1822–1901), 237, 267

Heyting, Arend (1898–1980), 203, 245n62; intuitionistic arithmetic (HA), 203

Hilbert, David (1862–1943), 108n23, 115–16, 154, 161, 173, 196, 200, 203, 215, 223, 233, 245, 277–78, 298, 300, 303, 313; on arithmetization, 82, 149; on geometry, 82, 118–19, 127; on ideal elements and existence, 155–56, 171; on problems, 31–32, 54; on real numbers, 92, 225, 243, 275, 306; on symbols and formulas, 49, 90

Hoyrup, Jens (1943–), 20–21, 50, 67, 152n52

Husserl, Edmund (1859–1938), 61n18, 64–65

hypothesis, 86, 161, 166, 173, 207, 247–48, 278, 280, 288; hypothetical conception, 62–64, 154–57. *See also* conjecture

ibn al-Haytham [Alhazen], Abu Ali al-Hasan (965–1040), 129–30

idealization: in ancient geometry, 113, 118, 139–40, 142–46, 165–66, 247; idealized theory, 52–53

infinity, 130, 155, 161, 167, 193n24, 199, 205, 214, 217, 239, 270; simple vs. arbitrary, 231–36, 250–53

intuition, 151, 154, 279, 298, 305; intellectual, 64, 79; in Kant's sense, 7, 109, 125–26, 132

intuitionism, 64–65, 159n9, 164, 200–05, 244–46, 271; intuitionistic set theory (IZF), 296, 304

Jacobi, Carl G J. (1804–1851), 149, 245; on pure math, 115–16, 242

Jané, Ignasi, 265n26, 268, 274n41, 289n14

Jones, Vaughn (1952–), 304–05

Kant, Immanuel (1724–1804) and Kantianism, 6–7, 10, 15, 63, 109, 112, 125–26, 131, 134, 146n47, 153, 154, 244, 312–13

al-Karaji, Abū Bakr ibn Muḥammad (c. 953–c. 1029), 53n7, 193n22 & 23

al-Kashi, Ghiyāth al-Dīn Jamshīd (1380–1429), 211, 213, 259n17; and the number continuum, 222

Kepler, Johannes (1571–1630), 204, 227n28; his conjecture, 25n12

al-Khwarizmi, Muḥammad ibn Mūsā (c. 780–c. 850), 117n5

Kitcher, Philip S. (1947–), 1, 6, 26–27, 34–35, 37, 45–46, 47; on "mathematical practices," 4, 28–29, 45–46, 51, 80

Klein, Felix (1849–1925), 55, 82, 85, 255n10, 298

Kline, Morris (1908–1992), 105, 214n17; on the Greeks, 208–09

knowledge, 11, 42, 62–63, 64; and learning by the individual, 10–11, 91, 104, 131, 148, 206, 296–98

Kolmogorov, Andrei N. (1903–1987): 18, 33

Kreisel, Georg (1923–2015), 9, 160

Kronecker, Leopold (1823–1891), 90, 116, 155, 171, 212n14, 267, 270, 299; against the completeness axiom, 164, 234, 235, 272, 308n40

Kuhn, Thomas (1922–1996), 13, 27, 28, 29, 34, 44, 46; on exemplars, 55–57; on normal science, 4–5, 35

Kummer, Ernst E. (1810–1893), 179

Lagrange, Joseph-Louis: (1736–1813) 89–90, 196, 217, 238; against diagrams, 49, 126, 151, 217

Lakatos, Imre (1922–1974), 28–29, 35, 44, 51, 159, 239, 250; on quasi-empiricism, 154

Lambert, J. H. (1728–1777), 218–20; on rigor and formalization, 126–27

Landau, Edmund (1877–1938), 298

Lautman, Albert (1908–1944), 170n20, 207; *mixtes*, 207–08

Lawvere, F. William (1937–), 36n22, 90n2

Lebesgue, Henri (1875–1941), 88, 212n14, 261, 264; integration and measure, 175, 180, 262, 278, 301; on sets, 263, 271–72, 273, 277

Legendre, Adrien-Marie (1752–1833), 179, 242

Leibniz, Gottfried Wilhelm (1646–1716), 6, 35n20, 91n4, 105, 112n1, 137, 237; and the calculus, 217, 238, 244

Liouville, Joseph (1809–1882), 86n59; and transcendental real numbers, 237

Littlewood, John E. (1885–1977), 178n27, 196n28

Liu Hui (3rd century CE), 18, 31n15, 51, 116n5, 120–21

Locke, John (1632–1704), 11, 75

logic, 36, 52–53, 56, 81, 94–99, 158–59, 201–05, 241, 249–50; classical vs. intuitionistic, 170–71, 200–05; first-order (FOL), 56, 174–76; Henkin semantics for SOL, 191n20, 289n14; metatheory, 92–93, 300; second-order (SOL), 94, 98–99, 191n20, 288–89

logicism, 155, 286

Luzin, Nikolai N. (1883–1950): 261, 277–78, 305; constructivism of, 264; on set theory, 168–69, 224, 277, 302

Mac Lane, Saunders (1909–2005), 90n2, 113n2, 314n50

Maddy, Penelope (1950–), 23–24, 155n5

Mahavira the Teacher (9th century), 210

Manders, Kenneth, 2, 25, 112, 130, 132–37, 142, 144n43, 282

mathematical instruments, 21, 38, 59–60, 79, 115, 123

mathematics: Arabic, 30, 210, 215; classical or "postulational" vs. constructivistic, 64–65, 172–73, 262–63, 270–73; elementary vs. advanced, 112–13, 163–64, 168–69; non-Western, 18–20, 22, 31n15, 50–51, 116, 119–20; pure vs. applied, 15–16, 115–16, 242–45; and science, 13–15, 35–36, 43, 241–43; as science of magnitudes, 35–36, 212; as science of structures, 36

Maxwell, James C. (1831–1879), 35n21, 311

measuring practices, 102–03, 265

Mill, John Stuart (1806–1873), 6, 10, 27

models, model theory, 92–94, 97–98, 198–200; and classical logic, 204; and the continuum, 244, 312–13; intended, 93; scientific models of phenomena, 6, 9–10, 15, 42–43, 163, 208; of ZFC set theory, 274, 278, 294–95, 302–03

Needham, Joseph (1900–1995), 120–21

Neoplatonism, 115, 227

Netz, Reviel, 132–33

neuroscience, 12, 65–66, 77, 79, 152

Newton, Sir Isaac (1643–1727), 14n17, 35n20, 43, 156, 208, 246; and the calculus, 217, 222, 238, 244; on geometry, 122–23, 165; on number and magnitude, 88, 215, 220; on quantities given by motion, 220, 224

Noether, Emmy (1882–1935), 33n19

number concept, 65–71; around 1800, 106; expansion of, 104–05, 209–11, 216, 227; Gauss's, 108–09

number systems: complex numbers, 104–10, 256; fractions (rational numbers) and approximation, 83, 85, 102, 116; natural numbers, 96–97, 182–83; quaternions,

number systems (*continued*)
57, 86, 105; real numbers, 92, 99, 105–06,
208 (and decimal fractions, 211, 213–14,
217–18, 221, 258); transfinite ordinals,
265–67
number theory, 14, 101, 115–16, 150, 182–
84, 196–97, 256; algebraic, 75n40, 179,
182–83; logical, 199–200; prime number
theorem, 183

objectivity, 9, 39, 159–63, 249–50, chapter 9
passim
objects, mathematical, 81, 173–76, 241; in
Euclidean geometry, 139–41; existence,
164–66, 256, 258–59; sets as ("objecthood" of sets), 250–51, 284
Ohm, Martin (1792–1872), 215n18, 227

paradoxes: Banach-Tarski, 159; Galileo's,
252; Richard's, 232; set-theoretic, 251n3,
266, 285, 294. *See also* Skolem, Thoralf
Parsons, Charles (1933–), 49n4, 124, 173–75
Pasch, Moritz (1843–1930), 118, 127, 133
Peirce, Charles S. (1839–1914), 49n4, 63; on
the continuum, 144, 223n23
Penrose, Roger (1931–), 207n1, 312
physics, 9, 204, 242–43, 245–46, 311–12
Pi (π), 86, 211; exact vs. approximate value,
30–31, 116; Lambert's proof of irrationality, 218–20; Lindemann's proof of being
a transcendental number, 237–38
Piaget, Jean (1896–1980), 11n11, 12, 13, 27
Plato (427–348 BCE), 6, 10, 115, 118, 124–
25, 146n47, 148n49, 170n20, 242
platonism, 7, 170–73, 224, 265n27, 275–76;
methodological platonism vs. realism,
170–71
Poincaré, J. Henri (1854–1912), 82, 116,
243n59, 263; on the continuum, 312; on
relations, 175; on sets, 267–68, 271
Popper, Karl (1902–1994), 11
practice, 28–29; mathematical, 28–34, 39–40;
multiplicity of, 35–40; scientific, 42–43,
161, 204, 242–44; technical, 20, 40–42,
113. *See also* counting practice; drawing
practices; measuring practices
predicativism, 176, 195, 234, 248, 311n44
Proclus Diadochus (412–485), 125, 127–28,
129–30, 142, 227n28
proof, 24–25, 57, 127, 150, 240; automated
or computer-assisted, 25, 59–60; diagram-
based, 131–36; multiple proofs, 24–25,

53–54, 107, 111, 240; purely arithmetical,
253–57; rejection of diagrams and visual
ingredients, 126–27, 149–50
Ptolemy of Alexandria (c. 90–c.168), 129–30,
210-11n8
Putnam, Hilary (1926–), 9, 63n21, 160, 282
Pythagorean (*gougu*) theorem, 30, 120–22,
134, 139, 151, 168

Quine, Willard Van O. (1908–2000), 6, 14,
37, 93, 154; on sets, 296, 299, 313–14

reductionism, 61n16, 101, 154; against, 5,
10, 35, 291
Riemann, Bernhard (1826–1866), 7, 33, 58,
63, 82, 90, 108n24, 110, 239, 243, 256;
antinomy, 312–13; on the continuum,
161, 166–67; on geometry, 85, 153–54,
156, 167n15; surfaces, 74, 149
Russell, Bertrand (1872–1970), 155, 157,
200, 205; on sets, 251, 258, 263; type
theory, 285–86

al-Samawa'l, Ibn Yaḥyā al-Maghribi (12th
century), 193, 211
Schrödinger, Erwin (1887–1961), 312n46,
244n60
Schwarz, H. A. (1843–1921), 82
Selberg, Atle (1917–2007), 183
sets, 250–51; arbitrary set, 262, 264; cumulative picture ("cone representation"),
290–92; descriptive set theory, 207n2,
234, 243, 263–64, 271, 277–78, 299, 301–
02, (Borel set) 271, 277, (projective set)
277–78; dichotomic, logicist conception
of, 286; iterative conception of, 283–87;
minimalist views, 295–96; NBG system,
299, 301; quasi-combinatorialism, 98,
233–34, 258–62, 264, 285. *See also* Axiom
of Choice; axioms of set theory; Zermelo-
Fraenkel axiom system
Shelah, Saharon (1945–), 265, 276n43
Sierpinski, Waclaw (1882–1969), 179–80, 277
Silver, Jack H. (1942–), 177n25, 303n33
Skolem, Thoralf (1887–1963), 92–93, 97,
198; Löwenheim-Skolem theorem, 92–
93; paradox, 92–94, 96
Solovay, Robert M. (1938–), 278, 302–03
space: Euclidean, 166, 266, 308; Polish,
234n43; space-time, 244. *See also under*
Baire, René L.; Hausdorff, Felix
Spelke, Elisabeth, 151

Steel, John R. (1948–), 279, 292
Stevin, Simon (1548–1620), 217, 237; on numbers, 207, 211–13, 226
Stifel, Michael (1487–1567), 213–14
structures in math, 36, 113, 199, 206–07, 241, 280, 297, 307; and morphisms, 173
systematic links, 37–39, 83–88, 96–97, 101, 147, 248–53, 297

Tait, William W. (1929–), 173, 266, 284
Tao, Terence, 184n6, 282n3
Tarski, Alfred (1901–1983), 93n5, 159, 288
Theon of Alexandria (4th century CE), 128, 144-45
topology, 88, 149–50, 225; of Baire space, 234n43, 259n19; of ℝ, 33n18 & 19, 235, 254. *See also* axioms: of Completeness

Ulrich, Georg Carl J. (1798–1879), 168–70
al-Uqlidisi, Abu'l Hasan (10th century), 211, 259n17

values, value judgment, 25, 31, 33, 51, 80, 114–15, 118–19, 122, 155, 201, 229, 246
van der Waerden, Bartel L. (1903–1996), 18, 33n19, 132n22
van Fraassen, Bas (1941–), 13n14, 176
visual thinking, 26, 74–75, 108–09. *See also* intuition
Voevodsky, Vladimir, 282n3

Wallis, John (1616–1703), 214–15
Weierstrass, Carl T.W. (1815–1897), 81–82, 108n24, 126, 231n37, 235, 239, 253, 254–55, 260, 299; doubts about set theory, 270; on numbers, 110, 227–28, 254. *See also* Bolzano-Weierstrass
Weil, André (1906–1998), 132n22, 150, 181n31, 282
Weyl, Hermann (1885–1955): 160, 311, 314; and AC, 263; his constructivism, 164, 171–72, 197, 202–04, 205, 235n47, 264, 272, 302
Wigner, Eugene (1902–1995), 15n19, 207n1
Wiles, Andrew J. (1953–), 150, 182; theorem (implying Fermat's last), 62n19, 101, 176, 183–84
Wittgenstein, Ludwig (1889–1951), 1n1, 12n13, 89, 93, 209n5
Woodin, Hugh (1955–), 276, 279, 303, 314

Zeno of Elea (5th century BCE), 166
Zermelo, Ernst (1871–1953), 111, 171, 224, 233, 252, 258, 261–63, 290, 306, 309n42; and the axiom of choice (AC), 172, 178–80, 272, 301; and the cumulative hierarchy, 74, 287, 294–95
Zermelo-Fraenkel axiom system (ZFC), 250–51, 282, 284, 294n22, 295, 297, 300, 302; alternatives to, 176, 234, 296, 299, 301; and CH, 274, 275, 281, 288; and the cumulative hierarchy, 287n11; and large cardinals, 231n38, 284, 288; metatheory and limits of, 96–98, 171, 277, 302–03; well-ordering theorem, 171–72, 263, 269
Zhao Shuang (3rd century), 120–21